MIC & MMIC
AMPLIFIER AND OSCILLATOR
CIRCUIT DESIGN

The Artech House Microwave Library

Analysis, Design, and Applications of Fin Lines, Bharathi Bhat and Shiban K. Koul
E-Plane Integrated Circuits, P. Bhartia and P. Pramanick, eds.
Filters with Helical and Folded Helical Resonators, Peter Vizmuller
GaAs MESFET Circuit Design, Robert A. Soares, ed.
Gallium Arsenide Processing Techniques, Ralph Williams
Handbook of Microwave Integrated Circuits, Reinmut K. Hoffmann
Handbook for the Mechanical Tolerancing of Waveguide Components, W.B.W. Alison
Handbook of Microwave Testing, Thomas S. Laverghetta
High Power Microwave Sources, Victor Granatstien and Igor Alexeff, eds.
Introduction to Microwaves, Fred E. Gardiol
LOSLIN: Lossy Line Calculation Software and User's Manual, Fred. E. Gardiol
Lossy Transmission Lines, Fred E. Gardiol
Materials Handbook for Hybrid Microelectronics, J.A. King, ed.
Microstrip Antenna Design, K.C. Gupta and A. Benalla, eds.
Microstrip Lines and Slotlines, K.C. Gupta, R. Garg, and I.J. Bahl
Microwave Engineer's Handbook: 2 volume set, Theodore Saad, ed.
Microwave Filters, Impedance Matching Networks, and Coupling Structures, G.L. Matthaei, L. Young and E.M.T. Jones
Microwave Integrated Circuits, Jeffrey Frey and Kul Bhasin, eds.
Microwaves Made Simple: Principles and Applications, Stephen W. Cheung, Frederick H. Levien, et al.
Microwave and Millimeter Wave Heterostructure Transistors and Applications, F. Ali, ed.
Microwave Mixers, Stephen A. Maas
Microwave Transition Design, Jamal S. Izadian and Shahin M. Izadian
Microwave Transmission Line Filters, J.A.G. Malherbe
Microwave Transmission Line Couplers, J.A.G. Malherbe
Microwave Tubes, A.S. Gilmour, Jr.
MMIC Design: GaAs FETs and HEMTs, Peter H. Ladbrooke
Modern Spectrum Analyzer Theory and Applications, Morris Engelson
Monolithic Microwave Integrated Circuits: Technology and Design, Ravender Goyal, et al.
Nonlinear Microwave Circuits, Stephen A. Maas
Terrestrial Digital Microwave Communications, Ferdo Ivanek, et al.

MIC & MMIC AMPLIFIER AND OSCILLATOR CIRCUIT DESIGN

Allen A. Sweet

Artech House
Boston • London

Library of Congress Cataloging-in-Publication Data

Sweet, Allen A., 1943-
  MIC and MMIC amplifier and oscillator circuit design / Allen A. Sweet.
    p.   cm.
  Includes bibliographical references.
  ISBN 0-89006-305-2
  1. Microwave amplifiers--Design and construction. 2. Oscillators, Microwave--Design and construction. I. Title.
TK7871.2.S94   1990   89-49100
621.381'325--dc 20   CIP

British Library Cataloguing in Publication Data

Sweet, Allen A., 1943-
  MIC and MMIC amplifier and oscillator circuit design.
  1. Electronic equipment. Microwave oscillators. Design. 2. Electronic equipment. Microwave amplifiers. Design
  I. Title.
  621.381323

  ISBN 0-89006-305-2

© 1990 ARTECH HOUSE, Inc.

685 Canton Street
Norwood, MA 02062

All rights reserved. Printed and bound in the United States of America. No part of this publication may be reproduced or utilized in any form or by any means, electronic or mechanical, including photocopying, recording, or by any information storage and retrieval system, without permission in writing from the publisher.

International Standard Book Number: 0-89006-305-2
Library of Congress Catalog Card Number: 89-49100

10   9   8   7   6   5   4   3   2   1

Contents

Preface	xi
Chapter 1 INTRODUCTION	1
Chapter 2 DEVICES	5
2.1 GaAs FETs	5
2.1.1 GaAs FET Material Overview	5
2.1.2 Basic Device Structure	6
2.1.3 Fabrication Technologies	6
2.1.4 GaAs FET Model and Equivalent Circuit	7
2.1.5 Large-Signal Effects	19
2.2 Bipolar Transistors	22
2.3 Varactor Diodes	25
2.3.1 Structure	25
2.3.2 Varactor Operation	26
2.4 PIN Diodes	32
2.4.1 Structure	32
2.4.2 Operation of a PIN Diode	32
2.5 YIG Resonators	36
2.5.1 Basics of Ferrimagnetism	36
2.5.2 Mode Behavior of the YIG Resonance	41
2.5.3 The Equivalent Circuit of a YIG Resonator	44
2.5.4 YIG Limiting	46
2.5.5 Anisotropic Effects in YIG	47
2.5.6 Electromagnet Design	50
2.6 Dielectric Resonators	52
2.6.1 Resonant Frequency of a Dielectric Resonator	54
2.6.2 Temperature-Dielectric Resonant Frequency Drift	56
2.6.3 Dielectric Resonator Tuning	57
2.6.4 Coupling to a Dielectric Resonator	57
Chapter 3 MICROSTRIP TRANSMISSION LINE FUNDAMENTALS	59
3.1 The Basic Equations of Microstrip Transmission Lines	59

3.2	Table of Z_0 and ϵ_e for Some Typical Substrates		62
3.3	The Smith Chart		62
3.4	Coupled Microstrip Transmission Lines		65
3.5	Using Microstrip Transmission Lines as Distributed Equivalents of Lumped Element Inductors and Capacitors		67
3.6	Lange Couplers		69
3.7	S-Parameters and Multiport Networks		70

Chapter 4 COMPUTER AIDED DESIGN TECHNIQUES 73

4.1	Linear CAD Analysis-Optimization Program		73
4.2	Transmission Line Parameter Calculation Programs		78
4.3	Nonlinear CAD Analysis-Optimization Programs		78
4.4	Layout and Graphics Programs		79
4.5	Integrated Design Workstations		80
4.6	HP-41C Microwave Design Programs		83
	4.6.1	Varactor Diode Capacitance—Program Listing	83
	4.6.2	GaAs FET Model Analysis—Program Listing	85
	4.6.3	The Inductance of a Wire—Program Listing	87
	4.6.4	Thermal Resistance of a Semiconductor Die—Program Listing	88
	4.6.5	Large-Signal Power, Efficiency, and Load Resistance of a GaAs FET—Program Listing	90
	4.6.6	Tee Attenuator—Program Listing	91
	4.6.7	Inductance of a Metal Strip—Program Listing	92
	4.6.8	Mutual Inductance of Two Metal Strips—Program Listing	93
	4.6.9	Two-Turn Spiral Inductors—Program Listing	94
	4.6.10	Three-Turn Spiral Inductors—Program Listing	97
	4.6.11	Printout for Three-Turn Spiral Inductors—Program Listing	101
	4.6.12	Schottky Barrier Diodes *I-V*—Program Listing	103
	4.6.13	Wound Coil Inductance—Program Listing	103
	4.6.14	Distributed Amplifier Basics—Program Listing	105

Chapter 5 AMPLIFIER CIRCUITS 107

5.1	The Basics		107
	5.1.1	Matching Techniques	107
	5.1.2	Gain Compensation	108
	5.1.3	Fano's Limit	109
	5.1.4	Stability	111
	5.1.5	Noise Match	112
	5.1.6	Power Match	112
	5.1.7	Harmonic Distortions and Intermodulation Products	112
	5.1.8	Gain Drift with Temperature	113

5.2	Amplifier Performance by Basic Type	114
	5.2.1 Reactively Matched Amplifiers	114
	5.2.2 Balanced Amplifiers	115
	5.2.3 Lossy Match Amplifiers	116
	5.2.4 Feedback Amplifier	117
	5.2.5 Distributed Amplifiers	120
5.3	Example 1: An MIC-Balanced Reactively Matched Amplifier for the 6 to 18 GHz Band	128
5.4	Example 2: An MIC Lossy-Match Amplifier for the 2 to 6 GHz Band	133
5.5	Example 3: An MMIC Feedback Amplifier for the 2 to 8 GHz Band	139
5.6	Example 4: An MIC Two-Stage Low-Noise Amplifier for the 4.5 to 5.0 GHz Band	145
5.7	Example 5: An MMIC Distributed Amplifier for the 2 to 20 GHz Band	152
5.8	Example 6: A One-Watt MIC Balance Power Amplifier for the 4 to 10 GHz Band	162
5.9	Example 7: A Balanced MIC PIN Attenuator for the 6 to 18 GHz Band	168
Chapter 6 OSCILLATOR CIRCUITS		175
6.1	Negative Resistance Concepts	175
6.2	Transistor Oscillator Basics	178
6.3	Feedback Techniques for Oscillator Transistors	180
6.4	Oscillator-Transistor Comparison	181
6.5	Frequency-Temperature Stability and Tuning Linearity	182
6.6	Oscillator Resonator Comparison	183
6.7	Oscillator Noise	183
6.8	Example 1: A 12 to 18 GHz FET Varactor-Tuned VCO	189
6.9	Example 2: A 4 to 8 GHz Bipolar Varactor VCO	194
6.10	Example 3: A YIG-Tuned 2 to 8 GHz Bipolar Oscillator	200
6.11	Example 4: A YIG-Tuned 6 to 18 GHz FET Oscillator	206
6.12	Example 5: An 8.4 GHz DRO FET Oscillator	211
Chapter 7 MIC LAYOUT AND FABRICATION		217
7.1	Layout Strategy	217
	7.1.1 Process Flowchart	217
	7.1.2 Design Fabrication Rules	217
	7.1.3 Low Parasitic Design	217
	7.1.4 Grounding	220
7.2	Parasitic Models	223
	7.2.1 Microstrip End Effects	223
	7.2.2 Microstrip Bends	226

| | | 7.2.3 | Symmetric Step Changes in Microstrip Width | 227 |

- 7.2.3 Symmetric Step Changes in Microstrip Width — 227
- 7.2.4 Microstrip Tee Junction — 227
- 7.2.5 Microstrip Cross Junction — 228
- 7.2.6 Bond Wires — 229
- 7.2.7 Thin-Film Resistor — 230
- 7.2.8 Grounds — 230
- 7.3 Transistors, Diodes, and Capacitor Chips — 231
- 7.4 Thermal Considerations — 233
- 7.5 Artwork — 236
  - 7.5.1 Rubyliths — 236
  - 7.5.2 Digitizing — 236
- 7.6 Mask-Making — 238
  - 7.6.1 Film or Chrome-Glass Masks — 238
  - 7.6.2 Step and Repeat — 238
- 7.7 Types of Substrates — 239
  - 7.7.1 Dielectric Choices — 239
  - 7.7.2 Types of Metal Systems — 240
- 7.8 Photolithography — 240
  - 7.8.1 Photoresist — 240
  - 7.8.2 Etching — 242
- 7.9 Resistor Stabilization — 244
- 7.10 Sawing — 245
- 7.11 Brazing Process — 245
- 7.12 Die Attach Process — 247
- 7.13 Wire Bonding — 248
  - 7.13.1 Wedge Bonding — 248
  - 7.13.2 Ball Bonding — 249

Chapter 8 FINAL LAYOUT OF MIC AMPLIFIER AND OSCILLATOR DESIGN EXAMPLES — 253

- 8.1 The 6 to 18 GHz Balanced Amplifier Final Layout — 254
- 8.2 The 2 to 6 GHz Two-Stage Lossy Match Amplifier Final Layout — 257
- 8.3 Two-Stage 4.5 to 5.0 GHz Low-Noise Amplifier Final Layout — 260
- 8.4 The 4 to 10 GHz One-Watt Amplifier Final Layout — 261
- 8.5 The 6 to 18 GHz Balanced PIN Attenuator Final Layout — 265
- 8.6 The 12 to 18 GHz FET Varactor VCO Final Layout — 270
- 8.7 The 4 to 8 GHz Bipolar Varactor VCO Final Layout — 273
- 8.8 The 2 to 8 GHz Bipolar YIG-Tuned Oscillator Final Layout — 276
- 8.9 The Final Layout of the 6 to 18 GHz FET YIG-Tuned Oscillator — 278
- 8.10 Dielectrically Tuned 8.4 GHz Oscillator Final Layout — 282

Chapter 9 MMIC LAYOUT AND FABRICATION — 287

- 9.1 MMIC Economics — 287
  - 9.1.1 Cost of Design Rules — 287

	9.1.2	Mask Set Costs	288
	9.1.3	Fabrication Costs	288
	9.1.4	Probing, Thinning, Dicing, and Packaging Costs	290
	9.1.5	Cost per Chip as a Function of Volume	291
	9.1.6	A Comparison between MIC and MMIC Costs	292
9.2	MMIC Process Description		294
	9.2.1	MMIC Structure	294
	9.2.2	GaAs Material	296
	9.2.3	Doping by Ion Implantation	296
	9.2.4	Ohmic Contact Metal	297
	9.2.5	Resistive Layers	298
	9.2.6	Gate Metal	299
	9.2.7	First Metal	300
	9.2.8	Dielectric	301
	9.2.9	Second-Level Metal	302
	9.2.10	Dielectric and Airbridge Vias	303
	9.2.11	Substrate Vias	304
	9.2.12	Final Wafer Process Steps	305
9.3	MMIC Design and Layout Strategy		305
	9.3.1	Foundry Design Rules	307
	9.3.2	Rough Cut Layout	318
	9.3.3	Circuit Reoptimization Including Element Models, Layout Parasitics, and Layout Modifications	324
	9.3.4	Sensitivity Analysis and Yield Predictions	325
	9.3.5	Final Layout	329

Chapter 10 FINAL LAYOUT OF THE MMIC AMPLIFIER EXAMPLES 331
10.1 MMIC Layout for the 2 to 8 GHz Two-Stage Feedback Amplifier 334
10.2 Final Layout of the 2 to 20 GHz Distributed Amplifier 349
Index 359
About the Author 367

Preface

The microwave electronics field, since its inception at the MIT Radiation Laboratory during WWII, has been in a continual state of evolution. No area of the field has seen greater changes than that of microwave amplifiers and oscillators. The active devices used in amplifiers and oscillators have evolved from vacuum tubes (klystrons and magnetrons) to two terminal solid state devices (Gunn, IMPATT, tunnel, and varactor diodes) and finally to three terminal solid state devices (GaAs FETs, and silicon bipolar transistors). At the same time, the transmission line media has changed from waveguides, to coaxial cable, and finally to microstrip lines. The analytic design tools available to the microwave circuits designer have gone through similar changes. Analytic calculations based on the application of Maxwell's equation to particular situations have given way to numerical solutions calculated by mainframe computers, which have largely been replaced by general transmission line optimizing software (Touchstone and SuperCompact) that runs on PC computers and workstations.

These three parallel paths have now evolved to such a point that, in most cases, all of the active, passive, and transmission line elements of an amplifier or oscillator are located on one or two semiconductor chips plus a ceramic circuit board, all of which are interconnected by hybrid assembly techniques (i.e., microwave integrated circuits, or MICs); or all the circuit elements are located on a single chip of semiconductor material (i.e., monolithic microwave integrated circuits, or MMICs). This is not to say that vacuum tubes and waveguide transmission lines are no longer important. They are very important in high power applications and at millimeter wavelengths, but most low and medium power microwave components and subsystems are constructed now with solid state devices and microstrip transmission lines. The improvements in size, weight, electrical performance, and cost of the new technologies compared to the early vacuum tube/waveguide amplifiers and oscillators is truly fantastic. In fact, the present generation of airborne radar, communications, and EW systems simply would not be possible without the newer technologies. These rapidly developing technologies have placed a heavy burden on the circuits designer, who must stay current with all of these new trends.

The purpose of this book is to help enable circuit designers to produce both cost effective and reliable amplifier and oscillator designs by using the latest device, circuit, and simulator technologies. Since the author strongly believes in learning through example, a dozen up-to-date design examples are presented which systematically "walk" the designer through the entire process from conception to completed art work and assembly drawings. A systematic design process emerges from these examples, and all of the necessary analytic tools are provided which enable the designer to use this process. A heavy emphasis is placed on the use of computer aided design tools. Similarly, a heavy emphasis is also placed on the importance of each designer cultivating his or her own creative design strategies.

Many good books are available which discuss the theory of operation of various microwave devices, and they are carefully referenced in this book. However, the intent of this book is to go beyond a theoretical discussion and present the strategies of design as a systematic process.

I am truly grateful for all the helpful suggestions and comments that have been offered over the past decade by colleagues and students, all of whom have helped to bring to light many of the issues addressed in this book. These individuals include students participating in courses I have taught for the University of California at Berkeley Extension Division and for Technology Services Corporation, many of whom repeatedly requested an "advanced design course." This book is the foundation of such a course.

I also wish to thank my many colleagues throughout the industry for their numerous helpful technical discussions in which I have been privileged to participate over the years. Of these, two are especially deserving of a special note of thanks. George Vendelin, friend, colleague, and associate, who for many years took on the lengthy task of reading the entire manuscript, offered many helpful suggestions. John Mezak, also a long time friend and colleague, read the YIG resonator section and offered many useful suggestions on this material.

Finally, I most gratefully acknowledge the long, countless hours devoted by my wife, Fran, to the word processing and editing of the entire manuscript. Her ability to arrange words properly in order to bring out the clarity of intended meaning, and her occasional use of the pointed question, "What are you really trying to say here?" at just the right time proved invaluable to the process of manuscript preparation. Without her tireless efforts and unyielding support, this book would not have been possible.

Chapter 1
Introduction

Most texts and reference books in the microwave electronics field focus on a general theoretical approach to design. My belief is that most of us learn best by first studying specific examples and then proceeding to a general understanding. Such a philosophy is the basis for this book. The material is structured to provide a definite process of design and is illustrated by numerous specific examples that have been taken from the author's design experience acquired since the mid-1970s.

The book had its beginnings in the numerous courses on microwaves and microwave amplifiers and oscillators taught by the author since 1981. These include courses for Technology Service Corporation of Silver Spring, Maryland, and the University of California, Berkeley, Extension Division. The content of these courses emphasized either basic operational principles or computer-aided design techniques, depending on the particular course. Over the years, students have asked for an advanced design course that would address all of these ideas simultaneously. The book is partially motivated by these requests. The author hopes that the book will simultaneously provide a text for an advanced amplifier and oscillator design course and a reference handbook for the practicing design engineer.

Certain important modifications have been made to the previous course materials in preparing this book. These modifications fall into three general areas:

- *Computer-aided design* (CAD) techniques are emphasized throughout the book. Because of the complexity of modern microwave circuits, to design without making use of CAD techniques is now virtually impossible. Chapter 4 is devoted to this topic, and all of the examples make use of CAD techniques. "Hand calculations" are discussed, not as an end product, but merely as a "first guess" for a CAD analysis and optimization, which then follows.

- A *monolithic microwave integrated circuit* (MMIC) design process with examples is included, along with the more standard MIC hybrid techniques. MMIC circuits, which are totally functional amplifiers or oscillators fabricated on a single chip of GaAs, are a very important new class of circuit that has grown swiftly in popularity in recent years. These circuits pose some unique design challenges in terms of economics, design precision, and manufacturing yield. Because of its uniqueness, MMIC design is treated as a separate area.

- Considerable detail on the fabrication techniques for both MIC and MMIC circuits is presented. This material is provided to familiarize the designer with how designs are fabricated so that he or she can become sensitive to the design issues affecting fabrication. Remember that *if a circuit cannot be economically fabricated, it is a failure, regardless of its performance.*

The paradox of microwave circuit design is that the basic topology of most microwave amplifiers and oscillators is very straightforward. Often the only components involved are an active device (GaAs FET or Si bipolar) plus four or five microstrip transmission line elements of various lengths and widths. Compared to a large analog IC chip or a digital microprocessor chip, such a microwave circuit seems very simple. However, strange as it may sound, the design of microwave amplifiers and oscillators poses one of the greatest technical challenges to circuit designers within the electronics industry today. In the past, there had been a heavy reliance on empirical techniques (tweaking, cut-and-try, scatter gunning) to make microwave circuits respond as expected. However, economic pressures and the advent of MMIC technology make the old empirical techniques very undesirable and sometimes impossible.

Why does a simple microwave circuit require a very complicated design process? The answer is threefold: (a) many layout "parasitic" elements complicate the circuit's basic topology; (b) the active devices are not completely reproducible and vary with a degree of randomness from circuit to circuit; and (c) circuit elements at microwave frequencies are distributed by nature, and care must be taken to consider the "distributed" aspect of many so-called lumped elements. The predicted performance of a circuit is only as accurate as the model used to make the prediction. Furthermore, modeling in the world of microwave circuits is a very complicated process (and a fine art), which often involves exacting measurements as well as applying appropriate simplifying assumptions. Because the modeling of active devices quite often is left to designers, they must become expert modelers. The modeling process is explored in depth throughout the book.

All of the numerous examples are practical circuits that, if constructed, will yield good performance. Each example is worked through a series of design steps, culminating in a final layout appropriate for fabrication.

Enough background material is given on the basics of semiconductor devices (GaAs FETs, bipolar transistors, varactor, and PIN diodes) and microstrip transmission lines so that the designer will have all of the important basic relationships at his or her fingertips.

The chapters describing layout (MIC and MMIC) tie it closely to fabrication techniques. A layout strategy is developed that from the beginning includes modeling of parasitic elements. The layout process is seen as interactive in the sense that changes in the layout always must be checked for how they affect performance. Often several iterations will be required before the designer simultaneously achieves the desired performance and a realistic layout.

The two final chapters deal exclusively with GaAs MMIC circuits. Because MMIC circuits cannot be experimentally modified (except with great difficulty), their calculated performance must be extremely accurate and, at the same time, forgiving enough to deliver the specified performance despite process variations of certain key device parameters. MMIC design requires very special strategies that combine the best modeling and layout techniques with a statistically based yield analysis.

The author hopes that designers at all levels of experience will profit from this book. Although the core material was derived from classroom lecture notes, the book is not intended to function only as a text, but rather as a reference and guide for both students and practicing design engineers.

Chapter 2
Devices

2.1 GaAs FETs

2.1.1 GaAs FET Material Overview

The principal component of all GaAs FET devices is the gallium arsenide material itself. GaAs has been recognized as a very special microwave material since the 1960s. The technological evolution of GaAs has made great strides over the past two decades, and today it rivals silicon in terms of purity, precision, and manufacturability.

The significance of gallium arsenide as a microwave material was first recognized by J. B. Gunn[1] of IBM, who discovered periodic high-frequency oscillations in a biased "resistor" of N-type GaAs in 1965. These mysterious oscillations were soon traced to a phenomenon called *negative differential mobility*, which is caused by an even more obscure phenomenon called *intervalley scattering*[2], which occurs within the energy band structure of GaAs. Negative differential mobility is visible in the velocity field diagram of GaAs shown in Figure 2.1. Many workers joined the gallium arsenide field in the mid-1960s, to both understand more about its material properties, and to apply GaAs as a material to potential applications, which turned out to be largely in the microwave region of the EM spectrum. This was a time of experimentation with various "diode" structures, including transit time, hybrid, LSA, "modes"[3] to name a few; all of which used gallium arsenide as their basic material.

Not until the 1970s was gallium arsenide applied to the field-effect transistor (FET). Because of the significantly higher electron drift velocity in gallium arsenide relative to silicon, the transit time delay in a GaAs FET is much less than it is in an equivalent silicon FET, resulting in a considerably higher frequency of operation for GaAs FETs than for silicon FETs. Today, GaAs FETs that operate up to 50 GHz are commercially available.

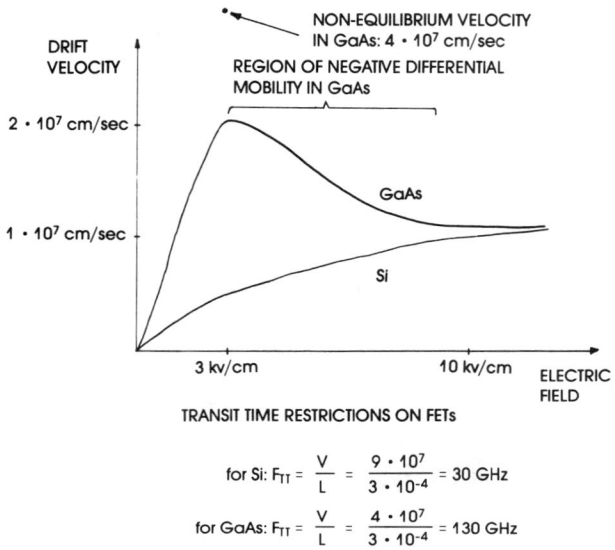

Fig. 2.1 The velocity field diagram of GaAs and Si showing the region of negative differential mobility in GaAs.

2.1.2 Basic Device Structure

The basic structure of a GaAs FET consists of a thin film of N-type gallium arsenide ($N_D \doteq 1.5 \cdot 10^{17} \text{cm}^{-3}$) with two ohmic contacts, called the *source* and the *drain* defining a current carrying channel. Deposited in the middle of this channel is a third contact, which is a rectifying Schottky barrier called the *gate*. The gate serves as a control element for the current that passes from the source to the drain through the channel. A cutaway diagram of this structure is shown in Figure 2.2.

The low-frequency equivalent circuit of a GaAs FET is a voltage controlled current source, called a *transconductance*. At microwave frequencies, various parasitic elements become significant, and must be included in any realistic equivalent circuit.

2.1.3 Fabrication Technologies

The active layer of gallium arsenide is deposited by one of four basic methods: vapor phase epitaxy, liquid phase epitaxy, molecular beam epitaxy (MBE), or ion implantation[4]. In the epitaxial processes, a thin film of gallium arsenide, containing the proper dopants, is deposited slowly to create a near-perfect crystal

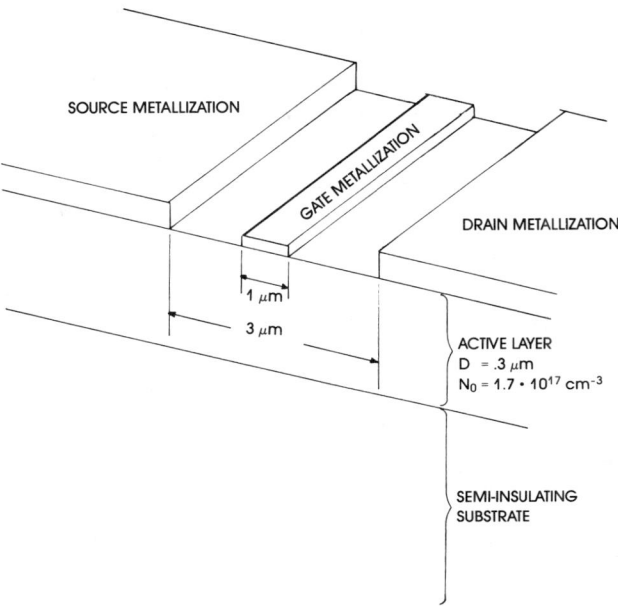

Fig. 2.2 FET channel details.

structure on the surface of a semi-insulating wafer of gallium arsenide. This epitaxial layer is on the order of .5 μm thick and its donor density is approximately $1.5 \cdot 10^{17}$ cm^{-3}. On top of the epitaxial gallium arsenide are deposited two ohmic metal contacts, the source and the drain, defining the FET's active channel. The gate metal is deposited in the center of the channel. A layout diagram showing the basic structure of a typical "tee"-gate GaAs FET is given in Figure 2.3. Gates are rectifying Schottky barrier contacts, composed of different metals than the ohmic contacts. Typical ohmic metals are gold and germanium, which later are alloyed into the gallium arsenide by a high-temperature braze process. Typical Schottky gate metals are aluminum or gold[5]. A variety of photolithographic processes such as lift-off and self-alignment are used to fabricate the extremely narrow gate metal[6]. The fabrication of GaAs FET devices is a fascinatingly complex blend of art and science that has made tremendous strides over the past two decades.

2.1.4 GaAs FET Model and Equivalent Circuit

In this section, we develop expressions for the elements of a simple GaAs FET equivalent circuit. Admittedly, this derivation is highly simplified, and the resulting

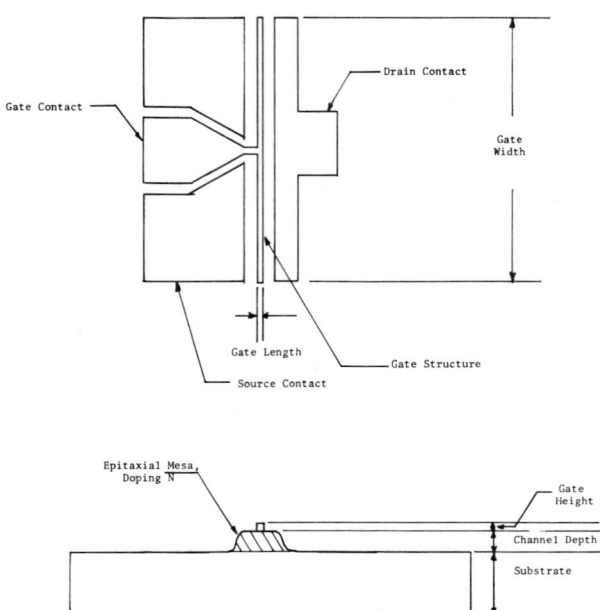

Fig. 2.3 The geometry of a GaAs FET with a t-gate.

expressions should be regarded as approximations. However, in spite of its simplicity, this basic model will offer invaluable insight into the characteristics of a wide variety of GaAs FET devices. Keep in mind that like any model, this simple model is no substitute for an experimentally derived device characterization.

2.1.4.1 Transconductance

The structure of a basic tee-gate GaAs FET is shown in Figure 2.3. Figure 2.4 shows the cross section of this FET's channel, exposing the placement of the active GaAs layer, the drain and the source contacts, and the gate. Transistor action occurs as a negative bias voltage is applied between the gate and the source, causing a depletion layer (a region of positive charge that cannot support electrical conduction) to be formed under the gate and extend more deeply into the active layer as the gate to source voltage is made more negative. Figure 2.5 gives a detailed diagram of the electric field, charge density, and electron drift velocity within the FET's channel. The essence of the FET's operation is that, as the gate voltage increases, the depletion layer extends deeper into the active layer, until it finally extends all the way through the active layer. Because a depletion layer supports no electrical conduction, charges traveling between the source and the

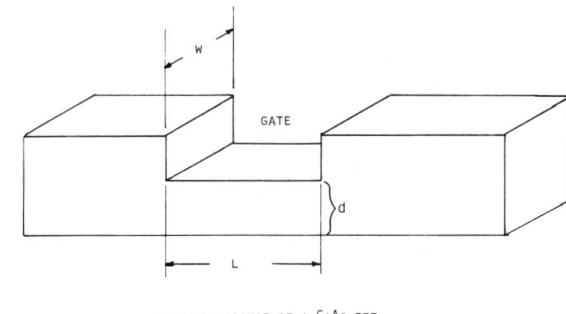

TRANCONDUCTANCE OF A GaAs FET

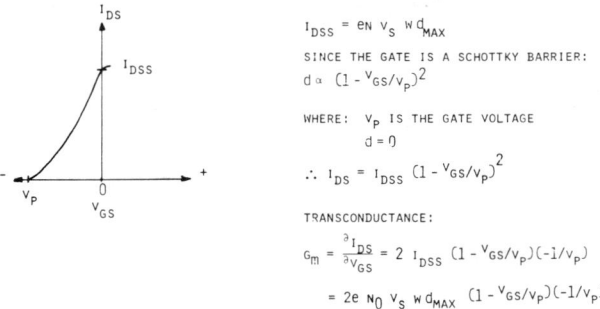

$I_{DSS} = eN\, v_s\, Wd_{MAX}$

SINCE THE GATE IS A SCHOTTKY BARRIER:
$d \propto (1 - V_{GS}/V_p)^2$

WHERE: V_p IS THE GATE VOLTAGE
$d = 0$

$\therefore I_{DS} = I_{DSS}\,(1 - V_{GS}/V_p)^2$

TRANSCONDUCTANCE:

$G_m = \dfrac{\partial I_{DS}}{\partial V_{GS}} = 2\, I_{DSS}\, (1 - V_{GS}/V_p)(-1/V_p)$

$= 2e\, N_0\, v_s\, W d_{MAX}\, (1 - V_{GS}/V_p)(-1/V_p)$

Fig. 2.4 The calculation of FET transconductance $= 2e\, N_0\, V_S\, W d_{MAX}\, (1 - V_{GS}/V_P)(-1/V_P)$.

drain are confined to an increasingly thinner layer of active material as the gate voltage increases, until all conduction stops at a critical point where the depletion layer extends completely through the active layer. This point is called *pinch-off*, and the corresponding gate to source voltage is called the *pinch-off voltage* (V_p).

The current flowing between the source and the drain contacts is controlled by the gate voltage from a maximum current (called I_{DSS}) for zero gate voltage, to zero current at a gate voltage equal to the pinch-off voltage (V_p). Calculated I–V characteristics of a GaAs FET are shown in Figure 2.6.

A measure of how the gate voltage controls the channel current is a parameter called *transconductance,* which is measured in siemens. Roughly, the transconductance of a GaAs FET, g_m, is equal to

$$g_m \doteq \frac{I_{DSS}}{V_p} \qquad (2.1)$$

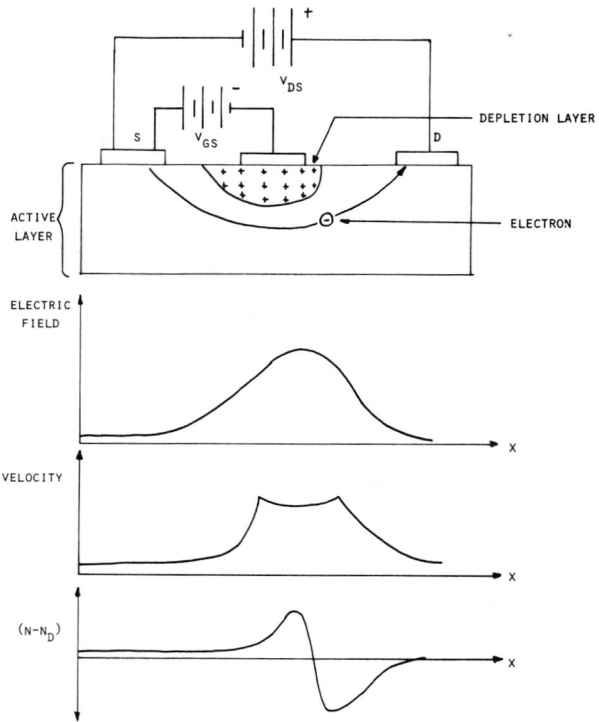

Fig. 2.5 Field effect transistor action.

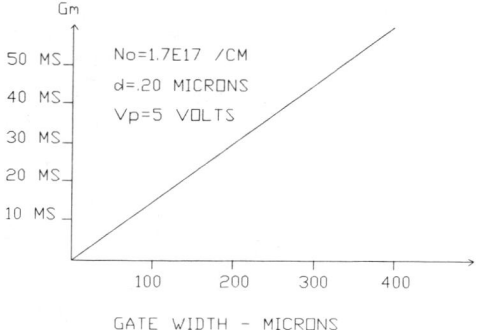

Fig. 2.6 The calculated transconductance of a GaAs FET.

A more accurate expression for transconductance can be developed by including the relationship for the depth of the depletion layer under the Schottky barrier gate. Referring to Figure 2.5, the maximum channel current, I_{DSS} is

$$I_{DSS} = en_0 v_s d_{max} \tag{2.2}$$

where
- e is the charge on an electron,
- n_0 is the active layer donor density,
- v_s is the electron drift velocity,
- w is the total width of the gate metal (the long direction),
- d_{max} is the total depth of the active layer.

The depth of the conducting channel under the gate is

$$d = d_{max} - \text{depletion depth}$$

For a Schottky gate, the relationship between d and gate voltage is[7]

$$d = d_{max}(1 - V_{GS}/V_p)^2 \tag{2.3}$$

Substituting this expression for the conducting channel depth into (2.2) yields an expression for drain current in terms of gate voltage:

$$I_{DS} = en_0 v_s w d_{max}(1 - V_{GS}/V_p)^2 \tag{2.4}$$

or

$$I_{DS} = I_{DSS}(1 - V_{GS}/V_p)^2 \tag{2.5}$$

The FET's transconductance may be found by taking the partial derivative of (2.5) with respect to V_{gs}

$$g_m = \frac{\partial I_{DS}}{\partial V_{GS}}$$

$$= 2I_{DSS}(1 - V_{GS}/V_p)\left(\frac{-1}{V_p}\right) \tag{2.6}$$

Notice the minus sign in the last term of (2.6). The minus sign indicates a 180° phase shift between gate voltage and drain current.

From (2.6) we see that the transconductance varies smoothly between $2I_{DSS}/V_p$ at $V_{gs} = 0$ to zero at $V_{GS} = V_p$. Compare this varying behavior to the constant value of transconductance predicted by (2.1). The transconductance of (2.1) is seen as the "average value" of the transconductance of (2.6). In practice, the transconductance of most FETs varies midway between the extremes of (2.1) and (2.6). In Figure 2.6, transconductance is plotted as a function of w for nominal conditions on n_0, V_p, and V_{gs}.

2.1.4.2 Gate Parasitic Elements

Ideally, a GaAs FET's equivalent circuit would be just a voltage-controlled drain current source, the value of which is the transconductance times the gate voltage. The ideal gate contact simply would be an open circuit, and the ideal drain output impedance also would be an open circuit. However, at microwave frequencies, real FETs behave somewhat differently. Parasitic elements such as gate capacitance and gate resistance dramatically alter the high-frequency performance of a GaAs FET. These elements must be taken into account if an equivalent circuit model is to be realistic. In Figure 2.7, a cutaway drawing of the active channel shows the regions associated with the parasitic elements.

Gate-to-source capacitance, C_{GS}, can be calculated in a straightforward way. Using the nomograph for uniformly doped GaAs Schottky barriers[8] shown in Figure 2.8, we can determine C', the gate capacitance per unit area (in pF/cm²)

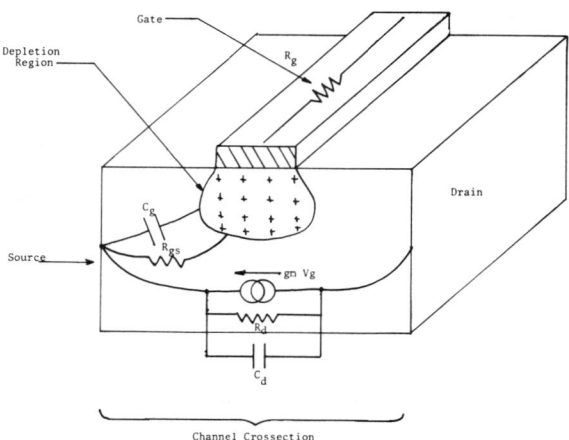

Fig. 2.7 Physical origins of the equivalent circuit of a GaAs FET.

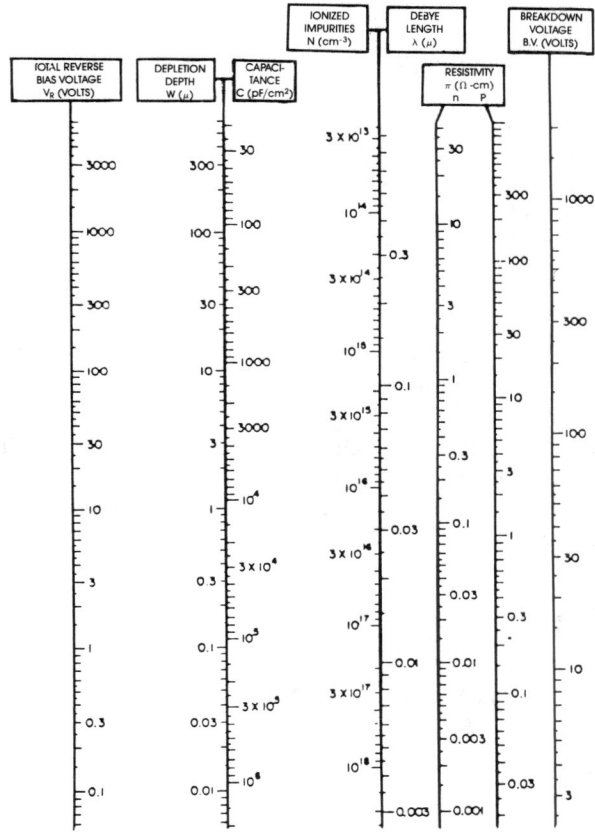

Fig. 2.8 Nomograph for a GaAs Schottky barrier. (From [8], p. 35. Reprinted with permission.)

by drawing a straight line between n_0, the ionized doping impurity density ($\sim 1.5 \cdot 10^{17}$ cm^{-2}) and the reverse bias voltage (~ 1.0 volts in normal operation). The gate capacitance then simply is the gate length, l, times the gate width, w, times C':

$$C_{GS} = l \cdot w \cdot C' \qquad (2.7)$$

In Figure 2.9, C_{GS} is plotted for various values of w and l for normal conditions on n_0 and V_{gs}.

Gate resistance is somewhat more difficult to calculate because it has two components. The first component is the resistance associated with the gate metal

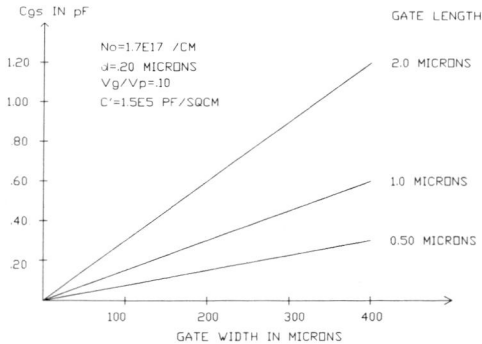

Fig. 2.9 Calculated plots of C_{gs} vs. gate width for FETs of various gate lengths.

itself, and the second component is the resistance of the undepleted GaAs between the gate and the source contacts. Let us consider the gate metal resistance first. Referring to Figure 2.10, the resistance of each half of a tee gate is modeled as that of an *R-C* ladder network:

$$R = \frac{\rho(w/2)}{3lh} = \frac{\rho w}{6lh}$$

where ρ is the resistivity of the gate metal, and h is the thickness of the gate metal.

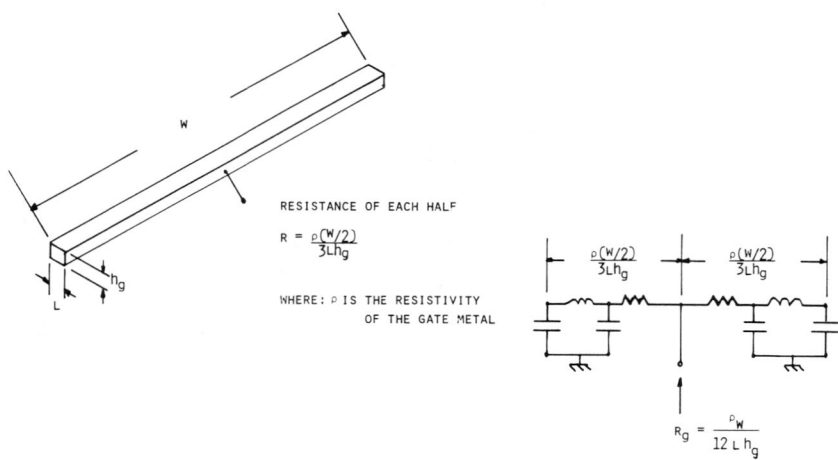

Fig. 2.10 Calculation of gate resistance.

By combining the two halves of the tee gate in parallel, the total gate metal resistance is

$$R_{G1} = \frac{\rho w}{12lh}$$

The component of the gate resistance associated with the undepleted GaAs is calculated, as shown in Figure 2.11, by assuming that the space between the ohmic source contact and the gate depletion region is one-half of the gate length, or $l/2$. The resistance associated with the undepleted material therefore is

$$R_{G2} = \frac{\rho_{GaAs}(l/2)}{wd_{max}}$$

Therefore, the total gate resistance is

$$R_G = R_{G1} + R_{G2} = \frac{\rho W}{12lh} + \frac{\rho_{GaAs}l}{2wd_{max}} \qquad (2.8)$$

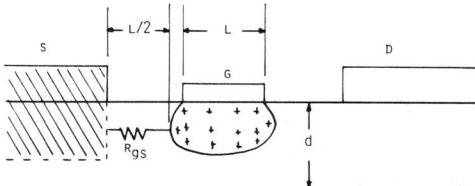

Fig. 2.11 The calculation of gate-to-source resistance in the undepleted GaAs.

Equation (2.8) shows that the gate resistance of a FET is composed of two factors, which vary in opposite ways in terms of w and l. Figure 2.12 is a graph of the behavior of R_G for various values of w and l, and nominal values of ρ, and ρ_{GaAs}, h, and d.

2.1.4.3 Drain Parasitic Elements

The drain region of a GaAs FET is modeled as a parallel resistor and capacitor combination. The resistor represents losses associated with the high but noninfinite output impedance of the FET, resulting from incomplete drain current saturation; and the capacitor represents the parallel plate capacitance between the source and the drain contacts.

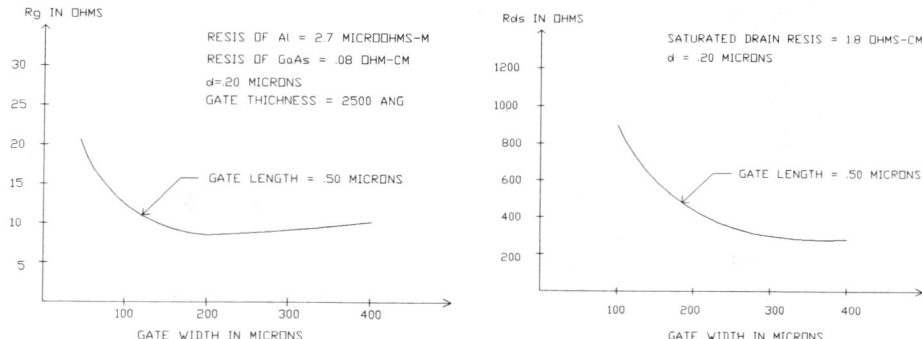

Fig. 2.12 Plots of the calculated values for R_g and R_{ds} of a Tee gate FET as a function of gate width, for a gate length of 50 microns.

Let us first consider the drain resistance. Referring to Figure 2.4, the resistance between the source and the drain for operation at $I_{DS} = .5L_{DSS}$ is

$$R_{DS} = \frac{\rho_s l}{w(d_{max}/2)} \tag{2.9}$$

where

ρ_s is the equivalent resistivity of the velocity saturated channel,
d_{max} is the total thickness of the active channel.

The value of ρ_s has been found experimentally to approximately equal 1.8 Ω-cm. This relationship is plotted in Figure 2.12, over a wide range of w. Although quite high in value, R_{DS}, nevertheless is a significant loss element in a FET's equivalent circuit and cannot be ignored.

The drain-to-source capacitance of an FET simply is the parallel plate capacitance between the drain and source contacts. This capacitance is

$$C_{DS} = \frac{\epsilon a}{L}$$

where

$\epsilon = \epsilon_R \epsilon_0,$
$\epsilon_R = 12.5$ in GaAs,
$\epsilon_0 = 8.85 \cdot 10^{-12}$ F/m,
$a = d_{max} w,$
$L \doteq 3l.$

Therefore, under these conditions, the drain-to-source capacitance is

$$C_{DS} = \frac{\epsilon_R \epsilon_0 d_{max} w}{3l} \tag{2.10}$$

Figure 2.13 is a plot of C_{DS} as a function of w.

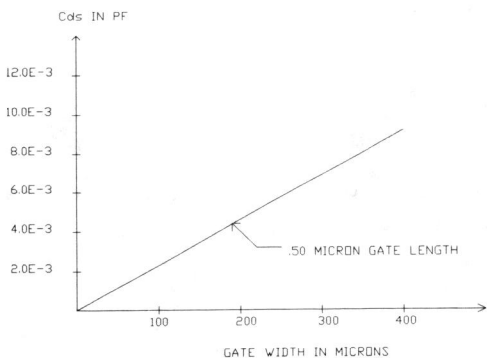

Fig. 2.13 A plot of the calculated values of C_{DS} vs. gate width for a FET with a .50 micron gate length.

2.1.4.4 Equivalent Circuit of a GaAs FET and Its Maximum Available Gain

The total equivalent circuit of a GaAs FET is shown in Figure 2.14. This circuit includes all of the elements derived in the last three sections. However, it is a simplified equivalent circuit and other parasitic elements such as drain, gate, source *contact* resistance, and gate-to-drain capacitance have been ignored. Even so, this simple equivalent circuit has proven to be very accurate in predicting the performance of GaAs FET devices operating in a wide variety of microwave amplifier and oscillator circuits.

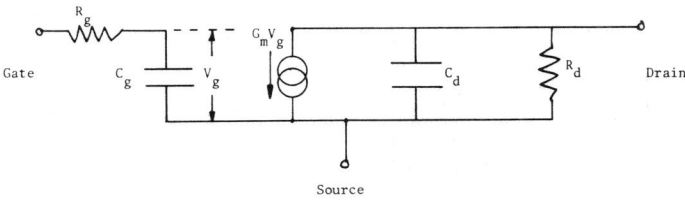

Fig. 2.14 The equivalent circuit of a GaAs FET.

An important performance parameter of a GaAs FET is its *maximum available gain* (MAG), which is the maximum power gain that can be achieved at each frequency assuming perfect input and output matching. MAG can be approximated by calculating the power gain of the simple equivalent circuit shown in Figure 2.14 at several frequencies, assuming that a perfect input and output match exists at each frequency. We use the simple equivalent circuit for this calculation, because a more complete equivalent circuit would lead to a mathematically intractable expression for MAG, adding little to our understanding. The result of this calculation is[8]

$$\text{MAG} = \frac{g_m^2}{(2\pi)^2} \left(\frac{R_{DS}}{R_G}\right)^{1/2} \left(\frac{1}{C_{GS}f}\right)^2 \tag{2.11}$$

To gain a feel for how a FET's geometry and material parameters affect MAG, we may substitute the expressions for g_m, R_{DS}, R_G, and C_{GS}, into (2.11):

$$\text{MAG} = \frac{[2I_{DSS}(1 - V_{gs}/V_p)]^2}{(2\pi)^2}$$

$$\left[\frac{\rho_s l/w(d_{max}/2)}{\left(\frac{\rho_w}{12lh} + \frac{\rho_{GaAs}l}{2wd_{max}}\right)}\right]^{1/2} \left(\frac{1}{l \cdot w \cdot C'f}\right)^2$$

where

$$I_{DSS} = en_0 v_s w d_{max}$$

Therefore,

$$\text{MAG} = \frac{\left[2en_0 v_s w d_{max}(1 - V_{GS}/V_p)\left(\frac{-1}{V_p}\right)\right]^2}{(2\pi)^2}$$

$$\left[\frac{\rho_s l/w(d_{max}/2)}{\left(\frac{\rho_w}{12lh} + \frac{\rho_{GaAs}l}{2wd_{max}}\right)}\right]^{1/2} \left(\frac{1}{lC'f}\right)^2 \tag{2.12}$$

As is clear from inspecting (2.12), to a good first approximation, MAG varies directly with the square of n_0 and inversely with the square of lf. This means that the easiest way to extend the frequency range of a GaAs FET is to simply reduce l.

Today, most GaAs FETs are designed with an l less than or equal to .5 μm, which is about the lower limit for optical photolithographic processing. By using alternate fabrication techniques, such as direct write electron-beam (E-beam) lithography, gates under .2 μm in length may be fabricated. A family of MAG curves calculated from (2.12) is shown in Figure 2.15. The MAG curve of a high performance microwave bipolar transistor has been added for comparison. Clearly, GaAs FETs have a significant power gain advantage over bipolar transistors at microwave frequencies.

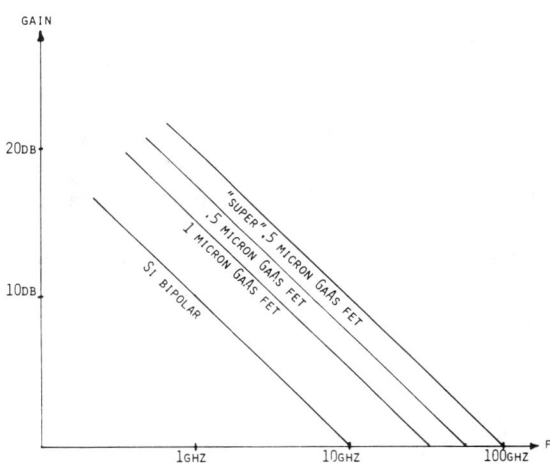

Fig. 2.15 MAG comparison for several devices.

2.1.5 Large-Signal Effects

GaAs FET devices operate according to the model developed here under only strict small-signal conditions. As signal levels rise, which is natural for power amplifiers and oscillators, the equivalent circuit elements will change their values, reflecting the nonlinearities that occur under large-signal conditions. To calculate certain important parameters such as power output and efficiency is impossible from a purely linear equivalent. For these reasons, to develop a simple large-signal "load line" FET model to facilitate the design of power amplifiers is very important.

Consider the family of characteristic I-V curves shown in Figure 2.16. Under large-signal conditions, the FET's voltage and current will describe a simple trajectory through this family of curves[9]. The maximum current, called I_m, is approximately 30 percent higher than I_{DSS}. This is because I_m occurs at positive

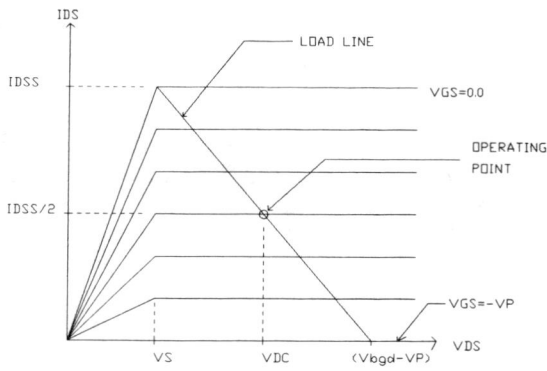

Fig. 2.16 Power, efficiency, and output resistance of a power FET calculated by the load line method.

gate voltage, completely eliminating the depletion layer and opening the channel fully.

The highest voltage that can exit between the drain and the source is the gate-to-drain breakdown voltage (V_{BGD}) minus the pinch-off voltage (V_p). This maximum voltage will be reached only as the drain current approaches zero. Current saturation occurs for only drain-to-source voltages above V_S. We assume that drain-to-source voltages below V_s are avoided because they imply high microwave losses in the nonsaturated region. Therefore, the FET's power generating ability is limited only by the parameters I_m, V_{BGD}, V_S, and V_p.

A load line analysis based on this family of characteristic curves can calculate the maximum RF drain-to-source voltage and the maximum RF drain current. By converting peak values to rms values and multiplying the maximum RF voltage by the maximum RF current, the maximum power output may be obtained.

The load line connects the point of maximum operating current (at $V_{DS} = V_S$ and $I_{DS} = I_m$) to the point of maximum voltage (at $V_{DS} = V_{BGD} - V_p$, and $I_{DS} = 0$), as shown in Figure 2.16. The power output is calculated as

$$P_{out} = \left(\frac{I_m}{2\sqrt{2}}\right)\left(\frac{V_{BGD} - V_p - V_S}{2\sqrt{2}}\right)$$

or

$$P_{out} = I_m\left(\frac{V_{BGD} - V_p - V_S}{8}\right) \tag{2.13}$$

where

I_m is the maximum drain current,
V_{BGD} is the gate-to-drain breakdown voltage,
V_p is the pinch-off voltage,
V_S is the drain saturation voltage.

The resistance of the load line is equal to the maximum RF voltage divided by the maximum RF current:

$$R_l = \frac{(V_{BGD} - V_p - V_S)}{I_m} \tag{2.14}$$

The dc power input, assuming the current is set at $I_m/2$, is

$$P_{dc} = \left(\frac{I_m}{2}\right)\left(\frac{V_{BGD} - V_p - V_S}{2} + V_S\right)$$

$$= \left(\frac{I_m}{2}\right)\left(\frac{V_{BGD} - V_p + V_S}{2}\right)$$

The power added efficiency is

$$\eta = \frac{\text{RF power output}}{\text{dc power input + RF power input}}$$

where RF power input is

$$\frac{1}{g}(P_{out}) = I_m \frac{(V_{BGD} - V_p - V_S)}{8g}$$

The term g is the amplifier's gain, and dc power input is

$$V_{dc} \times I_{dc} = \left[\frac{V_{BGD} - V_P + V_S}{2} + V_S\right]\frac{I_m}{2}$$

$$\eta = \frac{I_m(V_{BGD} - V_p - V_S)/8}{\left[\dfrac{V_{BGD} - V_p + V_S}{2}\right]\dfrac{I_m}{2} + \dfrac{I_m}{8g}(V_{BGD} - V_p - V_S)}$$

After some manipulation,

$$\eta = \frac{(V_{BGD} - V_p - V_S)(1 - 1/g)}{2(V_{BGD} - V_p + V_S)} \tag{2.15}$$

Equation (2.15) tells us that V_{BGD} is the major parameter controlling η. The term I_m does not enter into efficiency. Plots of P_{out} and η as a function of V_{BGD} are shown in Figure 2.17. Plots of P_{out} and R_l are shown in Figure 2.18. Notice that increasing I_m increases power but not efficiency. Increasing V_{BGD} increases both power and efficiency. For this reason, a chief objective when designing a high power, high efficiency FET is maximum V_{BGD}.

2.2 BIPOLAR TRANSISTORS

Since the invention of the transistor by William Schockley, Walter Brattain, and John Bardeen in 1947, processing technique improvements have led to progressively higher-frequency operation. By the 1970s, the use of ion implantation and electron-beam lithography had narrowed the base thickness to the point that true microwave frequency bipolar transistors became commercially available. Today, devices are readily available that oscillate to 10 GHz and beyond. These devices are most useful in oscillators, owing to their extremely low $1/f$ noise, and the relative ease with which feedback is applied to create the conditions for oscillation.

Today, most amplifier applications use GaAs FET devices, because of their inherently higher gain and higher frequency performance relative to bipolar transistors. However, for oscillator applications below 10 GHz, the bipolar transistor is a strong candidate because of its low oscillator noise and the relative ease with which it is configured as an oscillator.

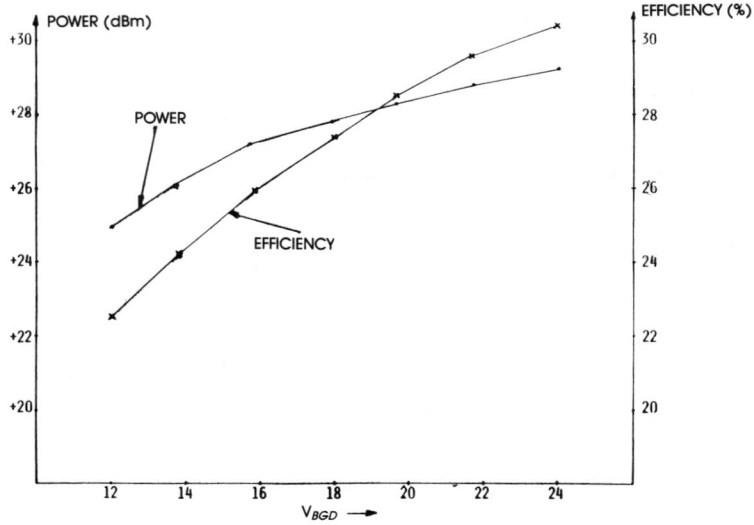

Fig. 2.17 Power output and efficiency *vs.* V_{BGD}, $I_{DSS} = 400\ MA$, $V_P = 4V$, $V_S = 2V$.

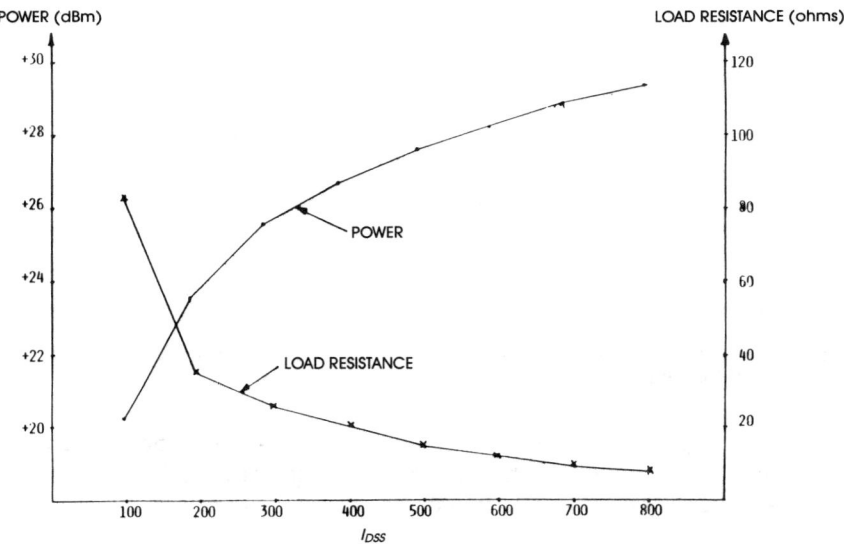

Fig. 2.18 Power and load resistance vs. I_{DSS} for $V_{BGD} = 15V$, $V_P = 4V$, $V_S = 2V$.

Most microwave frequency bipolar transistors are of the NPN type, and all are fabricated in silicon. The structure of a typical bipolar transistor is shown in Figure 2.19. Base and emitter contact fingers are interdigitated across the top of the device. The collector region is buried within the device and contacted at the back side of the chip. The base and emitter "fingers" are connected at common bonding pads as shown in Figure 2.20.

Bipolar transistors operate with a forward-biased emitter-base junction and a reverse-biased collector-base junction. Electrons from the emitter region are injected into the P-type base region. By diffusion, these electrons enter the collector region, where they are swept along through the base collector depletion zone to the collector region by the collector's positive potential. The current is amplified because almost all minority carriers (electrons) injected into the base region from

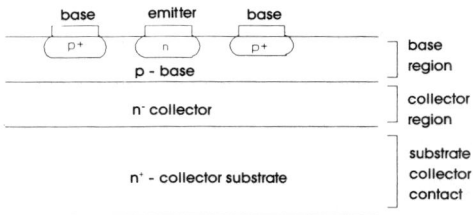

Fig. 2.19 The structure of a microwave bipolar transistor.

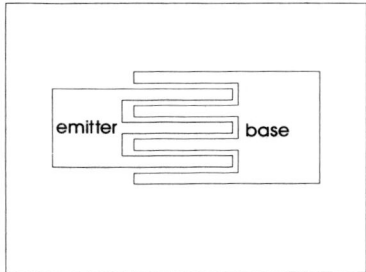

Fig. 2.20 Top side bonding pad pattern of a microwave bipolar transistor showing interdigitated base and emitter fingers.

the emitter region are transferred to the collector region. These electrons raise the collector's current quite high. The collector current is under the direct control of the base-to-emitter voltage, which means the collector current is controlled by the base current as the base-emitter junction is forward biased.

A simple equivalent circuit model for a microwave bipolar transistor[10] is shown in Figure 2.21. The elements in this circuit model are

r_b = base resistance,
r_e = emitter resistance,
C_e = emitter depletion capacitance,
C_c = collector depletion capacitance,
β = base-to-collector current gain.

The unity gain cut-off frequency of a bipolar transistor is given in terms of the various time delays[11]:

$$f_T = \frac{1}{2\pi\tau} \qquad (2.16)$$

where

$$\tau = \tau_e + \tau_b + \tau_{b,c} + \tau_c$$

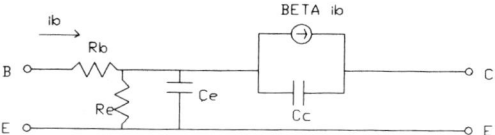

Fig. 2.21 A simple equivalent circuit of a microwave bipolar transistor.

and

τ_e = emitter-base junction charging time,
τ_b = base transit time,
$\tau_{b,c}$ = base-collector depletion zone transit time,
τ_c = collector-base junction charging time.

Microwave bipolar transistors are best modeled using S-parameter data based on actual measurements. The data usually can be obtained from the manufacturer of the transistor. If not, the data must be measured by the designer, using vector network analyzer techniques and a carefully designed test fixture to ensure accurate calibration and error correction.

2.3 VARACTOR DIODES

2.3.1 Structure

Varactor diodes are voltage controllable variable capacitance elements. Varactors are used to control the frequency of many types of oscillators and tunable filters. The basic structure of a varactor diode is shown in Figure 2.22. Varactor diodes are constructed from both silicon and gallium arsenide. In cross section, these diodes consist of a substrate (usually N^+ type material), an active layer of N type material and a thin P^+ layer followed by an ohmic contact. The active layer is in a necked portion of the diode called a *mesa,* whose purpose is to provide diode area control by selective etching.

Figure 2.23 shows the doping profile of an abrupt junction varactor diode.

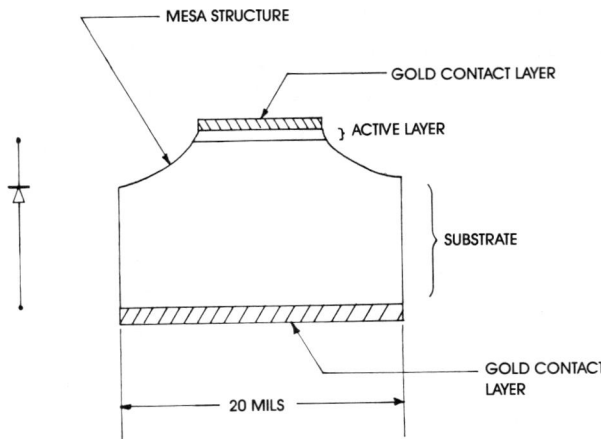

Fig. 2.22 Cross-sectional diagram of a varactor chip.

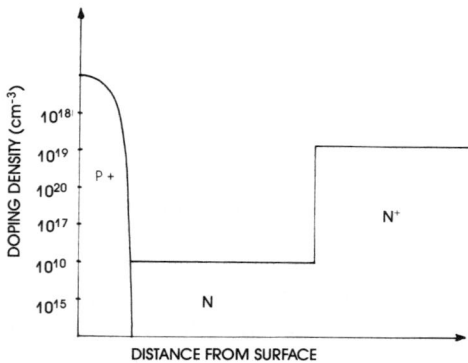

Fig. 2.23 Doping profile of an abrupt junction diode.

The heavily doped P region is closest to the top contact of the diode. The doping level changes abruptly at the junction to a region of flat N^- type doping and then changes abruptly to flat N^+ doping in the diode's substrate.

2.3.2 Varactor Operation

Figure 2.24 diagrams the internal electric field and space charge density within a varactor diode. A varactor is a junction diode that will support current flow when forward biased above its barrier potential. However, under reverse-bias conditions, a region of space charge, called a *depletion region,* forms within the N^- region of the diode. The width of this depletion region increases with increasing reverse bias, until it extends across the entire N region. This depletion region is positively charged and matched by an equal amount of negative charge (electrons) in the P region. As the reverse-bias voltage increases, these charge layers move apart, causing an effect similar to moving apart the conductive plates of a parallel

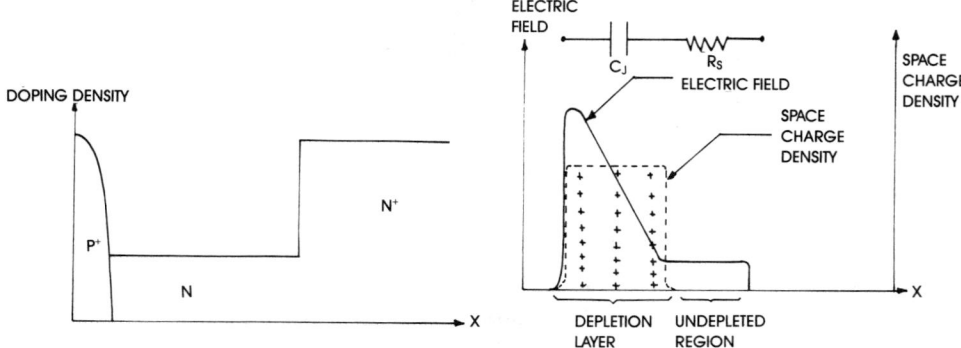

Fig. 2.24 Region of depletion within a varactor diode.

plate capacitor. As the charges are depleted, by increasing voltage, the capacitance of the diode decreases according to the relationship

$$C = \frac{\epsilon a}{d}$$

where d is the nominal separation between charges and a is their effective area.

At low reverse voltage, much of the N⁻ region is not depleted. This undepleted N-doped material determines the series resistance of the diode. As the reverse voltage increases, the depletion region extends over a greater portion of the N region, decreasing the diode's series resistance.

The relationships for the capacitance and resistance of an abrupt junction varactor diode are[12]

$$C(V) = a \left[\frac{\epsilon e n}{2(\phi - V)} \right]^{1/2} \tag{2.17}$$

$$C(0) = a \left[\frac{\epsilon e n}{2\phi} \right]^{1/2} \tag{2.18}$$

$$r_s(V) = \frac{l - w}{e n \mu a} \tag{2.19}$$

$$r_s(0) = \frac{l - \sqrt{2\epsilon\phi/en}}{e n \mu a} \tag{2.20}$$

$$w = \left[\frac{2\epsilon(\phi - V)}{en} \right]^{1/2} \tag{2.21}$$

$$w(0) = \left[\frac{2\epsilon\phi}{en} \right]^{1/2} \tag{2.22}$$

where
- a = junction area,
- ϵ = semiconductor dielectric constant,
- n = active region doping,
- V = bias voltage,
- ϕ = barrier potential,

e	=	electronic charge,
l	=	active region length,
w	=	depletion layer width,
μ	=	semiconductor mobility in the active region,
$C(V)$	=	varactor capacitance,
$r_s(V)$	=	series resistance.

The hyperabrupt junction varactor diodes have a nonuniform N^- region doping profile, which is tailored so that the capacitance changes more rapidly with bias voltage than in abrupt junction diodes. The profile of a hyperabrupt junction varactor is shown in Figure 2.25. Figure 2.26 compares the capacitance of a hyperabrupt diode with the capacitance of an abrupt diode. Notice that the hyperabrupt diode's capacitance changes more rapidly than that of an abrupt junction diode only over a limited range of bias voltage. However, the increased frequency tuning rate that can be achieved in this region could be very useful in some applications.

The capacitance of any varactor diode (either abrupt or hyperabrupt) as a function of voltage can be written as a simple generalized expression:

$$C(V) = aK\left(\frac{n}{V + \phi}\right)^\Gamma \qquad (2.23)$$

where

a is the cross-sectional area of the diode,
K is a constant,
n is the average doping of the active region,
ϕ is the built-in potential (approximately .70 V),
Γ is a slope parameter (about .50 for an abrupt junction and about 1.0 for a hyperabrupt junction diodes).

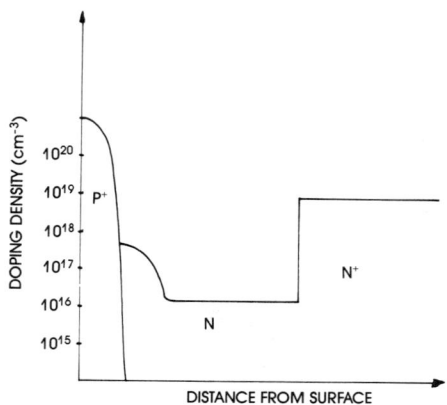

Fig. 2.25 Doping profile of a hyperabrupt junction diode.

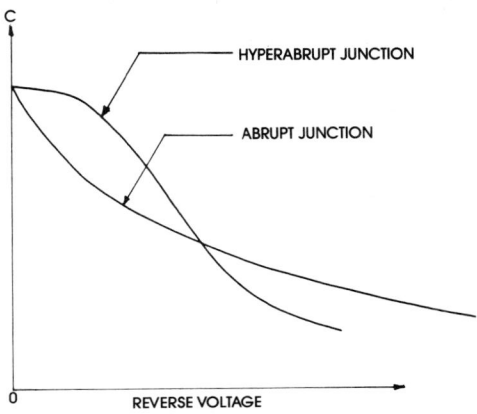

Fig. 2.26 Capacitance tuning characteristic for a general varactor.

Figure 2.27 compares the frequency tuning curve of an abrupt junction with that of a hyperabrupt junction varactor.

More difficult about which to generalize is the behavior the series resistance, R_S. However, mobility plays a key role in reducing R_S. High-mobility materials such as GaAs can significantly reduce R_S. We may calculate the difference in series

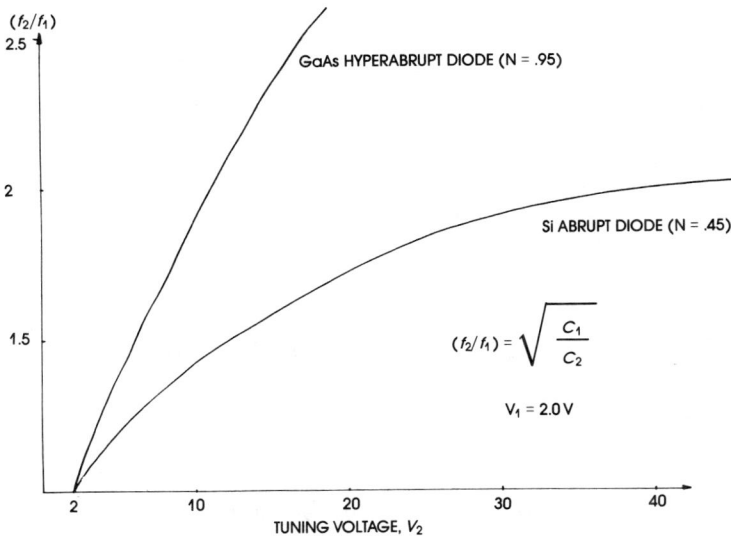

Fig. 2.27 A comparison of varactor tuning ratios.

resistance with Si and GaAs diodes based on mobility difference. In general, the resistance of the undepleted N material is

$$R_S = \frac{\rho l}{a}$$

where l is the length of the undepleted region, and

$$\rho = \frac{1}{en\mu}$$

so

$$R_S = \frac{l}{en\mu a}$$

GaAs mobility is roughly four times greater than Si mobility[13]:

$$\mu_{GaAs} \doteq 4\mu_{Si}$$

so

$$\frac{R_S(\text{silicon})}{R_S(\text{GaAs})} \doteq 4$$

The quality factor of a varactor diode is

$$Q = \frac{1}{2\pi f R_S C} \qquad (2.24)$$

Therefore,

$$Q_{GaAs} \doteq 4Q_{Si}$$

The cut-off frequency, f_C, is the frequency where the Q becomes 1.0:

$$f_C = \frac{1}{2\pi R_S C} \qquad (2.25)$$

Normally, varactor manufacturers specify Q at 50 MHz. The value of Q may be scaled to other frequencies by using the relationship

$$Q(f_1) = Q(f_2)\frac{f_2}{f_1}$$

Varactors are sensitive to RF power and must be operated below a critical RF "saturation" level to avoid the production of objectionable harmonic signals. This critical RF level can be calculated by recognizing that rectification and harmonic generation will occur if the RF voltage is allowed to swing above the forward knee voltage of the diode. Referring to Figure 2.28 and assuming that the dc voltage is zero (worst case) and the forward knee voltage, $V_f = .7$ V:

$$\text{maximum RF power} = \frac{V_f^2}{r}$$

$$= \frac{(.7/1.41)^2}{50}$$

$$= 5 \cdot 10^{-3} \text{ W or } + 7 \text{ dB/m}$$

Therefore, at zero bias voltage, the RF power impinging on a varactor diode must be less than +7 dBm to avoid harmonic generation.

The capacitance change with temperature of a varactor diode may be calculated from (2.23). Simplifying (2.23) slightly,

$$C(V) = \frac{C(0)}{(V + \phi)^\Gamma}$$

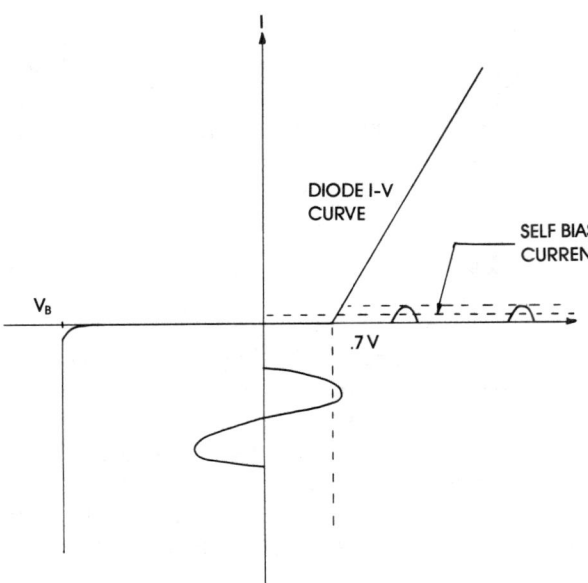

Fig. 2.28 Maximum input power to avoid self-bias effects, for tuning voltage equal to zero V.

Taking the partial derivative of capacitance with respect to temperature:

$$\frac{\partial C(V)}{\partial t} = \frac{-\Gamma C(0)}{(V + \phi)(V + \phi)^\Gamma} \left(\frac{\partial \phi}{\partial t}\right) \quad (2.26)$$

where

$$\frac{\partial \phi}{\partial t} = 2.3 \text{MV/}^\circ\text{C for silicon}$$

$$\phi = .70\text{V}$$

Some observations about the temperature slope are that it is directly proportionate to Γ and inversely proportional to the applied voltage. This means that hyperabrupt varactors biased at zero volts will have more temperature slope than abrupt junction varactors biased at high-reverse voltage. Applications where temperature-frequency stability is critical require a careful choice of varactor diode and bias point.

2.4 PIN DIODES

2.4.1 Structure

A PIN diode is composed of three semiconductor layers: a heavily doped P region; a high resistivity intrinsic, or I region; and a heavily doped N region. The I region of a PIN diode is typically 10 to 200 μm thick. Ohmic contacts on the P and N layers connect the diode to its circuit environment. Figure 2.29 shows the structure of a typical PIN diode.

2.4.2 Operation of a Pin Diode

PIN diodes are P-N junctions that contain a relatively thick I region in the middle of the P-N junction. PIN diodes are capable of passing forward dc current by carrier diffusion. Once sufficient forward voltage is applied to the diode to lower the built-in barrier potential, holes are injected into the I region from the P region and electrons are injected into the I region from the N region. These carriers diffuse across the I region in opposite directions. Because PIN diodes are quite thick, the transit time frequency associated with the carriers traveling across the I region is relatively low. Only at frequencies *below* the transit time frequency does the diode function like a normal P-N junction; for frequencies above the transit time fre-

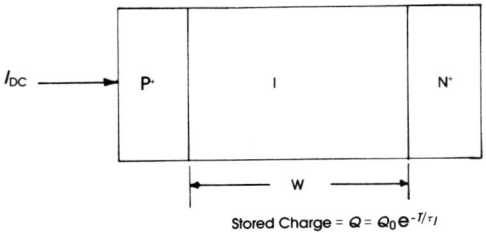

Fig. 2.29 A calculation of the high-frequency resistance of a PIN diode.

quency, a PIN diode's behavior will be quite different. Following White[14], the transit time for holes is given by

$$\tau_t = \frac{w^2}{D_P} \qquad (2.27)$$

where
    w is the width of the I region,
    D_P is the diffusion constant for holes.

Holes are the slower moving carriers in silicon, therefore their transit time frequency will be lower. From the Einstein relationship for diffusion constants,

$$D_P = \mu_P \frac{kT}{e}$$

where
    D_P is the hole diffusion constant,
    μ_P is the hole mobility,
    k is the Boltzmann's constant,
    T is temperature (kelvin),
    e is electronic charge.

Therefore,

$$f_t = 1/\tau_t = \frac{\mu_P kT/e}{w^2} \tag{2.28}$$

For silicon,

$$\mu_P = 500 \text{ cm}^2/\text{V} - \text{S}$$
$$T = 300 \text{ K}$$
$$f_t = \frac{1300}{w^2}$$

This works out to about 30 MHz for $w = 10$ μm.

Above f_t, a PIN diode's conductivity becomes very different from the classic diode behavior it exhibits at frequencies below f_t. The storage of charge within the I region largely is responsible for high-frequency conductivity. Holes and electrons injected into the I region recombine in a certain characteristic time, called the *lifetime*. Charge in the I region will reach a steady-state balance between charge decay from recombination and charge replenishment due to bias current. The hole and electron charge densities are given by

$$Q_P = I_0 \tau_P \tag{2.29}$$

$$Q_N = I_0 \tau_N \tag{2.30}$$

where

τ_P is the hole recombination lifetime,
τ_N is the electron recombination.

Note that the long lifetime does not imply slow switching speed. A properly designed current driver can remove the charge from the I region (switching the diode) in a time well under the lifetime. Lifetime is simply a measure of crystal purity because impurities in the material's crystal lattice provide recombination sites. Typical lifetimes are in the .1 to 20 μs range.

The high-frequency electrical conductivity of the I region (and the diode as a whole) can be calculated from the basic expression for the resistance of a three-dimensional (3D) cylinder:

$$R = \frac{w}{\sigma a}$$

where

    a is the cylinder's (the I region's) cross-sectional area:

$$\sigma = e(\mu_P P + \mu_N N)$$

    w is the cylinder's length (the I region's thickness)

Assuming that $P = N$ and $\tau = \tau_P = \tau_N$, meaning that equal hole and electron densities are injected into the I region, and because holes and electrons recombine only with one another, their lifetimes are functionally equal. The active resistance of a PIN diode is

$$R = \frac{w}{2ea\mu_{AP}P} \tag{2.31}$$

where

$$\mu_{AP} = \frac{2\mu_P \mu_N}{\mu_P + \mu_N}$$

$$\doteq 610 \text{ cm}^2/\text{V-S in silicon}$$

Using (2.29), we can calculate the hole charge density

$$Q_P = ePaw = I_0 \tau \tag{2.32}$$

Combining (2.31) and (2.32) yields

$$R = \frac{w^2}{2\mu_{AP}\tau I_0} \tag{2.33}$$

Notice that the active resistance depends only on lifetime, current, and thickness, not on area. Resistance is inversely proportional to bias current, I_0. This means that at low-bias current, a PIN diode is nearly an open circuit; whereas at high-bias current, a PIN diode is nearly a short circuit. This behavior makes the PIN diode an ideal microwave control element because its high-frequency resistance can be controlled conveniently over a *very* wide range with bias current as shown in Figure 2.30.

The equivalent circuit of a PIN diode is completed by adding a small value of fixed contact resistance, R_S, in series with the active I region of the diode. Also, the parallel plate capacitance of the I region shunts the active resistance in the equiva-

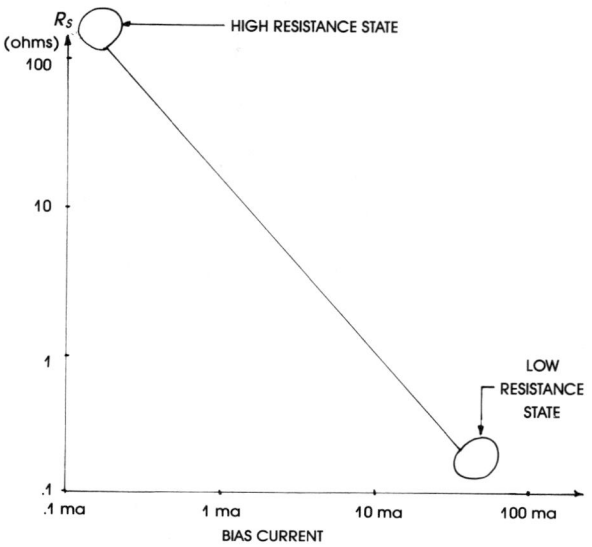

Fig. 2.30 Typical PIN resistance behavior with bias current.

lent circuit. This capacitance, c_j, may be calculated from the formula for the capacitance of a cylinder:

$$c_j = \frac{\epsilon_R \epsilon_0 \pi d_j^2}{4w} \quad (2.34)$$

where
ϵ_0 = 8.85 · 10^{-14} F/cm
ϵ_R = 11.8 for silicon
d_j = junction diameter
w = width of I region

The complete equivalent circuit of a PIN diode is shown in Figure 2.31.

2.5 YIG RESONATORS

2.5.1 Basics of Ferrimagnetism

Magnetically tunable microwave resonators for oscillator and filter applications can be constructed from ferrimagnetic materials such as *yittrium iron garnet* (YIG). Such resonators use the precessing motion of atomic spins to couple energy

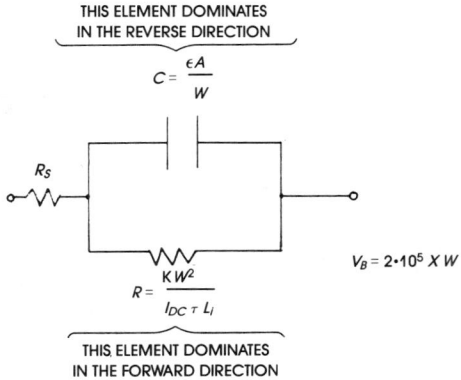

Fig. 2.31 The equivalent circuit of a PIN diode.

to and from a microwave frequency magnetic field. Spin is a purely quantum mechanical property of the electrons in orbits about the nucleus of an atom. Atomic electron spin can be visualized as similar to the rotation of planets in orbit about the sun. The spin precession frequency is determined precisely by a dc magnetic field. This means that a very carefully *controlled* microwave resonator can be constructed using these materials.

YIG, the chemical formula of which is $Y_2Fe_2(FeO_3)_3$, has a cubic crystal structure. Its Curie temperature is about 257°C, and its net magnetization is 5 Bohr magnetrons per formula unit[15]. The iron atoms within the atomic structure of YIG are responsible for its microwave resonance properties.

Consider for a moment the atomic structure of an iron atom, which is shown in Figure 2.32. The iron atom's magnetic properties result from a spin imbalance that occurs in the 3D orbital. Normally each orbital has an equal number of electrons with spin up and spin down. The imbalance in an iron atom's 3D orbital occurs in iron, because the 4S orbital happens to be at a lower energy than 3D, and consequently fills in first. According to Hund's rule[16], when an incomplete orbital (one that is filled out of sequence) does fill with electrons, it will do so in a way that maximizes spin imbalance. This process creates four unbalanced spins in 3D level of an iron atom. This net unbalanced spin leads to a natural magnetization within the ferrimagnetic material, such as iron.

A magnetic moment results from both net spin and the orbital motion of electrons. Because these motions are quantum mechanical in nature, the value of magnetic moment for an orbital electron is quantified according to the expression:

$$\mu_S = \mu_B N \qquad (2.35)$$

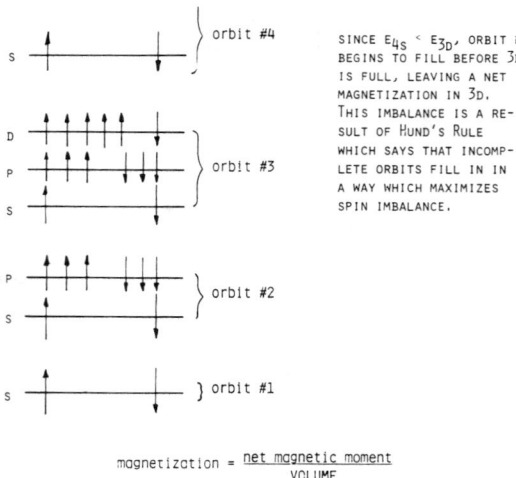

Fig. 2.32 Magnetization of (Fe^{26}).

where

$$\mu_B = \frac{eh}{4\pi mc} = \text{a Bohr magnetron},$$
N = a quantum number,
h = Planck's constant,
m = the mass of an electron,
c = the speed of light.

For iron, the value that N can assume is

$N = 2$ for orbital contributions

$N = 4$ for spin contributions

A very important motion, called *precession,* occurs when an atom with a net magnetic moment is subjected to both a dc and an ac magnetic field, which are at right angles to each other (see Figure 2.33). The magnetic moment within an iron atom can be thought of as a spinning top with a magnet imbedded in it. As the dc field is increased, the axis of the top is drawn into closer alignment with the dc field. As the magnetic moment is drawn closer to the axis of the dc magnetic field, its rotational velocity becomes higher. Increased rotational velocity means an increase in the frequency at which energy is coupled from the ac magnetic field to the

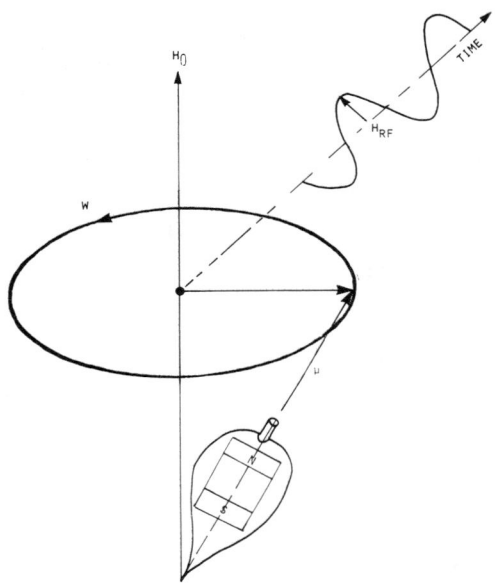

Fig. 2.33 Spin precession in YIG.

magnetic moment. This means the resonant frequency of the precessing magnetic moment is controlled by the dc magnetic field. See Figure 2.34 for a graphic depiction of this motion.

A convenient way to apply the dc and ac magnetic fields is to use a coupling loop around a sphere of YIG material. A dc magnetic field, applied in the plane of the coupling loop, and RF current, passed through the loop, will create an ac magnetic field at right angles to the dc magnetic field (see Figure 2.35).

The rate of precession, which is the resonant frequency, may be expressed as

$$f_R = \gamma \{H_O + H_A + (N_t - N_Z)M_S\} \tag{2.36}$$

where

γ = the gyromagnetic ratio = 2.8 MHz/Oe,
H_0 = dc magnetic field in Oe,
H_A = an anisotropic field, which depends on crystal orientation in Oe,
$(N_t - N_z)$ = a sample shape factor,
 $(N_t - N_Z)$ equals zero for a sphere,
M_s = net magnetization of the material.

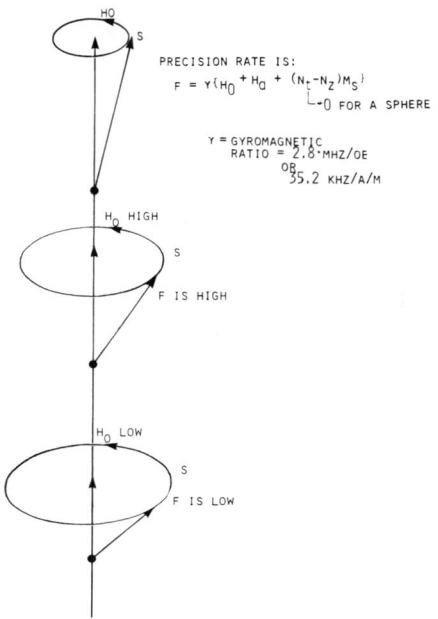

Fig. 2.34 Spin precession causes YIG resonance.

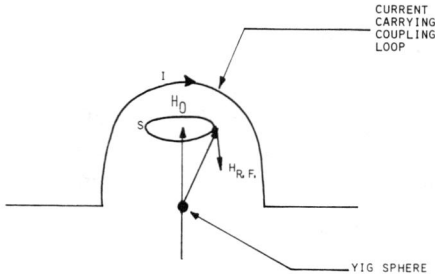

Fig. 2.35 Coupling to a YIG sphere.

Equation (2.36) predicts that the resonant frequency is very nearly equal to the dc magnetic field, H_0, times a universal constant γ. This means YIG resonators may be controlled very precisely with the magnetic field of an electromagnet.

2.5.2 Mode Behavior of the YIG Resonance

Iron atoms occur at regularly spaced locations within the lattice structure of a YIG crystal. The resonant behavior of a YIG sample depends on how all the magnetizations associated with the iron atoms move relative to each other in response to the magnetic field. Many relative magnetic motions of neighboring iron atoms are possible, these different motions are called *modes*. The simplest mode is the uniform precession mode, in which all the magnetizations of all the iron atoms precess in step (see Figure 2.36). Their magnetic moments are all aligned, and they precess in "locked step." The uniform precession mode, called the *110 mode*, is the mode of normal YIG resonator operation because its resonant frequency is given by (2.36). Higher-order modes, corresponding to more complicated relative motions have different resonant frequencies that tune with the dc magnetic field in more complicated ways than the 110 mode tunes.

A chart of many principal YIG "modes" is shown in Figure 2.37; this is the so-called magnetostatic mode chart[17]. Problems occur in practical YIG devices whenever a non-110 mode and the 110 mode occur at the same resonant frequency for a given magnetic field. The result is an oscillator that has small "glitches" in

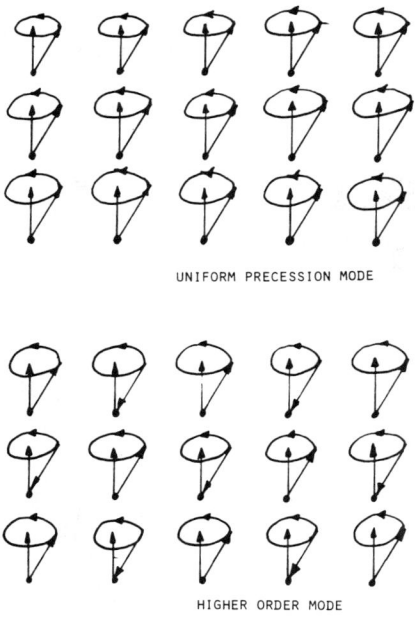

Fig. 2.36 Modes of a YIG sphere.

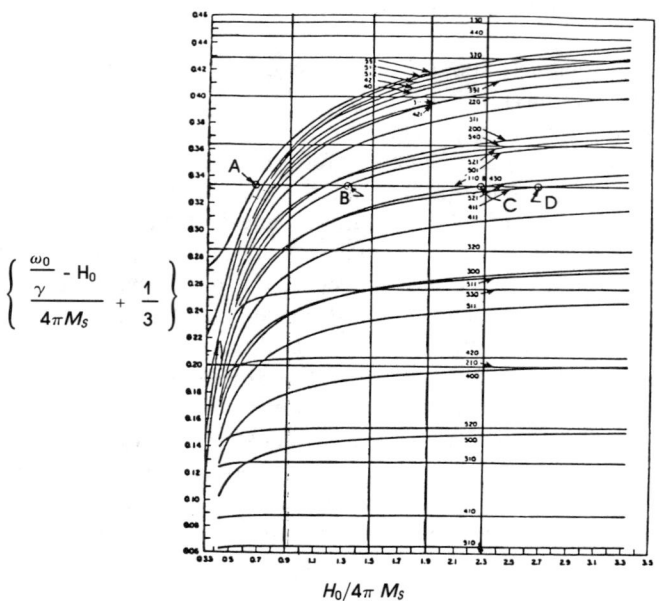

Fig. 2.37 Magnetostatic mode pattern ferrimagnetic resonance. (From [17], p. 695. Reprinted with permission.)

power as shown in Figure 2.38, and small discontinuities in frequency. This phenomenon is called a *crossing mode*.

Higher-order modes (non-110 modes) are a serious practical problem for YIG devices. These modes usually are excited by gradients occurring in the RF field profile that appears across the diameter of the YIG sphere. For instance, the shape of the net ac magnetization required to excite the 210 mode is shown in Figure 2.39.

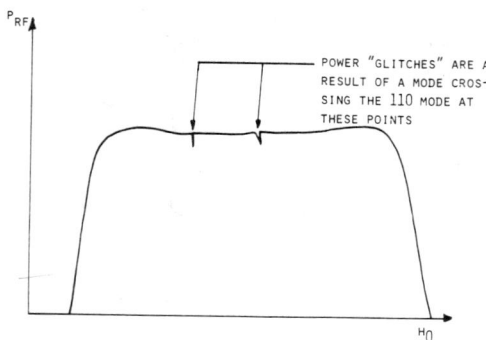

Fig. 2.38 The appearance of modes in a YIG oscillator's performance.

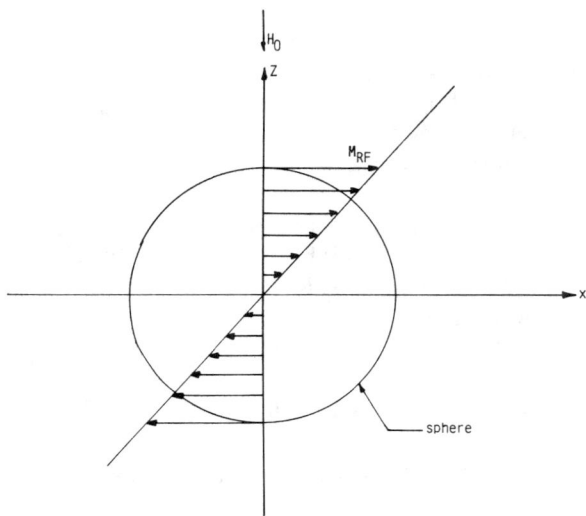

Fig. 2.39 Distribution of the RF magnetization within a YIG sphere during the 210 mode. This mode is excited by a gradient of the RF H field in the Z direction.

This mode can be minimized by placing the sphere in a region within the coupling loop where the ac magnetic field is *very* uniform spacially, as shown in Figure 2.40.

Sometimes higher-order modes can be eliminated by choosing a sphere with a particular YIG material magnetization, $4\pi M_s$. The term $4\pi M_s$ is a measure of the density of iron atoms within the YIG crystal. If a YIG crystal is doped with certain elements such as gallium, these doping atoms will replace the iron atoms at

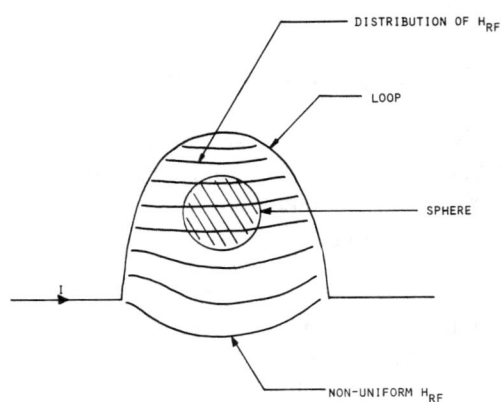

Fig. 2.40 Proper placement of sphere to avoid modes: Sphere *must* be placed in a region of uniform H_{RF} and H_0 to avoid modes other than 110.

occasional lattice sites. The doping process reduces the YIG material's $4\pi M_s$ by reducing the density of iron atoms. For pure YIG material, $4\pi M_s$ is 1780 gauss but may be reduced to as low as 200 gauss by heavy gallium doping. Assuming a selection of differently doped spheres is available, the tabulated data in Table 2.1 will enable the designer to move a troublesome coincidence frequency between the 110 mode and a higher-order mode either above or below the band of operation simply by changing the $4\pi M_s$ of the sphere. Assume that a 4 to 8 GHz YIG oscillator (using pure YIG material) has a glitch at 5.886 GHz due to a coincidence between the 110 and the 200 modes. Referring to Table 2.1, the 200 column tells us that if the value of $4\pi M_s$ is lowered to 1000 gauss, the coincidence frequency is lowered to 3.270 GHz, which is below the desired band of operation.

2.5.3 The Equivalent Circuit of a YIG Resonator

The equivalent circuit of a YIG resonator may be calculated from the following expressions[18]:

$$L_Y = \frac{(4\pi M_S) D_S^3}{42.77\pi^3 D_L^4 \omega} \left(\frac{11803}{\omega/2\pi}\right)^2 \left[\sin\frac{\omega L2}{11803} - \sin\frac{\omega L1}{11803}\right]^2$$

$$C_Y = \frac{1}{\omega^2 L_Y}$$

$$R_Y = \omega L_Y Q_\mu$$

$$Q_\mu = \left(\frac{\omega}{2\pi\gamma} - \frac{4\pi M_S}{3}\right) \Delta H$$

where
- L_Y is the YIG sphere's equivalent inductance in μH
- C_Y is the YIG sphere's equivalent capacitance in μF
- R_Y is the YIG sphere's equivalent resistance in ohms
- $4\pi M_S$ is the material's saturation magnetization, in gauss
- D_S is the YIG sphere's diameter, in inches
- D_L is the coupling loop's diameter, in inches
- L_1 is the length of the coupling loop, from the short circuited end to the beginning of coupling, in inches
- L_2 is the length of the coupling loop, from the short circuited end to the end of coupling, in inches
- Q_u is the unloaded Q, of the YIG sphere
- γ is the gyromagnetic ratio (2.8 MHz/Oe)
- ΔH is the YIG material's line width in Oe
- ω is the angular resonant frequency

Table 2.1 Higher-Order Mode Coincidences with the 110 Mode as a Function of Magnetization

	Higher-Order Magnetostatic Modes						
	$\overline{421}$	$\overline{311}$	521	200	501	$\overline{521}$	411
H4πM$_S$ for Resonance with Fundamental Mode	.8447	.9325	1.167	1.168	1.258	2.233	2.667
4πM (gauss)	Coincidence of Resonant Frequencies of Higher-Order Modes with Fundamental Mode as Function of 4πM$_S$						
1800	4257.3	4699.8	5881.7	5886.7	6340.3	11254.3	13441.9
1600	3868.7	4270.9	5344.9	5349.4	5761.6	10227.1	12214.9
1400	3226.8	3562.2	4457.9	4461.8	4805.6	8530.1	10187.9
1200	2838.2	3133.2	3921.1	3924.5	4226.9	7502.9	8961.9
1000	2365.2	2611.0	3267.6	3270.4	3522.4	6252.4	7467.6
900	2128.6	2349.9	2940.8	2943.4	3170.2	5627.2	6720.8
800	1892.1	2088.8	2614.1	2616.3	2817.9	5001.9	5974.9
700	1655.6	1827.7	2287.3	2289.3	2465.7	4376.7	5227.3
600	1419.1	1566.6	1960.6	1962.2	2113.4	3751.4	4480.6
500	1182.6	1305.5	1633.8	1635.2	1761.2	3126.2	3733.8
400	946.1	1044.4	1307.0	1308.2	1408.9	2501.0	2987.0
350	827.8	913.9	1143.7	1144.6	1232.8	2188.3	2613.3
300	709.5	783.3	980.3	981.1	1056.7	1875.7	2240.3
250	591.3	652.8	816.9	817.6	880.6	1563.1	1866.9
200	473.0	522.2	653.5	654.1	704.5	1250.5	1493.6

Note: For 4πMS values not tabulated, use
$F(MHZ) = 2.8 \underbrace{(4\pi MS)}_{Gauss} \underbrace{(H_0/4\pi MS)}_{\text{Mode number for resonance with 110 mode}}$

See Figure 2.41 for a diagram of loop and sphere dimensions. Quality factor is typically 500 to 2000 for YIG resonators.

The equivalent circuit of a YIG resonator, which is shown in Figure 2.42 simply is a parallel *R-L-C* resonator, in series with a fixed inductance (the loop inductance). The equivalent circuit of a particular YIG sphere may be measured, or it can be calculated by first calculating Q and then, from a knowledge of its resonant frequency, the values of L, R, and C are calculated. Because the sphere is magnetically tunable, each new value of dc magnetic field corresponds to new values of L, R, and C.

2.5.4 YIG Limiting

As the dc magnetic field is reduced, a YIG resonance becomes progressively lossy, and at sufficiently low frequencies, resonator Q disappears altogether. There is no corresponding high-frequency limit. This low frequency Q degradation phenomenon is called *YIG limiting*. Below a certain limiting frequency, normal resonator operation no longer is possible. Operation is possible in an intermediate region,

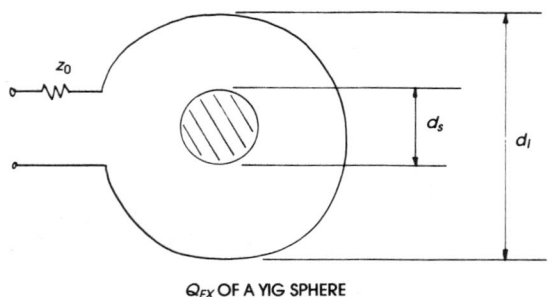

Q_{EX} OF A YIG SPHERE

$$Q_{EX} = \frac{3.4\, Z_0\, d_l^2}{N^2\, (4\pi M_S)\, d_s^3} \left[1 + \left(\frac{\omega L_S}{Z_0} \right)^2 \right]$$

WHERE: Z_0 is the characteristic impedance of transmission line

d_l is the loop diameter

d_s is the sphere diameter

$(4\pi M_S)$ is the saturation magnetization (Gauss)

N is the number of turns in the loop

L_S is the self inductance of the loop

ω is the angular frequency

Fig. 2.41 Calculating the q of a YIG sphere.

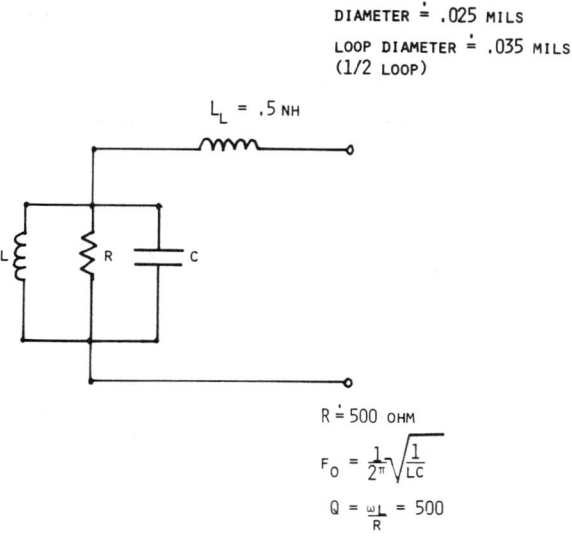

Fig. 2.42 The equivalent circuit of a YIG sphere.

called the *coincidence limiting region,* but just barely. For fields below this region, the loss becomes excessive for any operation. The two limiting regions are

$1/3(4\pi M_s) < H_0 < 2/3\,(4\pi M_s)$: coincidence limiting region

$H_0 < 1/3\,(4\pi M_s)$: high loss region

Temperature is an important factor in limiting. High-temperature operation is helpful because the limiting region moves to lower frequencies at higher temperatures. Table 2.2 shows the relationship between $4\pi M$ and limiting frequency at both room temperature and at high temperature (+90°C).

2.5.5 Anisotropic Effects in YIG

The resonant frequency of a YIG sphere is very nearly the gyromagnetic ratio times the dc magnetic field, but it depends on a small directional component related to an internal magnetic field, called *anisotropy*. YIG anisotropy is temperature dependent and can be used to advantage to provide temperature compensation for the natural negative temperature-frequency drift of an oscillator's electromagnetic shell.

Table 2.2 Limiting Frequency as a Function of Gallium Doping Level

210 Mode Separator (MHz)	Limiting Hot (MHz) (+90°C)	$4\pi M_S$ (gauss)	Limiting Room Temp. (MHz)
0– 40	0– 140	0– 101	0– 260
40– 50	140– 180	101– 126	260– 307
50– 60	180– 220	126– 151	307– 354
60– 70	220– 260	151– 176	354– 401
70– 80	260– 310	176– 201	401– 449
80– 90	310– 360	201– 226	449– 496
90–100	360– 405	226– 251	496– 544
100–110	405– 450	251– 277	544– 591
110–120	450– 500	277– 302	591– 633
120–130	500– 545	302– 327	633– 686
130–140	545– 590	327– 352	686– 733
140–150	590– 635	352– 377	733– 780
150–160	635– 680	377– 402	780– 827
160–170	680– 725	402– 427	827– 874
170–180	725– 770	427– 453	874– 921
180–190	770– 810	453– 478	921– 968
190–210	810– 900	478– 528	968–1063
210–230	900– 985	528– 578	1063–1158
230–250	985–1070	578– 629	1158–1253
250–270	1070–1160	629– 679	1253–1348
270–290	1160–1250	679– 729	1348–1442
290–310	1250–1330	729– 779	1442–1537
310–330	1330–1420	779– 830	1537–1631
330–350	1420–1510	830– 880	1631–1726
350–370	1510–1590	880– 930	1726–1820
370–390	1590–1680	930– 980	1820–1915
390–430	1680–1860	980–1081	1915–2104
430–470	1860–2050	1081–1182	2104–2293
470–510	2050–2230	1182–1282	2293–2482
510–590	2230–2600	1282–1483	2482–2860
590–670	2600–2970	1483–1684	2860–3239
710	3150	1785	3239–3620

A YIG resonator's frequency, including anisotropy is

$$f_R = 2.8 \left\{ H_0 + \left[2 - \left(\frac{5}{2}\right) \sin^2\theta - \left(\frac{15}{8}\right) \sin^2\theta \right] \frac{K_1}{M_S} \right\} \quad (2.37)$$

where

H_0 is the dc magnetic field,

θ is a rotational angle about an axis normal to the 111 crystal plane of the YIG sphere,

M_s is the temperature dependent saturation magnetization,

K_1 is a constant

As shown in Figure 2.43, as the YIG sphere is rotated about the proper axis, its resonant frequency varies periodically with angle in a nearly sinusoidal manner. When the frequency is above the nominal frequency, the sphere exhibits a large positive temperature-frequency drift; when the frequency is below the nominal frequency, the sphere exhibits a large negative temperature-frequency drift. With experience, a technician can align the angle of a YIG sphere to provide nearly perfect temperature compensation over a wide range of temperatures and frequencies.

Often a YIG sphere is mounted on a heating element (called a *heater*) to maintain the sphere at a constant high temperature. Heaters are used to avoid limiting and improve an oscillator's temperature-frequency stability. See Figure 2.44 for an example of a YIG heater.

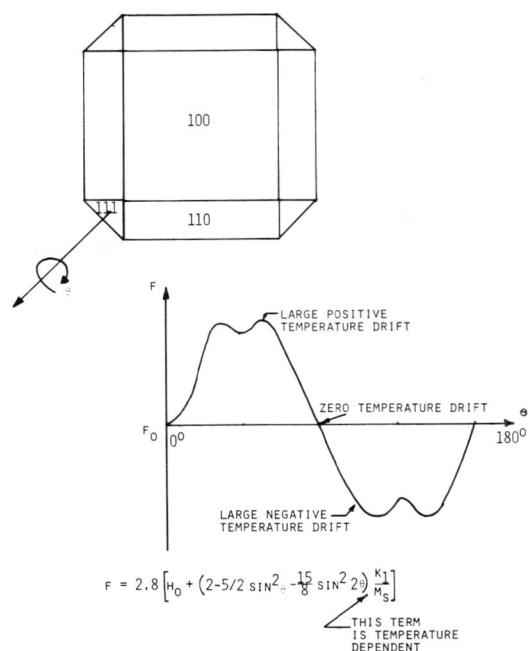

Fig. 2.43 Anisotropy of YIG.

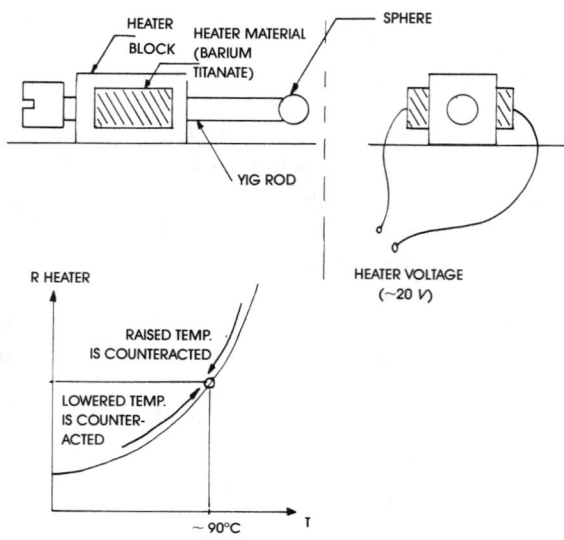

Fig. 2.44 A YIG heater.

2.5.6 Electromagnet Design

All YIG devices are tuned by electromagnets. The electromagnet's outer shell forms the mechanical housing for the electronics inside the YIG oscillator or YIG filter and, therefore, is an important part of any YIG oscillator design effort.

A cross-sectional drawing of the basic electromagnet design is shown in Figure 2.45. This magnet consists of a coil of wire wound around a core of special

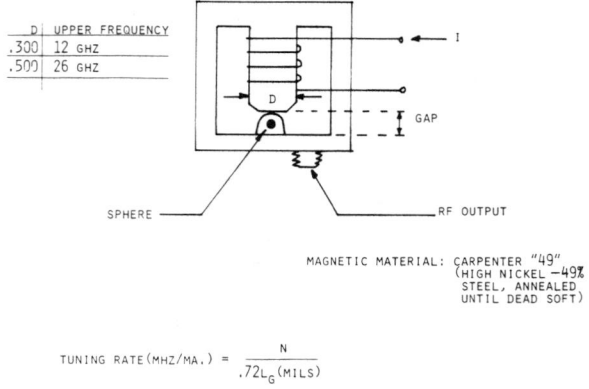

Fig. 2.45 Magnetic structure design relationships.

high-permeability steel (Carpenter 49). The YIG sphere is located in the center of an air gap, between the core's pole piece and the magnetic return path supplied by the magnet's outer shell. Both the pole piece and the outer shell are made of Carpenter 49 steel.

The frequency tuning rate of this structure is calculated approximately from the empirical expression

$$S(\text{MHz/mA}) = \frac{N}{.72 L_g(\text{mils})} \qquad (2.38)$$

where
    N is the number of turns in the coil,
    L_g is the length of the air gap in mils.

Direct current magnetic field and frequency are proportional in YIG devices. At some high magnetic field, the magnetic material (Carpenter 49) in the electromagnet will saturate and linear tuning will not be possible at higher frequencies. To prevent magnetic saturation in higher frequency operation, the pole piece diameter must be increased. The relationship between the upper frequency limit and pole piece diameter is given in Table 2.3.

Table 2.3 Frequency Limits as a Function of Pole Piece Diameter

Upper Frequency Limit	Pole Piece Diameter
12 GHz	.300 in
26 GHz	.500 in
40 GHz	.700 in

A cross section of an electromagnet's coil is shown in Figure 2.46. The basic relationships for designing the electromagnet's coil are

$$R = \text{number of rows} = \frac{H}{d} \qquad (2.39)$$

$$L = \text{number of layers} = \frac{t}{d} \qquad (2.40)$$

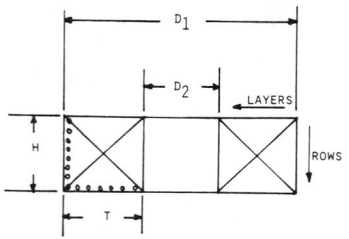

Fig. 2.46 Coil design relationships.

$$t = (6.463L + 1)(.134d) \tag{2.41}$$

where
    L = number of layers,
    R = number of rows,
    d = diameter of wire,
    H = coil height,
    t = coil thickness,
    N = (number of rows)(number of layers) = $R \times L$ (2.42)

Two common electromagnet housing designs used in YIG-tuned oscillators are the symmetric E-frame structure and the unsymmetric cup–end-cap structure. Both structures work well, although the cup–end-cap structure has become more popular due to its ease of assembly and manufacture. Figures 2.47 and 2.48 show the rough dimensions of these structures for various frequency ranges. In all cases, the material is Carpenter 49.

2.6 DIELECTRIC RESONATORS

Traditionally, microwave resonators have consisted of air dielectric, metal-walled, waveguide cavities. Such resonators suffer from a variety of practical limitations such as large size, low Q due to metal wall losses, temperature-frequency drift due

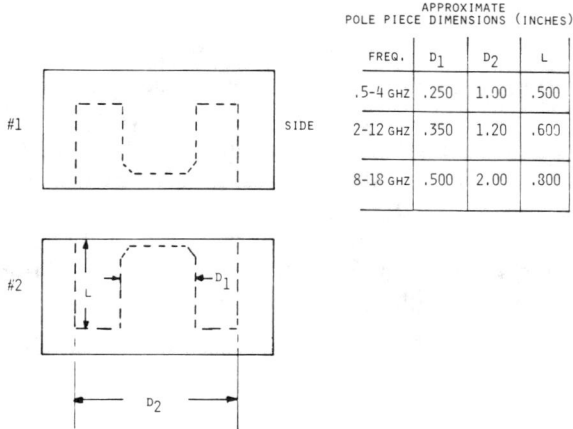

Fig. 2.47 E-frame magnetic structure.

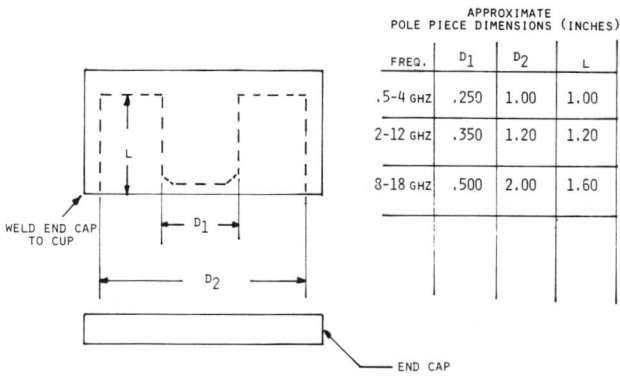

Fig. 2.48 Cup–end-cap magnetic structure.

to metal expansion, and long-term drift due to the gradual relaxation of built-in metal stresses. All of these resonator limitations translate directly to performance limitations for any oscillator constructed with such a resonator. Dielectric resonators overcome most of these limitations. Dielectric resonators [19] may be constructed for resonant frequencies from 1 GHz to over 30 GHz, with quality factors of 1,000 to 40,000 or more. The Q factor depends on the kind of dielectric material and on frequency. Dielectric resonators are very stable with temperature, which means that they are ideal resonators for high-stability low-noise oscillators for communications and radar applications.

2.6.1 Resonant Frequency of a Dielectric Resonator

Many of the limitations of an air dielectric metal-walled cavity can be avoided by using a cavity filled with a high dielectric material. The metal walls in fact are unnecessary in such a resonator because the boundary between the high-dielectric material and the surrounding low-dielectric air forms the cavity boundaries. These dielectric resonators, such as the one shown in Figure 2.49, consist of simply a cylindrical "puck" of high dielectric material such as barium titanium oxide ($\epsilon = 38$), whose resonant frequency is determined by the condition that the puck's diameter equals one wavelength in the high-dielectric material. This means

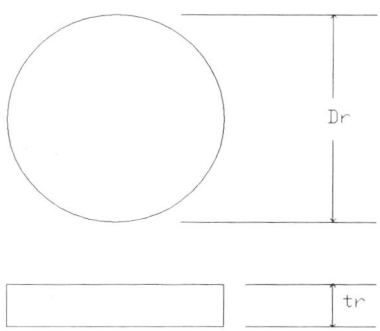

Fig. 2.49 The basic configuration of a dielectric resonator puck.

$$f_R = \frac{c}{\sqrt{\epsilon} d_R} \qquad (2.43)$$

where
- c is the speed of light in a vacuum ($3 \cdot 10^{10}$ cm/s),
- ϵ is the resonator puck's dielectric constant,
- d_R is the puck's diameter, and
- f_R is the resonant frequency.

The most common mode of operation of a dielectric resonator is called the $TE_{01\delta}$ mode[18]. The pattern of the electric and magnetic fields within the puck is shown in Figure 2.50.

It also is possible to excite spurious modes within a dielectric resonator. To avoid such moding, certain restrictions must be observed on the thickness, t_r, of a resonator puck. A common criterion for avoiding spurious mode is

$$.35 d_R \leq t_r \leq .45 d_R$$

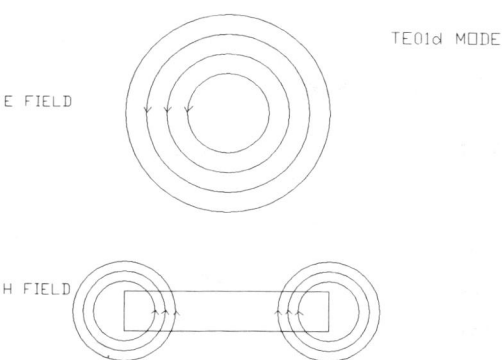

Fig. 2.50 The electric and magnetic field patterns within a dielectric resonator puck.

To account for the effect of the resonator puck's thickness is important in determining the resonator's fundamental frequency[20]. The effect of thickness is included in the following expression for the resonator's frequency:

$$f_0 = \frac{8.55}{\sqrt{\epsilon}(\pi/4)^{1/3}(d_R^2 t_r)^{1/3}} \tag{2.44}$$

A designer may arrive at the actual dimensions of a resonator puck to produce a particular resonant frequency in an experimental circuit by using the following iteration process:

1. Use (2.44) to make a first estimate of the diameter and thickness of the resonator puck.
2. Using the resonator chosen in step 1, *measure* its actual resonant frequency operating in the circuit of interest.
3. Calculate

$$K_{\text{new}} = \sqrt{\epsilon}\left(\frac{\pi}{4}\right)^{1/3}(d_r^2 t_r)^{1/3} f_{0,\text{ actual}}$$

4. Calculate a new resonator thickness from

$$t_{r,\text{new}} = \frac{K_{\text{new}}}{(\epsilon)^{3/2}\left(\dfrac{\pi}{4}\right) d_r^2 f_0}$$

5. If $t_{r,new}$ is within the range $.35d_r \leq t_r \leq .45d_r$, the new thickness and the original diameter define the actual resonator puck dimensions. If t_r is outside this range, start the process again with a slightly different resonator diameter.

2.6.2 Temperature-Dielectric Resonant Frequency Drift

There are three contributions to the temperature-frequency drift of a dielectric resonator[21]. They are

- Temperature coefficient of the dielectric constant:

$$\tau_\epsilon = \frac{1}{\epsilon} \frac{\partial \epsilon}{\partial t}$$

- Temperature coefficient of expansion of the dielectric resonator:

$$\alpha_l = \frac{1}{d} \frac{\partial d}{\partial t}$$

- Temperature coefficient of the metal in the outer housing:

$$\tau_l = \frac{1}{l} \frac{\partial l}{\partial t}$$

The total frequency-temperature coefficient of an oscillator built with a dielectric resonator is

$$\tau_f = \frac{1}{f_0} \frac{\partial f_0}{\partial T} = -(A\tau_\epsilon + B\alpha_l + C\tau_L) \tag{2.45}$$

where

$A \doteq \frac{1}{2}$, $B \doteq 1.0$, and C is in the range of .05 to 1.0, which depends on the position of the resonator relative to the metal housing.

With a careful choice of dielectric material and housing configuration, reducing τ_f to an extremely small number is possible; τ_f on the order of 4ppm/°C is possible with practical design. Since the high Q resonator's temperature stability dominates that of all other components in a practical oscillator, sources with stability on the order of 4 ppm/°C are practical using resonators.

2.6.3 Dielectric Resonator Tuning

A dielectric resonator may be tuned by a number of techniques for perturbing its fringing fields. These techniques include various forms of metal and dielectric screws, plungers, and disks, which are attached to the housing wall and adjusted mechanically in position relative to the resonator puck. The unperturbed resonant frequency of a dielectric puck may be tuned by as much as 20 percent using one of these techniques. Figure 2.51 shows a scheme for tuning a dielectric resonator that uses a metal disk.

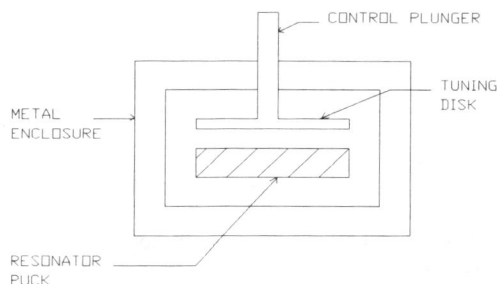

Fig. 2.51 A tuning scheme for a dielectric resonator using a metal disk.

2.6.4 Coupling to a Dielectric Resonator

The strong exterior magnetic field of the $TE_{01\delta}$ mode suggests the use of a loop coupling technique. However, because most dielectric resonators will be used in conjunction with microstrip circuits, it is most desirable to couple the resonator directly to a microstrip line. This can be done easily by placing the resonator puck in close proximity to a microstrip line so that the resonator's external magnetic field surrounds the microstrip line, as shown in Figure 2.52. For maximum coupling, the resonator puck may slightly overlap the microstrip line. The condition for maximum coupling, which is also shown in Figure 2.51, is that the length l, which is the distance from the center of the resonator puck to the center axis of the microstrip line be $.35\ D_R$, where D_R is the resonator puck's diameter. Some separation between the resonator puck and the surface of the microstrip line or dielectric material may be necessary to maintain high external Q under coupled operation.

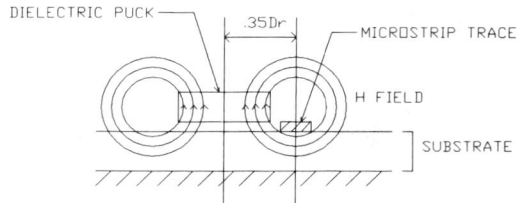

Fig. 2.52 Offset between the center line of a dielectric resonator and the center line of a microstrip trace for maximum coupling.

REFERENCES

1. J. B. Gunn, "Microwave Oscillations of Current in II–IV Semiconductors," *Solid State Comm.,* Vol. 1, September 1963, pp. 88–91.
2. D. E. McCumber, and A. G. Chynoweth, "Theory of Negative-Conductance Amplification and Gunn Instabilities in 'Two Valley' Semiconductors," *IEEE Trans. Electron Devices,* Vol. ED-13, January 1966.
3. B. K. Ridley, "The Inhibition of Negative Resistance Dipole Waves and Domains in N-GaAs," *IEEE Trans. Electron Devices,* Vol. Ed-13, January 1966.
4. P. Wolf, *IBM J. Res. Development,* Vol. 14, 1970, p. 125.
5. R. E. Williams, *Gallium Arsenide Processing Techniques,* Artech House, Dedham, MA, 1984.
6. Williams, *ibid.*
7. Williams, *ibid.*
8. A. G. Milnes, *Semiconductor Devices and Integrated Electronics,* Van Nostrand Reinhold, New York, 1980, p. 35.
9. J. V. Dilorenzo and W. R. Wiseman, "GaAs Power MESFETs: Design, Fabrication, and Performance," *IEEE Trans. Microwave Theory Tech.,* Vol. MTT-27, May 1979.
10. G. V. Vendelin, *Design of Amplifiers and Oscillators by the S-Parameter Method,* John Wiley & Sons, New York, 1982, pp. 44–45.
11. A. G. Milnes, *Semiconductor Devices and Integrated Electronics, op. cit.,* pp. 230–232.
12. *Ibid.,* pp. 143–145.
13. O. Madelung and D. Meyerhofer, *Physics of II-IV Compounds,* John Wiley and Sons, New York, 1964, pp. 133–141.
14. J. White, *Semiconductor Control,* Artech House, Dedham, MA, 1974, pp. 39–66.
15. M. L. Keith and R. Roy, "Structural Relations among Double Oxides of Trivalent Elements," *American Mineralogist,* Vol. 39, P1, January 1954.
16. C. Kittel, *Introduction to Solid State Physics,* John Wiley and Sons, New York, 1967, p. 437.
17. P. C. Fletcher, and R. B. Bell, "Ferrimagnetic Resonance Modes in Spheres," *J. Applied Physics,* Vol. 30, 1959, pp. 687–698.
18. P.S. Carter, "Equivalent Circuit of Orthagonal-Loop-Coupled Magnetic Resonance Filters and Bandwidth Narrowing due to Coupling Inductance," *IEEE Trans. on MTT,* Vol. MTT-18, No. 2, February 1970, pp. 100–105.
19. K. A. Zaki and A. E. Atia, "Modes in Dielectric Wave Guide and Resonators," *IEEE Trans. Microwave Theory Techn.,* Vol. MTT-31, December 1985, pp. 1039–1045.
20. "A Designer's Guide to Microwave Dielectric, Ceramics," Trans-Tech, Inc., March 1988.
21. "Temperature Coefficients of Dielectric Resonators," *Tech. Brief No. 831,* Trans-Tech, Inc.

Chapter 3
Microstrip Transmission Line Fundamentals

3.1 THE BASIC EQUATIONS OF MICROSTRIP TRANSMISSION LINES

Both MIC and MMIC circuits rely on microstrip lines to realize many common circuit elements. Because microstrip lines are so important and so basic, this chapter is devoted to their characteristics.

The physical structure of a microstrip transmission line is shown in Figure 3.1. Microstrip is a quasi-TEM transmission media consisting of a conductive metal strip of width, w, in contact with a dielectric slab of relative dielectric constant ϵ_r, and thickness, t, the bottom side of which is contacted by a ground plane. Microstrip transmission line is characterized by a characteristic impedance Z_0, and an effective dielectric constant ϵ_e. The basic equation for wave length on a microstrip line is

$$f\lambda_g = \frac{c}{\sqrt{\epsilon_e}} \tag{3.1}$$

where
 λ_g is the wavelength in microstrip,
 c is the speed of light in a vacuum ($3 \cdot 10^{10}$ cm/s),
 ϵ_e is the effective dielectric constant,
 f is the frequency.

The phase velocity of a wave propagating along a microstrip line is

$$v_p = \frac{c}{\sqrt{\epsilon_e}} \tag{3.2}$$

The characteristic impedance of a microstrip line in its most basic form is

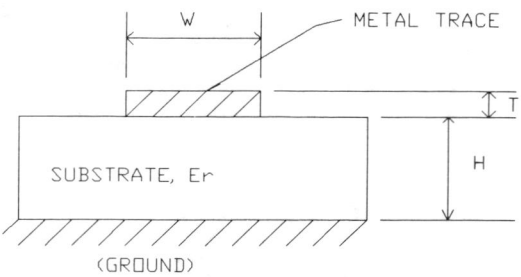

Fig. 3.1 The basic physical structure of a microstrip line.

$$Z_0 = \frac{1}{v_p C} \tag{3.3}$$

where C is capacitance/length.

Everything a designer needs to know about a microstrip line can be summed up in a knowledge of Z_0 and ϵ_e.

Closed form expressions for Z_0 and ϵ_e that are accurate to within 2 percent have been given by Wheeler [1] and Schneider [2], as follows.
For $w/t < 1$:

$$Z_0 = \frac{60}{\sqrt{\epsilon_R}} \ln\left[8t/w_e + .25w_e/t\right] \tag{3.4}$$

$$\epsilon_e = \frac{\epsilon_R + 1}{2} + \frac{\epsilon_R - 1}{2}\left[\left(1 + \frac{12t}{w_e}\right)^{-1/2} + .04\left(1 - \frac{w_e}{t}\right)^2\right] \tag{3.5}$$

For $w/t > 1$:

$$Z_0 = \frac{377}{\sqrt{\epsilon_e}\,[w_e/t = 1.393 + .667 \ln(w_e/t + 1.444)]} \tag{3.6}$$

$$\epsilon_e = \frac{\epsilon_R + 1}{2} + \frac{\epsilon_R - 1}{2}(1 + 12t/w_e)^{1/2} \tag{3.7}$$

where

$$w_e = w + \frac{t'}{\pi}\left(\ln\frac{2t_e}{t'} + 1\right) = \text{effective microstrip width}$$

$t_e = t - 2t'$ = effective substrate thickness,
w = physical line width,
t = substrate dielectric thickness,
ϵ_R = substrate relative dielectric constant,
t' = metal thickness.

Microstrip transmission lines are by nature dispersive; that is, ϵ_e is a function of frequency. Getsinger [3] has given an expression for the frequency dependence of ϵ_e:

$$\epsilon_e(f) = \epsilon_r - \frac{\epsilon_r - \epsilon_e}{1 + f^2(2\mu_0 t/Z_0)^2 \left\{ \pi^2/12 \left[(\epsilon_e - 1) + \alpha^2 \frac{(\epsilon_R - \epsilon_e)}{\epsilon_R} - 1 \right] \left[\frac{(\epsilon_R - 1)(\epsilon_R - \epsilon_e)}{\epsilon_e(\epsilon_R - 1)} \right] \right\}} \quad (3.8)$$

where
$\alpha \doteq 3.0$
$\mu_0 = 4\pi \times 10^{-7}$ H/M

The approximate shape of $\epsilon_e(f)$ is shown in Figure 3.2

The two mechanisms for electrical loss in a microstrip transmission line are conductor loss and dielectric loss. Normally conductor loss is dominant; however, with some materials, such as alumina, dielectric loss can become significant at sufficiently high frequencies.

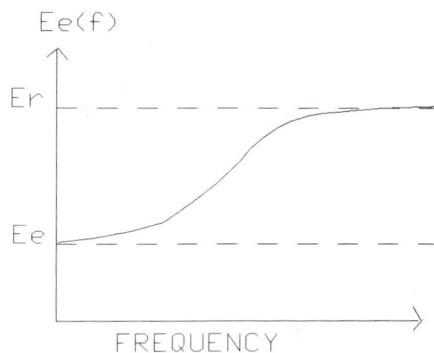

Fig. 3.2 The dispersive behavior of $\epsilon_e(f)$ in a microstrip.

Consider first the conductor loss contribution. The following expression for conductor loss was calculated by Hammerstad and Bekkadal [4]:

$$\alpha_c = .072 \frac{\sqrt{f}\lambda_g}{w\, Z_0} \text{ dB/wavelength} \tag{3.9}$$

This expression has been found to be optimistic because it ignores conductor surface roughness, which can be a significant fraction of a skin depth in real microstrip lines. Surface roughness can increase conductor loss by as much as 60%.

Notice that α_c varies inversely with w, the microstrip line's width. Low ϵ_R substrates, such as duroid, which have large values of w for a given Z_0, have relatively low conductor loss.

Hammerstad and Bakkadal [4] also have given an expression for dielectric loss:

$$\alpha_d = 27.3 \frac{\epsilon_R(\epsilon_e - 1)\tan\delta}{\epsilon_e(\epsilon_R - 1)} \text{ dB/wavelength} \tag{3.10}$$

where $\tan\delta$ is the material's loss tangent.

For many microstrip lines, $\epsilon_R \doteq \epsilon_e$, in which case (3.10) becomes $\alpha_d \doteq 27.3 \tan\delta$. For most practical microstrip lines, α_d is 5 to 10 times less than α_c, which means that α_d can often be safely ignored.

3.2 TABLES OF Z_0 AND ϵ_e FOR SOME TYPICAL SUBSTRATES

This section contains tabulations of microstrip characteristic impedance and effective dielectric constant for several popular substrate materials of standard thicknesses. Designs performed with these specific substrates can be carried out on a "first cut" basis, using only the information contained in the tables. Of course, final designs should make use of a generalized computer program, such as Touchstone or SuperCompact to include the effects of dispersion and loss. These tables were calculated using the line calculation program (see Section 4.2) Dimstrip. Tables 3.1 and 3.2 are for alumina substrates of two different thicknesses. Table 3.3 is for a duroid (soft board) substrate; and Table 3.4 is for a GaAs substrate.

3.3 THE SMITH CHART

The Smith chart [5] is an important tool for performing many types of general transmission line calculation. The Smith chart works with all types of transmission lines, including microstrip lines. Figure 3.3 shows a Smith chart, with a transmission line circuit used as an example. The Smith chart is a graphical transmission

Table 3.1 Tabulated Microstrip Data For Alumina, $\epsilon_R = 9.9$, $t = 10$ mils, $f = 12$ GHz

w (mils)	Z_0 (Ω)	ϵ_e
44.	20	8.05
23.9	30	7.47
14.4	40	7.04
9.3	50	6.74
6.05	60	6.51
4.10	70	6.36
2.65	80	6.22
1.73	90	6.11
1.15	100	6.01

Table 3.2 Tabulated Microstrip Data For Alumina, $\epsilon_R = 9.9$, $t = 25$ mils, $f = 12$ GHz

w (mils)	Z_0 (Ω)	ϵ_e
102.5	20	8.33
57.3	30	7.73
36.3	40	7.31
23.8	50	6.98
15.9	60	6.72
10.5	70	6.54
7.19	80	6.40
4.66	90	6.27
3.15	100	6.18

Table 3.3 Tabulated Microstrip Data For Duroid, $\epsilon_R = 2.1$, $t = 10$ mils, $f = 12$ GHz

w (mils)	Z_0 (Ω)	ϵ_e
99.9	20	1.94
61.0	30	1.89
42.9	40	1.85
31.0	50	1.81
23.3	60	1.78
17.8	70	1.75
14.0	80	1.73
11.0	90	1.71
8.8	100	1.69

Table 3.4 Tabulated Microstrip Data For GaAs, $\epsilon_R = 12.5$, $t = 5$ mils, $f = 12$ GHz

w (mils)	Z_0 (Ω)	ϵ_e
16.7	20	9.52
9.85	30	8.93
5.77	40	8.43
3.68	50	8.10
2.37	60	7.85
1.49	70	7.66
.93	80	7.50
.61	90	7.39
.35	100	7.26

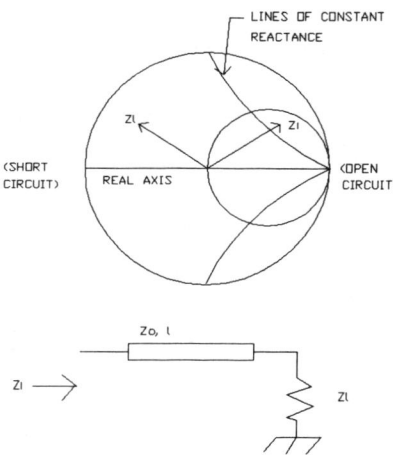

Fig. 3.3 The Smith chart analyzing a basic transmission line circuit.

line impedance calculator because it calculates the impedance of any length of transmission line of characteristic impedance Z_0, of length, l, expressed in fractions of a wavelength or in degrees of phase and terminated by an impedance Z_l. The Smith chart graphically performs the complex transformation of impedances done by the transmission line equations, solving these equations in a simple, straightforward way. The process for using the Smith chart is as follows:

- Normalize Z_l to Z_0.
- Locate the normalized Z_l (i.e., Z_l/Z_0) on the Smith chart.

- Calculate the transmission line's length in terms of λ_g, (i.e., calculate l/λ_g).
- "Rotate" Z_l along a circle of constant radius centered on the chart's origin ($\Gamma = 0.0$) an "angle" l/λ_g from the load.
- The new point, Z_i, will be the impedance of the line section terminated in an impedance, Z_l. The reflection coefficient of Z_l will be a vector drawn from the origin to Z_i. Scales are available on the Smith chart that make it possible to read the magnitude and the angle of the reflective coefficient, and the real and imaginary components of normalized impedance.

The Smith chart is a powerful "thinking tool" for expressing the impedance of microwave devices in a transmission line environment. Many nontransmission line microwave devices are characterized as a locus of points in frequency on a Smith chart; for example, varactor and PIN diodes, bipolar transistors, and GaAs FETs.

3.4 COUPLED MICROSTRIP TRANSMISSION LINES

Microstrip lines running parallel to each other on a common dielectric substrate (as shown in Figure 3.4) will cross couple microwave energy. This coupling is the basic operational principle behind various types of directional coupler designs. Coupling also is the mechanism for undesirable crosstalk between transmission lines located close together on a common substrate. Such coupling always is detrimental to performance.

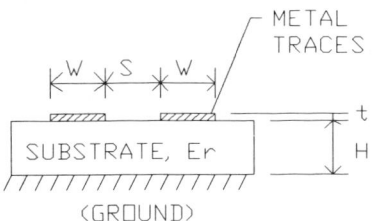

Fig. 3.4 Coupled microstrip lines.

Coupled lines exhibit both even and odd mode excitations of the types shown in Figure 3.5. Each mode has its own characteristic impedance (Z_{0o}) and (Z_{0e}) and its own characteristic phase velocity (v_{po}) and (v_{pe}).

Following Edwards [6], let C be the capacitance per unit length of each microstrip line, and let C_1 be the capacitance per unit length of the two strips without the substrate. The effective dielectric constant, ϵ_e, may be defined in terms of these capacitances as

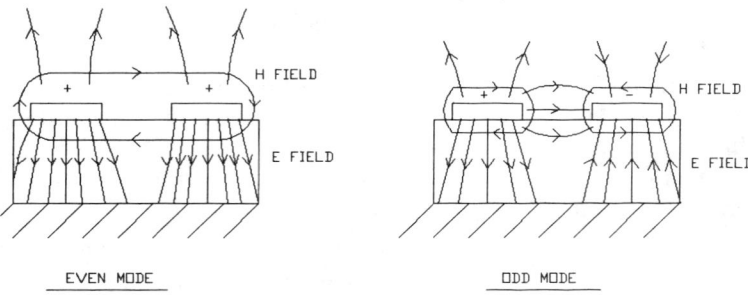

Fig. 3.5 Even and odd mode excitations of coupled microstrip lines.

$$\epsilon_e = \frac{C}{C_1} \tag{3.11}$$

From (3.2) and (3.3), a characteristic impedance may be calculated in terms of C and C_1 as

$$Z_0 = \frac{1}{cC_1\sqrt{\epsilon_e}} \tag{3.12}$$

Dividing this impedance into even and odd mode impedances yields

$$Z_{0e} = \frac{1}{cC_{1e}\sqrt{\epsilon_{ee}}} \tag{3.13}$$

$$Z_{0o} = \frac{1}{cC_{1o}\sqrt{\epsilon_{ro}}} \tag{3.14}$$

and

$$v_{pe} = \frac{c}{\sqrt{\epsilon_{ee}}} \tag{3.15}$$

$$v_{Po} = \frac{c}{\sqrt{\epsilon_{eo}}} \tag{3.16}$$

where c is the speed of light in a vacuum.

A common relationship exists between the even and odd mode impedances of coupled lines and the characteristic impedance of a single transmission line:

$$Z_0 = \sqrt{Z_{0E}Z_{0O}} \tag{3.17}$$

This is only approximately true for microstrip because the even and odd mode velocities are unequal, which changes the phase relationship between modes, modifying (3.15). Nevertheless, (3.17) is a useful approximation.

The coupling factor k'', in dB between coupled microstrip lines is given as

$$k = 20 \log \frac{Z_{0E} - Z_{0O}}{Z_{0E} + Z_{0O}} \tag{3.18}$$

The exact analysis of the even and odd mode impedances and the effective dielectric constant is best done by computer, using any one of a number of fundamental methods such as those described by Bryant and Weiss [7]. A number of excellent commercial software programs, which are discussed in Chapter 4, accurately calculate all the parameters of coupled microstrip lines.

3.5 USING MICROSTRIP TRANSMISSION LINES AS DISTRIBUTED EQUIVALENTS OF LUMPED ELEMENT INDUCTORS AND CAPACITORS

Often microwave circuit design begins with topologies containing only lumped elements: capacitors, inductors, and resistors. To realize such a circuit on a microstrip substrate, the lumped elements must be replaced by their microstrip equivalents. Typically, inductors become narrow microstrip lines, and shunt capacitors become wide microstrip lines (or stubs). However, it is important for the designer to recognize that microstrip elements are never pure lumped elements, because they always contain some parasitic elements. These parasitics must be considered in any overall circuit model because they often have a significant effect on performance. The equivalent circuits (including parasitic elements) of both narrow and wide microstrip lines are shown in Figure 3.6.

The high-impedance "inductive" line is modeled as a π section, with parasitic shunt capacitors located at either end of the inductor. The low-impedance "capacitance" line is modeled as a T section with parasitic series inductors.

Consider first the high Z_0 inductive line. Any transmission line has a reactance (assuming a short at one end) of

$$\omega L = X_L = Z_0 \sin\left(\frac{2\pi l}{\lambda g}\right) \tag{3.19}$$

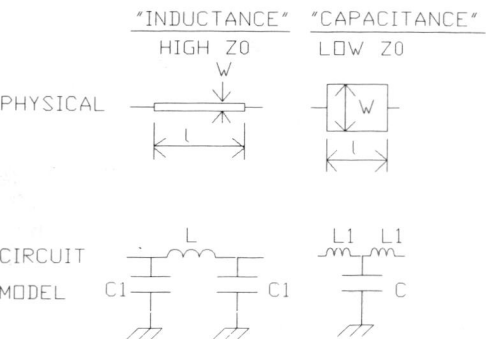

Fig. 3.6 Inductive and capacitive microstrip lines.

Solving (3.19) for line length, *l*, yields an expression for the length of the microstrip in terms of its inductance:

$$l = \frac{\lambda g}{2\pi} \sin^{-1}\left(\frac{\omega L}{Z_0}\right) \tag{3.20}$$

However, this line also has a shunt capacitive susceptance equal to

$$\omega C = \frac{1}{Z_0} \tan\left(\frac{\pi l}{\lambda g}\right) \tag{3.21}$$

or the shunt parasitic capacitances, C_1, at either end of the line section are

$$C_1 = \frac{C}{2} = \frac{1}{2\omega Z_0} \tan\left(\frac{\pi l}{\lambda g}\right) \tag{3.22}$$

For short lines, $\tan X \doteq X$, and

$$C_1 \doteq \frac{l}{2fZ_0\lambda g} \tag{3.23}$$

The low Z_0, capacitive line has a shunt capacitance equal to

$$\omega C = B = \frac{1}{Z_0} \sin\left(\frac{2\pi l}{\lambda g}\right) \tag{3.24}$$

Solving for the line length, l, in terms of shunt capacitance

$$l = \frac{\lambda g}{2\pi} \sin^{-1}(\omega C Z_0) \qquad (3.25)$$

The parasitic series inductance is given as

$$L_1 = \frac{L}{2} = \frac{Z_0}{2\omega} \sin\left(\frac{2\pi l}{\lambda g}\right) \qquad (3.26)$$

For short lines, $\sin X \doteq X$. So,

$$L_1 \doteq \frac{Z_0 l}{Zf\lambda g} \qquad (3.27)$$

3.6 LANGE COUPLERS

A very important class of microstrip couplers is the Lange coupler, named for its inventor, Julius Lange [8]. The Lange coupler, which is shown in Figure 3.7, is an interdigitated, microstrip, quadrature phase 3 dB coupler. Lange couplers are capable of true quadrature phase coupling over two octaves of bandwidth because they compensate for the even and odd mode phase velocity differences. The direct and coupled ports of a Lange coupler are able to track each other by 90° ± 5° even over two or more octaves.

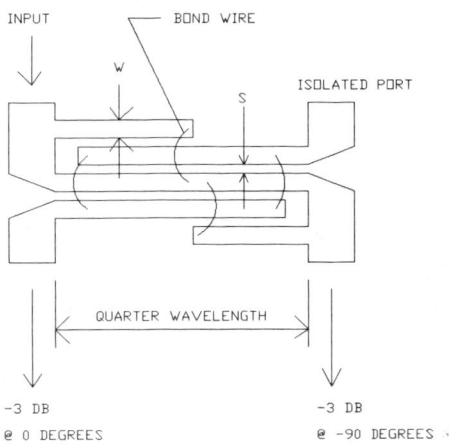

Fig. 3.7 The layout of a Lange coupler.

Lange couplers consist of three main fingers and two half-length fingers. These fingers are connected in the proper configuration by very short (low-inductance) bond wires. All coupler fingers are spaced by a gap, S. The total length of the coupler is $\lambda_g/4$ at the *center* of its band. All lines leading to and from a Lange coupler must be of 50 Ω characteristic impedance.

Both Touchstone® and SuperCompact® contain Lange coupler models. Lange couplers normally are designed using one of these programs. However, a good first guess for an octave band Lange coupler on 20 mil alumina is $w = 1.5$ mils and $S = 1.0$ mils. For a two-octave coupler, a good first guess is $w = 1.0$ mils and $S = .60$ mils. The reduction in S is necessary to slightly overcouple the coupler to expand its bandwidth.

Lange couplers are invaluable building blocks for many microwave circuits, including balanced amplifiers and attenuators, and phase shifters. Some applications of Lange couplers are discussed in Chapter 5.

3.7 S-PARAMETERS AND MULTIPORT NETWORKS

Often the behavior of a microwave device or network can be summed up by a set of S-parameters, which are a tabulation of the magnitudes and angles of their reflection and transmission coefficients at discrete frequencies. S-parameter concepts and conventions are very important when working with Touchstone and SuperCompact files. All S-parameters are commonly referred to a specific transmission line characteristic impedance, which usually is 50 Ω, because the reflection and transmission coefficients have meaning only in a transmission line environment. This approach is used quite often by the manufacturers of diodes, GaAs FETs, and bipolar transistors to communicate the electrical characteristics of these devices to the circuit designer who can then insert the S-parameters of a given device directly into a Touchstone or SuperCompact file, and analyze the behavior of any circuit that contains the device. No model or equivalent circuit is necessary, because the S-parameters contain *all* the necessary small-signal information to completely characterize the device.

S-parameters may be created for a network with an arbitrary number of ports. However, from a practical standpoint, amplifier and oscillator designers rarely have occasion to use networks with more than four ports. Most diodes effectively are one-port devices, GaAs FETs and bipolar transistors are two-port devices, and Lange couplers are four-port devices. Amplifiers and oscillators themselves usually are viewed as two-port devices, although exceptions do exist, such as an amplifier with both a direct and an attenuated output.

The numbering system for S-parameters is as follows: "diagonal" S-parameters, $S_{11}, S_{22}, S_{33}, S_{44}$, are the magnitude and angle of the reflection coefficients at ports 1,2,3, and 4, respectively. The return loss of each port of the network is equal to

$$RL_i = 20 \log |S_{ii}| \qquad (3.28)$$

The "nondiagonal" S-parameters, such as S_{21}, S_{12}, S_{31}, are the magnitudes and angles of the transmission coefficients between pairs of ports. In general, the transmission S-parameters have the form S_{ij}, where i is the output port and j is the input port, giving a sense of direction to the transmission coefficient. The gain, or loss, along the transmission path from j to i is

$$\text{gain/loss} = 20 \log |S_{ij}| \qquad (3.29)$$

The phase shift between ports j and i is $\angle S_{ij}$. Figure 3.8 graphically demonstrates how to interpret the S-parameters of 1-, 2-, 3-, and 4-port networks.

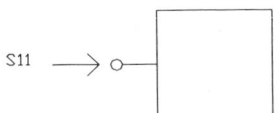

One-port network

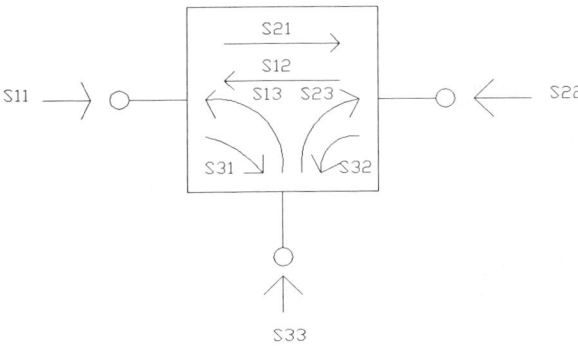

Three-port network

Fig. 3.8 Graphical representations of the S = parameters for one-, two-, three-, and four-port networks.

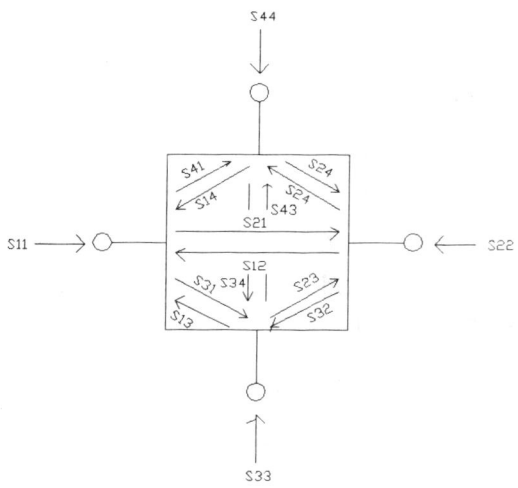

Four-port network

Fig. 3.8 (cont'd).

REFERENCES

1. H. A. Wheeler, "Transmission-Line Properties of Parallel Strips Separated by a Dielectric Sheet," *IEEE Trans. Microwave Theory Tech.*, Vol. MTT-13, March 1965, pp. 172–185.
2. M. V. Schneider, "Microstrip Lines for Microwave Integrated Circuits," *Bell Systems Tech. J.*, Vol. 48, May–June 1969, pp. 1421–1444.
3. W. J. Getsinger, "Microstrip Dispersion Model," *IEEE Trans. Microwave Theory Tech.*, MTT-21, January 1973, pp. 34–39.
4. Hammerstad, E., and F. Bekkadal, "A Microstrip Handbook," *ELAB Report*, STR 44 A 74169, University of Trondheim, Norway, 1975, pp. 98–110.
5. P. H. Smith, *Electronic Applications of the Smith Chart,* McGraw-Hill, New York, 1969.
6. T. C. Edwards, *Foundations for Microstrip Circuit Design,* John Wiley and Sons, New York, 1983, pp. 131–140.
7. T. Bryant and J. Weiss, "Parameters of Microstrip Transmission Lines and of Coupled Pairs of Microstrip Lines," *IEEE Trans. Microwave Theory Tech.*, Vol. MTT-16, December 1968, pp. 1021–1027.
8. J. Lange, "Interdigitated Strip-Line Quadratives Hybrid," *IEEE Trans. Microwave Theory Tech.*, Vol. MTT-17, December 1969, pp. 1150–1151.

Chapter 4
Computer Aided Design Techniques

4.1 LINEAR CAD ANALYSIS-OPTIMIZATION PROGRAM

Only a few computer programs perform analysis and circuit optimization of general *linear* microwave networks. These very specialized programs allow a designer to analyze the gain, phase, return loss, delay, impedance, and admittance of a general microwave network, and optimize the network's element values to achieve a *user defined set* of performance goals. Such programs are an invaluable tool for any designer, and it is a necessity that every designer have access to at least one. Most linear analysis-optimization programs are very expensive, and they often run on expensive computers. However, these programs are worth the money because they perform the task of circuit optimization, which forms the cornerstone of almost every microwave amplifier and oscillator design effort. The two major linear analysis and optimization programs are *Touchstone*® by EEsof, Inc., Westlake Village, California, and *SuperCompact*® by Compact Software, Inc., Patterson, New Jersey.

In terms of hardware requirements, versions of both Touchstone and SuperCompact are available for IBM compatible personal computers, for main frame computers, and for a new class of desk top computers called workstations. Before purchasing any computer hardware, be sure to contact the software program's publisher regarding the specific hardware requirements for any and all programs you intend to use.

Linear analysis and optimization programs are a lot like languages in that even though each deals with the same information, the grammar and syntax of one is markedly different from the other. Just as with a language, it takes the user considerable time and effort to become fluent in the analysis-optimization program of his or her choice. Although Touchstone and SuperCompact, in a basic sense, are almost identical programs (and may be directly translated one to the other), a user who has become comfortable with the syntax, organizational structure, and editor of one program, is very likely to stick with that particular program. For the past five years, I have used Touchstone almost exclusively (because of availability), achieving a high level of fluency and comfort with it. For this reason, the examples in this book are worked using Touchstone. However, you must recognize that any and all examples just as easily could be run on SuperCompact. Regrettably, SuperCompact users must first "translate" these Touchstone examples into SuperCompact to run them on their own computer—a direct result of the software "tower of Babble" with which we must all live on a daily basis.

Let us now briefly investigate the organization of a linear analysis-optimization circuit file. Touchstone files are broken down into a number of constituent parts, called *blocks*. There are six main blocks: VAR, CKT, FREQ, OUT, GRID, and OPT. Several other specialized blocks also may be used from time to time. The function of each of the five basic blocks is

    VAR   The variable block is used to describe common variables that appear in more than one element in the CKT block.

    CKT   Describes all circuit elements within the network and specifies their interconnection according to a user-defined system of nodes. Up to four total input or output ports may be defined.

    FREQ   Specifies starting frequency, stopping frequency, and frequency steps.

    OUT   The output block allows the user to specify which output parameters are to be calculated. The choices are any or all of the following: S-parameters, power gain, voltage gain, return loss, phase, time delay, impedance, and admittance. The output block also specifies on what grid (or screen) a parameter is to be displayed.

    GRID   Defines the scale of the X and Y axes of the graphics display of each grid.

    OPT   Allows the user to define the performance goals the circuit is to achieve. When commanded to optimize, the optimizer will change the variables within the elements which have been designated (with a backslash replacing the equal sign) as variables of the optimizer, until the best possible compromise is achieved between the requested goals. It is left to the user to determine if these optimized parameters are "physically realizable." This means, for example, that a line width of .001 mils may produce the desired performance on the computer, but be unrealizable in practice.

By way of example, let us write a Touchstone file for the simple FET amplifier circuit in Figure 4.1. Note that any comments, which are made to the right of the exclamation (!) symbol, will not be read into the Touchstone file. We will make use of this "comment" system to describe each element. Due to the simplicity of this circuit, no VAR block will be necessary.

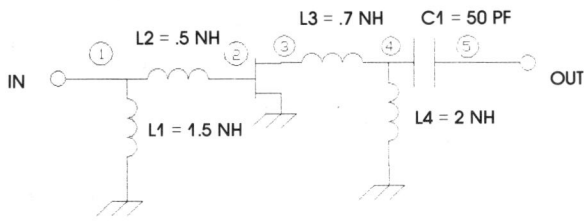

Fig. 4.1 A simple GaAs FET amplifier circuit used as an example of a circuit that is analyzed and optimized by a computer-aided program such as Touchstone or SuperCompact.

```
CKT
    IND    1    0    L\    1.5        ! L1 is a 1.5 NH inductor connected between node 1 and node 0 (ground).
    IND    1    2    L\    .5         ! L2 is a .5 NH inductor connected between node 1 and node 2.
    S2PA   2    3    0    NE71000    ! The active device is an NEC NE71000 GaAs FET whose separate S-parameter file is read by Touchstone into a two-port network, S2PA. The FET's input is connected to node 2, and its output is connected to node 3; its common port is grounded.
    IND    3    4    L\    .7         ! L3 is a .7 NH inductor connected from node 3 to node 4.
    IND    4    0    L\    2.0        ! L4 is a 2.0 NH inductor connected from node 4 to ground.
```

		CAP	4	5	C=	50	! C1 is a 50 pf capacitor connected from node 4 to node 5.
		DEF2P	1	5		AMP	! The total amplifier is defined as a two-port network named AMP, which is connected from node 1 to node 5.
FREQ							
		SWEEP	2	8	1		! The frequency range of the Touchstone analysis will be 2GHz to 8 GHz in 1 GHz steps.
OUT							
		AMP	dB	[S21]	GR1		! The forward gain, S21. in dBs of AMP is presented on output grid 1.
		AMP	dB	[S11]	GR2		! The input return loss. S11, in dBs is presented on output grid 2.
		AMP	dB	[S22]	GR3		! The output return loss. S22, in dBs is presented on output grid 3.
GRID							
		RANGE	2	8	1		! The X-axis of all grids will be 2 GHz to 8 GHz in 1 GHz steps.
		GR1	0	10	1		! The Y-axis of grid 1 will run from 0 dB to +10 dB in 1 dB steps.
		GR2	−20	0	5		! The Y-axis of grid 2 will run from −20 dB to 0 dB in 5 dB steps.

	GR3	−20	0	5	! The Y-axis of grid 3 will run from −20 dB to 0 dB in 5 dB steps.
OPT					
	RANGE	2	6		! The frequency range for optimization will be 2 to 6 GHz.
	AMP	dB	[S21]	=7.0	! The optimizer goal for forward gain is *exactly* 7.0 dB.
	AMP	dB	[S11]	<−15	! The optimizer goal for input return loss is *less than* (more negative than) −15 dB.
	AMP	dB	[S22]	<−15	! The optimizer goal for output return loss is *less than* (more negative than) −15 dB.

Once the file is entered into Touchstone and run in the optimizer mode, the values of L1, L2, L3, and L4 will be varied until the performance of the network is as close to the optimizer goals as is allowed by the flexibility of the chosen topology. The time required for optimization is a complicated function of the complexity of the circuit to be optimized and the speed of the computer. In general, simple circuits running on fast computers, optimize quickly; whereas complicated circuits running on slow computers, optimize slowly.

If one goal is to be favored over another, weighting factors may be entered into the OPT block to specify the relative value of the goals. As the optimization process proceeds, Touchstone automatically updates both the circuit file and the output grids. After each iteration of the optimizer, an error function is calculated and displayed. This error function is a relative measure of how far the circuit's performance deviates from the goals.

Both Touchstone and SuperCompact have more than 100 elements in their circuit element libraries. These elements include nearly all of the active and passive elements normally encountered in the design of microwave amplifiers and oscillators. The values of all these elements may be made variables of the optimizer. Many examples of Touchstone files written for actual amplifier and oscillator circuits are presented in Chapters 5 and 6.

4.2 TRANSMISSION LINE PARAMETER CALCULATION PROGRAMS

Sometimes it is important to simply obtain the exact operating parameters (Z_0, ϵ_e) of a transmission line system. Several commercially available programs perform this task easily and quickly. These programs calculate the parameters of microstrip lines of various thicknesses, relative dielectric constants, and line widths. They also calculate the parameters of strip lines, coplaner wave guides, and in some cases, coaxial lines. Usually both single and coupled pairs of lines may be analyzed. Very often these programs give useful results that are used to determine the "starting values" in Touchstone and SuperCompact files.

Two such transmission line programs and their publishers are *Dimstrip*®, published by Jan Van Craeynest of Palo Alto, California, and *LineCalc*® from EEsof, Inc., Westlake Village, California.

I have had extensive personal experience with both and found working with each program to be fast, accurate, and easy. All designers should have at least one of these programs available to them, to allow quick analysis of the basic parameters of a wide variety of transmission line structures.

4.3 NONLINEAR CAD ANALYSIS-OPTIMIZATION PROGRAMS

There long has been an urgent need for programs that will analyze and optimize nonlinear circuits in the same way that Touchstone and SuperCompact analyze and optimize linear circuits. This need grows out of the inherently nonlinear character of most amplifiers and *all* oscillators. Important operating parameters, such as power compression-saturation, harmonic generation, mixing products, dc to RF conversion efficiency, and bias effects simply cannot be addressed by conventional linear analysis-optimization programs. However, at the time of this writing, the situation is beginning to change and important programs are becoming available for nonlinear analysis of microwave networks. In the past, the only available nonlinear analysis software package was the popular SPICE family of programs, including a specially configured microwave version called *Microwave SPICE*®, published by EEsof. SPICE is a powerful program for nonlinear device analysis; however, its major drawback in the microwave field is its time domain nature. This means that a frequency domain calculation must be handled by SPICE one point at a time, which is very time consuming. Nevertheless, at this time, Microwave SPICE offers the only fully nonlinear design tool for oscillators, and should be carefully considered by designers who need to predict more than simple start oscillation conditions.

Two new programs, *Libra*®, published by EEsof and *Microwave Harmonica*®, published by Compact Software, solve some of the drawbacks of SPICE by providing a true nonlinear analysis in the frequency domain. Both programs use a harmonic balance technique that sums the total current at all *harmonic frequencies* (up to a user specified highest harmonic) to zero at each node, thereby satisfying Kirchhoff's laws.

Both Libra and Microwave Harmonica use nonlinear models for bipolar transistors, GaAs FETs, and diodes, which are deducible on theoretical grounds and also may be measured in the laboratory.

The harmonic balance programs offer the designer a new window on circuit analysis. The following strictly nonlinear performance parameters may be calculated using Libra and Microwave Harmonica:

- Power output
- Gain compression
- dc to RF conversion efficiency
- Harmonic generation
- Intermodulation product levels
- Bias effects
- Effects of two or more input signals
- Mixing
- Rectification

At this time, it is too early to predict what impact these nonlinear programs will have on the microwave industry. No doubt, nonlinear programs have the potential to enable designers to make a quantum leap in predicting circuit performance. It seems very likely that the ability to analyze, rather than guess at, the nonlinear properties of a microwave circuit will significantly advance the state of the designer's art, once these programs come into general use and acceptance. Also, as the popularity of the nonlinear programs increases, we expect a substantial library of nonlinear device parameters to become generally available, just as the success of Touchstone and SuperCompact have been accompanied by the availability of a large number of *linear* device parameters.

4.4 LAYOUT AND GRAPHICS PROGRAMS

The computer-aided design revolution has profoundly affected the drafting and layout field, just as programs such as Touchstone and SuperCompact have revolutionized the art and science of the circuit designing. A number of layout software tools now are available to the designer. Although not an absolute necessity in the design process, layout software has the potential to accelerate the overall design process and add significantly to its accuracy, precision, and convenience. Many designers now consider layout software a necessity.

Layout programs are available both as stand-alone tools and companions to circuit analysis-optimization programs. Several of the generic drafting programs, such as AutoCAD® by Autodesk, Inc., of Sausalito, California, are very useful for microwave circuit layout. In fact, most of the graphics in this book have been drawn using AutoCAD. The companion programs, such as EEsof's MICAD, a companion to Touchstone, offer the important advantage of being able to create a layout directly from a circuit analysis file. The ability to create a layout directly can save considerable time, as the manual design step of converting the elements of a circuit diagram into layout graphics, is eliminated. Table 4.1 lists the leading layout and graphics programs including their publishers, companion programs, if any, and a brief description.

Table 4.1 Layout and Graphics Programs

Program	Publisher	Circuit Analysis Companion	Description
AutoCAD®	AutoDesk	—	Generic drafting program * multi-levels * many options
Anvil®	MCS, Inc.	—	Generic drafting program * multilevel * many options
CALMA®, GDSII	Calma/G.E.	—	Multi-level layout package * ideal for MMICs * direct mask making capability
MICAD®	EEsof, Inc.	Touchstone	Layout companion to Touchstone. Ability to make rubyliths on plotter
Autoart®	Compact Software, Inc.	SuperCompact	Layout companion to SuperCompact

4.5 INTEGRATED DESIGN WORKSTATIONS

The microwave industry is following the lead of the silicon IC industry by moving toward integrated design software packages that are resident in workstation-level computers. These computers usually use one of the UNIX operating systems and offer superior speed and display graphics over that normally available with personal computers. The hardware for these integrated systems currently is manufactured by Apollo, Sun, and Hewlett-Packard. Normally, the hardware and the software come as a complete package, fully configured by the software publisher prior to shipment as a "turn-key" system.

Workstations have tremendous potential, matched in kind by their high price. Most users will find the enormous capability and expense of an integrated workstation justifiable only for the more complicated MMIC designs. Workstations offer the MMIC designer some unique and labor saving features.

- *Schematic capture,* a feature that allows the user to enter circuit elements as graphical entities on a schematic diagram. Element values appear graphically next to the element's symbol. Node numbering is automatic, once the designer has "drawn the schematic," making a separate circuit file unnecessary.
- *Foundry standard cell element library.* This library of standard elements, which the GaAs MMIC foundries have made available to the software publisher, allows the user to call up both circuit models and layout cells for many standard elements. See Chapter 9 for more details on standard cells.
- *Linear and nonlinear analysis software* is available simultaneously on most integrated workstations.
- *Pop-up windows* allow the user to simultaneously view the circuit's schematic diagram, electrical performance, and layout. As changes are made in either the schematic or the layout, performance parameters are updated automatically.
- Display graphics are superior to those normally available on PC computers.
- *Final output* is available directly in GDSII, and may be used to control mask making machines.

Table 4.2 summarizes the integrated workstations currently available to the microwave industry. More packages are in development at the time of this writing, so the industry may expect to see a greater variety and selection of this important class of design tools in the future.

One integrated design program is very reasonable in cost. This program is called *PUFF* (named for Peter Yarrow and Leonard Lipton's magic dragon of folk song fame). PUFF was developed by David Rutledge of the Department of Electrical Engineering, California Institute of Technology in Pasadena, and Richard Compton of the Department of Electrical Engineering, Cornell Unviersity, Ithaca, New York.

PUFF is used as a teaching tool at Cal Tech and Cornell Universities for quickly introducing students to the computer-aided designing of microwave circuits. It is not intended for the design of the highly complicated microwave circuits normally encountered in industrial applications. However, PUFF does have many of the same features as other integrated design systems, such as a schematic

Table 4.2 Integrated Design Workstations

Program	Publisher	Linear Analysis	Nonlinear Analysis	Graphics/ Layout	Output Format	Computer Platform
MMIC Design Workstation	EEsof	Touchstone	Libra	MICAD, CADANCE	GDSII	Apollo DN 3000/4000 or Sun 3
GaAs Station	Compact Software	Super Compact MIC	Microwave Harmonica	AutoArt		Apollo, Sun, various mainframes
HP85150A Microwave Design System	Hewlett-Packard	MLS	—	MAG	HP EGS ME-10/ME-30 GDS II	HP 9000
PUFF	Cal Tech/ Cornell	Yes (No optimizer)	—	Yes	Printer graphics	IBM compatible PC
Academy	EEsof	Touchstone	Libra	MICAD	CDSII, etc.	Apollo, Sun, PC 386 compatibles (coming in late 1989)

capture data entry system, graphical performance parameter output both on rectangular coordinates and on a Smith chart, and a graphics-layout display. All graphic output is displayed simultaneously in separate "windows." The layouts produced by PUFF may be printed or plotted directly on most popular printers. Also, by setting the scale factors appropriately, these layouts may be used as artwork to make photomasks.

A software package that bridges the gap between standard simulation software, such as Touchstone and SuperCompact, and integrated design workstations, is Academy® from EEsof. Academy has the ability to link Touchstone or Libra to the MICAD layout program. Data entry is by schematic capture. Like the integrated design workstations, Academy is capable of simultaneously displaying a circuit's schematic, its linear or nonlinear (with Libra) analysis, and its layout; each in its own window. Changes in either the layout or the schematic automatically updates the other as well as the simulation.

Academy is moderately priced, and runs on Sun or Apollo platforms. The publisher is planning to release a copy of Academy that will run on high-performance 386 PC platforms using the OS/2 operating system. Since 386 PC "clones" are now becoming available at low to moderate cost, Academy, running on one of these machines, soon may offer designers most of the advantages of an integrated design workstation, at an affordable cost.

4.6 HP-41C MICROWAVE DESIGN PROGRAMS

Many simple design calculations may be handled quickly and efficiently using programmable calculators such as the HP-41C. Once written, programs may be stored on magnetic cards for convenient recall. Such programs are at their best performing analytic calculations that can be coded in 300 lines or less. The following is a collection of fourteen useful HP-41C programs for microwave designs that I have developed over the last decade. Each program is listed first, followed by an example. To use any of these programs, you need only enter the program into your HP41C series calculator and run the example to gain a feel for the program's use. It is very convenient to store each program on a magnetic card for future use.

4.6.1 Varactor Diode Capacitance—Program Listing

```
01  LBL "VARACTR"           08  "V1"
02  "C0,PICOF"              09  PROMPT
03  PROMPT                  10  STO 03
04  STO 01                  11  "V2"
05  "N"                     12  PROMPT
06  PROMPT                  13  STO 04
07  STO 02                  14  RCL 04
```

15 ENTER ↑
16 .7
17 +
18 STO 05
19 RCL 03
20 ENTER ↑
21 .7
22 +
23 STO 06
24 RCL 05
25 ENTER ↑
26 RCL 06
27 /
28 ENTER ↑
29 RCL 02
30 Y ↑ X
31 STO 07
32 "C1/C2 RATIO="
33 ARCL X
34 AVIEW
35 RCL 07
36 SQRT
37 "F1/F2 RATIO="
38 ARCL X
39 AVIEW
40 RCL 03
41 ENTER ↑
42 .7
43 +
44 STO 08
45 .7

46 ENTER ↑
47 RCL 08
48 /
49 ENTER ↑
50 RCL 02
51 Y ↑ X
52 ENTER ↑
53 RCL 01
54 *
55 "C1,PICOF="
56 ARCL X
57 AVIEW
58 RCL 04
59 ENTER ↑
60 .7
61 +
62 STO 09
63 .7
64 ENTER ↑
65 RCL 09
66 /
67 ENTER ↑
68 RCL 02
69 Y ↑ X
70 ENTER ↑
71 RCL 01
72 *
73 "C2,PICOF="
74 ARCL X
75 AVIEW
76 END

Varactor Diode Capacitance—Example

```
              XEQ "VARACTR"
C0,PICOF
              2.0000    RUN
N
              1.0000    RUN
V1
              1.0000    RUN
```

V2
        10.0000   RUN
C1/C2 RATIO=6.2941
F1/F2 RATIO=2.5088
C1,   PICOF=0.8235
C2,PICOF=0.1308

4.6.2. GaAs FET Model Analysis—Program Listing

01 LBL "FET ANA"	33 ENTER ↑	65 *
02 "GATE LENGTH"	34 .7 E7	66 RCL 09
03 PROMPT	35 *	67 /
04 STO 01	36 RCL 03	68 "GM,.2IdSS="
05 "GATE WIDTH"	37 *	69 ARCL X
06 PROMPT	38 RCL 02	70 AVIEW
07 STO 02	39 *	71 RCL 08
08 "DOPING"	40 RCL 04	72 ENTER ↑
09 PROMPT	41 *	73 RCL 01
10 STO 03	42 1 E-5	74 *
11 "DEBTH, CHANNEL"	43 *	75 RCL 02
12 PROMPT	44 STO 11	76 *
13 STO 04	45 "IdSS="	77 1 E-8
14 "GATE METAL RESI"	46 ARCL X	78 *
15 PROMPT	47 AVIEW	79 STO 14
16 STO 05	48 RCL 11	80 "C, GATE="
17 "GATE METAL HEIG"	49 ENTER ↑	81 ARCL X
18 PROMPT	50 2	82 AVIEW
19 STO 06	51 *	83 RCL 05
20 "FREQUENCY"	52 .707	84 ENTER ↑
21 PROMPT	53 *	85 RCL 02
22 STO 07	54 RCL 09	86 *
23 "GATE CAP/AREA"	55 /	87 12
24 PROMPT	56 STO 13	88 /
25 STO 08	57 "GM, .5IdSS="	89 RCL 01
26 "PINCH OFF VOLTS"	58 ARCL X	90 /
27 PROMPT	59 AVIEW	91 RCL 06
28 STO 09	60 RCL 11	92 /
29 "DRAIN VOLTAGE"	61 ENTER ↑	93 1 E4
30 PROMPT	62 2	94 *
31 STO 10	63 *	95 STO 12
32 1.602 E-19	64 .447	96 8 E-2

97 ENTER ↑	131 "C, DRAIN="	165 ENTER ↑
98 RCL 01	132 ARCL X	166 RCL 14
99 *	133 AVIEW	167 *
100 RCL 02	134 1.8	168 1/X
101 /	135 ENTER ↑	169 X ↑ 2
102 2	136 RCL 01	170 RCL 18
103 /	137 *	171 *
104 RCL 04	138 RCL 02	172 RCL 19
105 /	139 /	173 *
106 1 E4	140 RCL 04	174 1 E18
107 *	141 /	175 *
108 RCL 12	142 2	176 LOG
109 +	143 *	177 10
110 STO 15	144 1 E4	178 *
111 "R, GATE="	145 *	179 "MAG="
112 ARCL X	146 STO 17	180 ARCL X
113 AVIEW	147 "R, DRAIN="	181 AVIEW
114 13	148 ARCL X	182 RCL 11
115 ENTER ↑	149 AVIEW	183 ENTER ↑
116 8.85	150 RCL 13	184 2
117 *	151 ENTER ↑	185 /
118 RCL 04	152 2	186 RCL 10
119 *	153 /	187 *
120 RCL 02	154 PI	188 .3
121 *	155 /	189 *
122 3	156 X ↑ 2	190 LOG
123 /	157 STO 18	191 10
124 RCL 01	158 RCL 17	192 *
125 /	159 ENTER ↑	193 "POWER="
126 1 E-4	160 RCL 15	194 ARCL X
127 *	161 /	195 AVIEW
128 2	162 SQRT	196 "END"
129 /	163 STO 19	197 END
130 STO 16	164 RCL 07	

GaAs FET Model Analysis—Example

```
        XEQ "FET ANA"
GATE   LENGTH   (mi-
crons)
          1.0000    RUN
```

GATE WIDTH (microns)
          500.0000   RUN
DOPING (cm^{-3})
          1.2+17   RUN
DEBTH, CHANNEL (microns)
          .2000   RUN
GATE METAL RESI (ohms-cm)
          2.7-06   RUN
GATE METAL HEIG (microns)
          .2000   RUN
FREQUENCY (Hz)
          13+09   RUN
GATE CAP/AREA (pF/cm^2)
          7+04   RUN
PINCH OFF VOLTS (volts)
          .0000   CLX
          4.0000   RUN
DRAIN VOLTAGE (V)
          6.0000   RUN
IdSS=134.5680 (mA)
GM, .5IdSS=47.5698 (ms)
GM, .2IdSS=30.0759 (ms)
C, GATE=0.3500 (pF)
R, GATE=9.6250 (ohms)
C, DRAIN=0.1918 (pF)
R, DRAIN=360.0000 (ohms)
MAG=12.2873 (dB)
POWER=20.8318 (dBm)

4.6.3 The Inductance of a Wire—Program Listing

01 LBL "WIREIND"
02 RCL 02
03 ENTER ↑
04 RCL 01
05 /

06 4
07 *
08 LN
09 2
10 *
11 1.5
12 −
13 RCL 02
14 *
15 .00254
16 *
17 "IND="
18 ARCL X
19 AVIEW
20 "END"
21 END

The Inductance of a Wire—Example

D (mils) =   .7000 STO 01
L (mils) = 25.0000 STO 02
         XEQ "WIREIND"
IND=0.5349 nH

4.6.4. Thermal Resistance of a Semiconductor Die—Program Listing

01 LBL "THERM R"	16 ARCL X	31 AVIEW
02 RCL 12	17 AVIEW	32 RCL 06
03 ENTER ↑	18 RCL 13	33 ENTER ↑
04 2.54 E-3	19 ENTER ↑	34 RCL 07
05 /	20 2.54 E-3	35 +
06 RCL 06	21 /	36 1/X
07 /	22 RCL 06	37 1 E3
08 RCL 07	23 /	38 *
09 /	24 RCL 07	39 2.54
10 RCL 01	25 /	40 /
11 /	26 RCL 02	41 RCL 03
12 RCL 12	27 /	42 /
13 *	28 STO 18	43 STO 19
14 STO 17	29 "R2="	44 "R3="
15 "R1="	30 ARCL X	45 ARCL X

46 AVIEW	66 1/X	86 +
47 RCL 08	67 1 E3	87 STO 22
48 ENTER ↑	68 *	88 "RT="
49 RCL 09	69 2.54	89 ARCL X
50 +	70 /	90 AVIEW
51 1/X	71 RCL 05	91 1
52 1 E3	72 /	92 ENTER ↑
53 *	73 STO 21	93 RCL 16
54 2.54	74 "R5="	94 −
55 /	75 ARCL X	95 RCL 15
56 RCL 04	76 AVIEW	96 *
57 /	77 RCL 17	97 RCL 22
58 STO 20	78 ENTER ↑	98 *
59 "R4="	79 RCL 18	99 RCL 14
60 ARCL X	80 +	100 +
61 AVIEW	81 RCL 19	101 "TJ="
62 RCL 10	82 +	102 ARCL X
63 ENTER ↑	83 RCL 20	103 AVIEW
64 RCL 11	84 +	104 "END"
65 +	85 RCL 21	105 END

Thermal Resistance of a Semiconductor Die—Example

K_{DIE}(W-cm^{-1}-°C^{-1})	1.5000	STO 01
$K_{DIE\ ATTACH}$	= 3.0000	STO 02
K_{RIB}	=3.90000	STO 03
$K_{CARRIER}$	= .1700	STO 04
$K_{HOUSING}$	= .9700	STO 05
DIE W × L (mils)	15.0000	STO 06
	15.0000	STO 07
RIB W × L (mils)	25.0000	STO 08
	200.0000	STO 09
CARRIER W × L (mils)	500.0000	STO 10
	500.0000	STO 11
DIE THICKNESS (mils) =	5.0000	STO 12
DIE ATTACH THICK =	1.0000	STO 13
T_A (°C)	= 85.0000	STO 14
P_{dc} (W)	= .3500	STO 15
P_{RF} (W)	= .0100	STO 16
	XEQ "THERM R"	
R1 =29.1630 :	CHIP (°C/W)	

R2 = 0.5833 : DIE ATTACH (°C/W)
R3 = 3.3650 : RIB (°C/W)
R4 = 10.2928 : CARRIER (°C/W)
R5 = 0.4059 : HOUSING (°C/W)
RT = 43.8100 : TOTAL (°C/W)
TJ = 100.1801: JUNCTION TEMP. (°C)

4.6.5 Large-Signal Power, Efficiency, and Load Resistance of a GaAs FET— Program Listing

01 LBL "PWCALC"
02 "IdSSM ?"
03 PROMPT
04 STO 01
05 "VbGD ?"
06 PROMPT
07 STO 02
08 "VP ?"
09 PROMPT
10 STO 03
11 "VS ?"
12 PROMPT
13 STO 04
14 "GAIN ?"
15 PROMPT
16 ENTER ↑
17 10
18 /
19 10 ↑ X
20 STO 05
21 RCL 02
22 ENTER ↑
23 RCL 03
24 −
25 RCL 04
26 −
27 STO 06
28 RCL 01
29 *
30 8
31 /
32 LOG
33 10
34 *
35 "PO="
36 ARCL X
37 AVIEW
38 RCL 06
39 ENTER ↑
40 RCL 04
41 +
42 RCL 04
43 +
44 STO 07
45 RCL 05
46 1/X
47 CHS
48 ENTER ↑
49 1.0
50 +
51 RCL 06
52 *
53 2
54 /
55 RCL 07
56 /
57 100
58 *
59 "EFF="
60 ARCL X
61 AVIEW
62 RCL 06

63 ENTER ↑
64 RCL 01
65 /
66 1000
67 *

68 "RL="
69 ARCL X
70 AVIEW
71 "END"
72 END

Large-Signal Power, Efficiency, and Load Resistance of a GaAs FET—Example

```
         XEQ "PWCALC"
IdSSM ? (mA)
         250.0000      RUN
VbGD ? (volts)
          15.0000      RUN
VP ? (volts)
           3.0000      RUN
VS ? (volts)
           1.5000      RUN
GAIN ? (dB)
           6.0000      RUN
PO=25.1604 (dBm)
EFF=29.1204 (%)
RL=42.0000 (Ω)
```

4.6.6 Tee Attenuator—Program Listing

01 LBL "PAD"
02 "R2?"
03 PROMPT
04 STO 00
05 SCL 00
06 X ↑ 2
07 ENTER ↑
08 2500
09 +
10 SQRT
11 RCL 00
12 −
13 STO 01
14 "R1="
15 ARCL X
16 AVIEW

17 RCL 01
18 ENTER ↑
19 RCL 00
20 +
21 50
22 +
23 RCL 01
24 *
25 STO 02
26 RCL 01
27 ENTER ↑
28 50
29 +
30 RCL 00
31 *
32 RCL 02

33 +
34 STO 03
35 50
36 ENTER ↑
37 RCL 00
38 *
39 RCL 03
40 /

41 LOG
42 20
43 *
44 "ATTEN,dB="
45 ARCL X
46 AVIEW
47 "END"
48 END

Tee Attenuator—Example

$$\text{XEQ ''PAD''}$$
R2? : SHUNT RESISTOR-Ω
140.0000   RUN
R1=8.6607 : SERIES RESISTORS-Ω
ATTEN,dB=−3.0397

4.6.7 Inductance of a Metal Strip—Program Listing

01 LBL "WIRE"
02 RCL 00
03 ENTER ↑
04 2
05 *
06 STO 05
07 RCL 01
08 ENTER ↑
09 RCL 02
10 +
11 STO 06
12 RCL 05
13 ENTER ↑
14 RCL 06
15 /
16 LN
17 STO 03
18 RCL 01
19 ENTER ↑
20 RCL 02
21 +
22 RCL 00

23 /
24 .2235
25 *
26 STO 04
27 RCL 03
28 ENTER ↑
29 .500
30 +
31 RCL 04
32 +
33 2 E-4
34 *
35 RCL 00
36 *
37 STO 07
38 "END"
39 END

Inductance of a Metal Strip—Example

LENGTH (microns)   = 250.0000 STO 00
WIDTH (microns)   = 20.0000 STO 01
THICKNESS (microns) =   5.0000 STO 02
                       XEQ "WIRE"
                       0.1759 nH***

4.6.8 Mutual Inductance of Two Metal Strips—Program Listing

01 LBL "MUT"
02 RCL 00
03 ENTER ↑
04 RCL 01
05 /
06 STO 02
07 RCL 02
08 LN
09 STO 03
10 RCL 02
11 1/X
12 X↑2
13 ENTER ↑

14 .25
15 *
16 −1
17 *
18 RCL 02
19 1/X
20 +
21 1.0
22 −
23 RCL 03
24 +
25 RCL 00
26 *
27 2 E-4
28 *
29 STO 04
30 "END"
31 END

Mutual Inductance of Two Metal Strips—Example

LENGTH (microns) = 250.0000 STO 00
SPACING (microns) =   5.0000 STO 01
                  XEQ "MUT"
                  0.1466 nH***

4.6.9 Two-Turn Spiral Inductors (see Figure 4.2)—Program Listing

01 LBL "2TURN"	14 "T?"	27 PROMPT
02 "A?"	15 PROMPT	28 STO 08
03 PROMPT	16 STO 04	29 RCL 02
04 STO 00	17 "RS?"	30 ENTER ↑
05 "B?"	18 PROMPT	31 −4.0
06 PROMPT	19 STO 05	32 *
07 STO 01	20 "LB?"	33 RCL 00
08 "W?"	21 PROMPT	34 +
09 PROMPT	22 STO 06	35 RCL 01
10 STO 02	23 "LENG B?"	36 +
11 "S?"	24 PROMPT	37 4.0
12 PROMPT	25 STO 07	38 *
13 STO 03	26 "Ee?"	39 STO 27

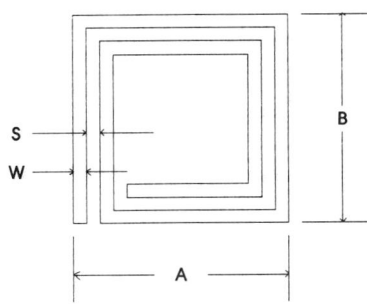

Fig. 4.2 The layout of a two-turn spiral inductor. Dimensions in microns.

40 RCL 03	68 RCL 09	96 RCL 02
41 ENTER ↑	69 +	97 ENTER ↑
42 −9.0	70 1/X	98 RCL 04
43 *	71 ENTER ↑	99 +
44 RCL 27	72 3 E5	100 RCL 09
45 +	73 *	101 /
46 STO 09	74 4.0	102 .2235
47 RCL 05	75 /	103 *
48 ENTER ↑	76 RCL 08	104 STO 29
49 RCL 09	77 SQRT	105 RCL 28
50 *	78 /	106 ENTER ↑
51 RCL 02	79 STO 13	107 .50
52 /	80 RCL 09	108 +
53 STO 10	81 ENTER ↑	109 RCL 29
54 RCL 10	82 2.0	110 +
55 ENTER ↑	83 *	111 2.0 E-4
56 7.0	84 STO 30	112 *
57 *	85 RCL 02	113 RCL 09
58 STO 11	86 ENTER ↑	114 *
59 RCL 11	87 RCL 04	115 1000
60 ENTER ↑	88 +	116 *
61 6.0	89 STO 31	117 STO 14
62 /	90 RCL 30	118 RCL 02
63 STO 12	91 ENTER ↑	119 ENTER ↑
64 RCL 07	92 RCL 31	120 RCL 03
65 ENTER ↑	93 /	121 +
66 7.0	94 LN	122 STO 35
67 *	95 STO 28	123 RCL 01

124 ENTER ↑	166 XEQ "MUT"	208 *
125 RCL 35	167 RCL 32	209 RCL 06
126 −	168 STO 22	210 ENTER ↑
127 STO 15	169 LBL "MUT"	211 7000
128 RCL 35	170 RCL 26	212 *
129 ENTER ↑	171 ENTER ↑	213 +
130 −2.0	172 RCL 03	214 RCL 14
131 *	173 /	215 +
132 RCL 00	174 STO 33	216 STO 23
133 +	175 RCL 33	217 RCL 23
134 STO 16	176 LN	218 ENTER ↑
135 RCL 35	177 STO 34	219 .20
136 ENTER ↑	178 RCL 33	220 *
137 −2.0	179 1/X	221 STO 24
138 *	180 X↑2	222 RCL 13
139 RCL 01	181 ENTER ↑	223 ENTER ↑
140 +	182 .25	224 2.0
141 STO 17	183 *	225 *
142 RCL 35	184 −1	226 PI
143 ENTER ↑	185 *	227 *
144 −3.0	186 RCL 33	228 X↑2
145 *	187 1/X	229 RCL 23
146 RCL 00	188 +	230 *
147 +	189 1.0	231 1.2
148 STO 18	190 −	232 *
149 RCL 15	191 RCL 34	233 1/X
150 STO 26	192 +	234 1 E9
151 XEQ "MUT"	193 RCL 26	235 *
152 RCL 32	194 *	236 STO 25
153 STO 19	195 2.0 E-4	237 RCL 09
154 RCL 16	196 *	238 "LENGTH="
155 STO 26	197 STO 32	239 ARCL X
156 XEQ "MUT"	198 "RTN"	240 AVIEW
157 RCL 32	199 RCL 19	241 RCL 10
158 STO 20	200 ENTER ↑	242 "RDC="
159 RCL 17	201 RCL 20	243 ARCL X
160 STO 26	202 +	244 AVIEW
161 XEQ "MUT"	203 RCL 21	245 RCL 11
162 RCL 32	204 +	246 "RI="
163 STO 21	205 RCL 22	247 ARCL X
164 RCL 18	206 +	248 AVIEW
165 STO 26	207 2000	249 RCL 12

```
250 "R2="              257 RCL 24           264 AVIEW
251 ARCL X             258 "L2="            265 RCL 13
252 AVIEW              259 ARCL X           266 "FR="
253 RCL 23             260 AVIEW            267 ARCL X
254 "L1="              261 RCL 25           268 AVIEW
255 ARCL X             262 "CP="            269 STOP
256 AVIEW              263 ARCL X           270 END
```

Two-Turn Spiral Inductors (See Figure 4.2)—Example

                     XEQ "2TURN"

A? (microns)
                   250.0000   RUN
B? (microns)
                   250.0000   RUN
W? (microns)
                     20.0000   RUN
S? (microns)
                     20.0000   RUN
T? (microns)
                      5.0000   RUN
RS? (metal resistivity, Ω/square)
                      .0200   RUN
LB? (bend inductance, nH)
                   −.0600   RUN
LENG B? (equiv. bend length, microns)
                 −40.0000   RUN
Ee? (effective dielectric constant)
                    8.5000   RUN
LENGTH=1,500.0000 microns
RDC=1.5000 ⎫
R1=10.5000 ⎬   ohms
R2=1.7500 ⎭
L1=1,167.3650 pH
L2=233.4730 pH
CP=40.6695 fF
FR=21.0859 GHz.

4.6.10 Three-Turn Spiral Inductors (see Figure 4.3)—Program Listing

01 ◆LBL "3TURN"   02 "A?"   03 PROMPT

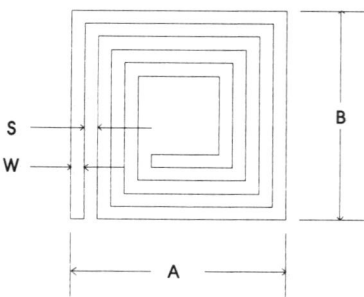

Fig. 4.3 The layout of a three-turn spiral inductor. Dimensions in microns.

04 STO 00	32 +	60 *
05 "B?"	33 6.0	61 STO 11
06 PROMPT	34 *	62 RCL 11
07 STO 01	35 STO 27	63 ENTER ↑
08 "W?"	36 RCL 02	64 6.0
09 PROMPT	37 ENTER ↑	65 /
10 STO 02	38 35.0	66 STO 12
11 "S?"	39 *	67 RCL 07
12 PROMPT	40 STO 36	68 ENTER ↑
13 STO 03	41 RCL 03	69 7.0
14 "T?"	42 ENTER ↑	70 *
15 PROMPT	43 −25.6	71 RCL 09
16 STO 04	44 *	72 +
17 "RS?"	45 RCL 36	73 1/X
18 PROMPT	46 −	74 ENTER ↑
19 STO 05	47 RCL 27	75 3 E5
20 "LB?"	48 +	76 *
21 PROMPT	49 STO 09	77 4.0
22 STO 06	50 RCL 05	78 /
23 "LENG B?"	51 ENTER ↑	79 RCL 08
24 PROMPT	52 RCL 09	80 SQRT
25 STO 07	53 *	81 /
26 "Ee?"	54 RCL 02	82 STO 13
27 PROMPT	55 /	83 RCL 09
28 STO 08	56 STO 10	84 ENTER ↑
29 RCL 00	57 RCL 10	85 2.0
30 ENTER ↑	58 ENTER ↑	86 *
31 RCL 01	59 7.0	87 STO 30

```
 88 RCL 02
 89 ENTER ↑
 90 RCL 04
 91 +
 92 STO 31
 93 RCL 30
 94 ENTER ↑
 95 RCL 31
 96 /
 97 LN
 98 STO 28
 99 RCL 02
100 ENTER ↑
101 RCL 04
102 +
103 RCL 09
104 /
105 .2235
106 *
107 STO 29
108 RCL 28
109 ENTER ↑
110 2.0 E-4
111 *
112 RCL 09
113 *
114 1000
115 *
116 STO 14
118 RCL 02
118 ENTER ↑
119 RCL 03
120 +
121 STO 35
122 RCL 01
123 ENTER ↑
124 RCL 35
125 -
126 STO 15
127 RCL 35
128 ENTER ↑
129 -2.0

130 *
131 RCL 00
132 +
133 STO 16
134 RCL 35
135 ENTER ↑
136 -2.0
137 *
138 RCL 01
139 +
140 STO 17
141 RCL 35
142 ENTER ↑
147 -3.0
144 *
145 RCL 00
146 +
147 STO 18
148 RCL 35
149 ENTER ↑
150 -3.0
151 *
152 RCL 01
153 +
154 STO 37
155 RCL 35
156 ENTER ↑
157 -4.0
158 *
159 RCL 00
160 +
161 STO 38
162 RCL 35
163 ENTER ↑
164 -4.0
165 *
166 RCL 01
167 +
168 STO 39
169 RCL 35
170 ENTER ↑
171 -5.0

172 *
173 RCL 00
174 +
175 STO 40
176 RCL 15
177 STO 26
178 XEQ "MUT"
179 RCL 32
180 STO 19
181 RCL 16
182 STO 26
183 XEQ "MUT"
184 RCL 32
185 STO 20
186 RCL 17
187 STO 26
188 SEQ "MUT"
189 RCL 32
190 STO 21
191 RCL 18
192 STO 26
193 SEQ "MUT"
194 RCL 32
195 STO 22
196 RCL 37
197 STO 26
198 XEQ "MUT"
199 RCL 32
200 STO 41
201 RCL 38
202 STO 26
203 XEQ "MUT"
204 RCL 32
205 STO 42
206 RCL 39
207 STO 26
208 XEQ "MUT"
209 RCL 32
210 STO 43
211 RCL 40
212 STO 26
213 XEQ "MUT"
```

214 RCL 32	241 *	268 +
215 STO 44	242 2.0 E-4	269 RCL 14
216 ◆LBL "MUT"	243 *	270 +
217 RCL 26	244 STO 32	271 STO 23
218 ENTER ↑	245 "RTN"	272 RCL 23
219 RCL 03	246 RCL 19	273 ENTER ↑
220 /	247 ENTER ↑	274 .20
221 STO 33	248 RCL 20	275 *
222 RCL 33	249 +	276 STO 24
223 LN	250 RCL 21	277 RCL 13
224 STO 34	251 +	278 ENTER ↑
225 RCL 33	252 RCL 22	279 2.0
226 1/X	253 +	280 *
227 X ↑ 2	254 RCL 41	281 PI
228 ENTER ↑	255 +	282 *
229 .25	256 RCL 42	283 X ↑ 2
230 *	257 +	284 RCL 23
231 −1	258 RCL 43	285 *
232 *	259 +	286 1.2
233 RCL 33	260 RCL 44	287 *
234 1/X	261 +	288 1/X
235 +	262 2000	289 1 E9
236 1.0	263 *	290 *
237 −	264 RCL 06	291 STO 25
238 RCL 34	265 ENTER ↑	292 RCL 09
239 +	266 11000	293 END
240 RCL 26	267 *	

Three-Turn Spiral Inductors (See Figure 4.3)—Example

	XEQ "3TURN"	
A? (microns)		
	250.0000	RUN
B? (microns)		
	250.0000	RUN
W? (microns)		
	5.0000	RUN
S? (microns)		
	5.0000	RUN
T? (microns)		
	5.0000	RUN

RS? (metal resistivity, ohm/square)
                        .0200   RUN
LB? (bend inductance, nH)
                       −.0600   RUN
LENG B? (equiv. bend length, microns)
                     −40.0000   RUN
Ee? (effective dielectric constant)
                       8.5000   RUN
                              XEQ "PRINT"
LENGTH=2,700.0000 (microns)
RDC=10.8000 ⎤
R1=75.6000   ⎬   (ohm)
R2=12.6000   ⎦
L1=4,714.5177 pH
L2=942.9035 pH
CP=39.6231 fF
FR=10.6301 GHz

4.6.11 Printout for Three-Turn Spiral Inductors—Program Listing

```
01 ◆LBL "PRINT"
02 "LENGTH="
03 ARCL X
04 AVIEW
05 RCL 10
06 "RDC="
07 ARCL X
08 AVIEW
09 RCL 11
10 "R1="
11 ARCL X
12 AVIEW
13 RCL 12
14 "R2="
15 ARCL X
16 AVIEW
17 RCL 23
18 "L1="
19 ARCL X
20 AVIEW
21 RCL 24
```

22 "L2="
23 ARCL X
24 AVIEW
25 RCL 25
26 "CP="
27 ARCL X
28 AVIEW
29 RCL 13
30 "FR="
31 ARCL X
32 AVIEW
33 "STOP"
34 "END"
35 END

Printout for Three-Turn Spiral Inductors—Example

	XEQ "3TURN"	
A? (microns)		
	250.0000	RUN
B? (microns)		
	250.0000	RUN
W? (microns)		
	5.0000	RUN
S? (microns)		
	5.0000	RUN
T? (microns)		
	5.0000	RUN
RS? (metal resistivity, ohm/square)		
	.0200	RUN
LB? (bend inductance, nH)		
	−.0600	RUN
LENG B? (equiv. bend length, microns)		
	−40.0000	RUN
Ee? (effective dielectric constant)		
	8.5000	RUN
	XEQ "PRINT"	

LENGTH=2,700.0000 (microns)
RDC=10.8000
R1=75.6000   (ohm)
R2=12.6000

L1=4,714.5177 pH
L2=942.9035 pH
CP=39.6231 fF
FR=10.6301 GHz

4.6.12 Schottky Barrier Doides I-V—Program Listing

```
01 ◆LBL "SCHOT"
02 RCL 02
03 ENTER ↑
04 RCL 03
05 /
06 E ↑ X
07 1.0
08 -
09 RCL 01
10 *
11 "END"
12 END
```

Schottky Barrier Diodes I-V—Example

I_s (amps) = 5-15 STO 01
V (volts) = .7500 STO 02
kt/e (volts) = .0260 STO 03
　　　　XEQ "SCHOT"
　　　　0.0169 AMPS*
　　　　1.0000 STO 02
　　　　　　RUN
　　252.6992   ***
　　　　.8000 STO 02
　　　　　　RUN
　　　0.1153   ***
　　　　.8500 STO 02
　　　　　　RUN
　　　0.7890   ***

4.6.13 Wound Coil Inductance—Program Listing

```
01 ◆LBL "COIL"
```

02 RCL 00
03 ENTER ↑
04 9
05 *
06 STO 03
07 RCL 01
08 ENTER ↑
09 10
10 *
11 STO 04
12 RCL 03
13 ENTER ↑
14 RCL 04
15 +
16 STO 05
17 RCL 00
18 ENTER ↑
19 RCL 02
20 *
21 X↑2
22 RCL 05
23 /
24 "END"
25 END

Wound Coil Inductance—Example

DIAMETER (mils) = 20.0000 STO 00
LENGTH (mils)   = 30.0000 STO 01
TURNS   = 5.0000 STO 02
                  XEQ "COIL"
                  20.8333 nH***

4.6.14 Distributed Amplifier Basics—Program Listing

01 ◆LBL "DISTAMP"
02 RCL 01
03 ENTER ↑
04 RCL 02
05 *
06 2
07 /
08 PI
09 /
10 RCL 03
11 /
12 RCL 00

13 /
14 LOG
15 20
16 *
17 STO 05
18 50
19 ENTER ↑
20 PI
21 *
22 RCL 03
23 *
24 RCL 00
25 *
26 1/X
27 1000
28 *
29 STO 06
30 50
31 ENTER ↑
32 PI
33 /
34 RCL 00
35 /
36 STO 07
37 RCL 05
38 ENTER ↑
39 20

40 /
41 10 ↑ X
42 RCL 04
43 *
44 50
45 /
46 RCL 02
47 /
48 STO 08
49 RCL 05
50 "AO="
51 ARCL X
52 AVIEW
53 RCL 06
54 "W="
55 ARCL X
56 AVIEW
57 RCL 07
58 "L="
59 ARCL X
60 AVIEW
61 RCL 08
62 "IDC="
63 ARCL X
64 AVIEW
65 "END"
66 END

Distributed Amplifier Basics—Example

f_c (GHz) = 30.0000 STO 00
G_M' (ms/mm) = 150.0000 STO 02
N = 5.0000 STO 01
C'_{gs} (pF/mm) = 1.0000 STO 03
I'_{DSS} (mA/mm) = 250.0000 STO 04
XEQ "DISTAMP"
AO=11.9952 GAIN (dB)
W=0.2122 GATE WIDTH (mm)
L=0.5305 LINE INDUCTANCE (nH)
IDC=0.1326 AMPS (dc)

Chapter 5
Amplifier Circuits

5.1 THE BASICS

5.1.1 Matching Techniques

All microwave transistor amplifiers, except distributed amplifiers, rely on some form of matching to achieve flat gain over a desired frequency range. We mathematically define a conjugate match as $\Gamma_l = \Gamma_g^*$ [1], where Γ_g is the transistor's reflection coefficient, and Γ_l is the load's reflection coefficient. As was discussed in Subsection 2.1.4.4, FETs have a maximum available gain that is a decreasing function of frequency. Any amplifier designed to operate from f_l to f_h, will have no more gain at f_h than MAG(f_h). In fact, if an amplifier is designed to have flat gain (as is usually the case), it will have no more gain than MAG(f_h) over the entire band f_l to f_h. Matching for flat gain becomes a matter of providing the best possible match into and out of the FET at the frequency f_h to achieve a gain close to MAG(f_h) and also selectively to mismatch the FET between f_l and f_h to "throw away" gain, in order to maintain flat gain over f_l to f_h. Mismatching must compensate for the FET's increasing MAG as frequency decreases. Refer to Figure 5.1 for a graphic representation of this concept.

If an amplifier is to operate over a narrow bandwidth, the matching problem becomes a simple matter of achieving a good match at f_h. But if a broad bandwidth is desired, the selective mismatch concept must be employed. Mismatch loss, M_c, is defined as [2] the ratio of the power available from a generator to the power delivered to a load

$$M_C = \frac{|1 - \Gamma_L \Gamma_g|^2}{(1 - |\Gamma_L|^2)(1 - |\Gamma_g|^2)} \tag{5.1}$$

where Γ_l is the load's reflection coefficient and Γ_g is the generator's reflection coefficient. Mismatch may be applied either in the FET's input, or its output.

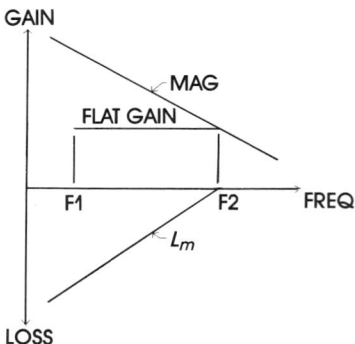

Fig. 5.1 Amplifier gain equalization with the introduction of selective mismatch loss, L_M.

5.1.2 Gain Compensation

The selective mismatch concept can be applied either to the circuit between the generator and the input of the FET or to the circuit between the FET and its load. It is common practice, however, to provide a selective mismatch only at the input of the FET and to conjugately match the FET's output to maximize power transfer to the load. Figure 5.2 shows an example of a typical Smith chart display of an amplifier's input and output matching network, with selective mismatch applied to the FET's input and an optimum match at its output. Note how the FET's input reflection coefficient matches only the conjugate of the input matching circuit's reflection coefficient at f_h, whereas the FET's output is closely matched to its output circuit over the entire frequency range.

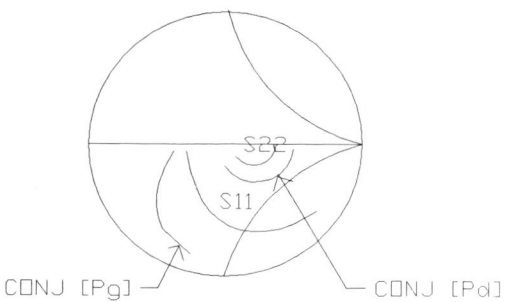

Fig. 5.2 A Smith chart display of the matching process within a transistor amplifier.

Some examples of typical matching networks are given in Figure 5.3. In their simple lumped element forms, these networks are either of a low-pass or a high-pass type. Matching networks often are natural extensions of the elements within the FET's own equivalent circuit. Realistic circuits will use microstrip elements to play the role of these ideal lumped elements.

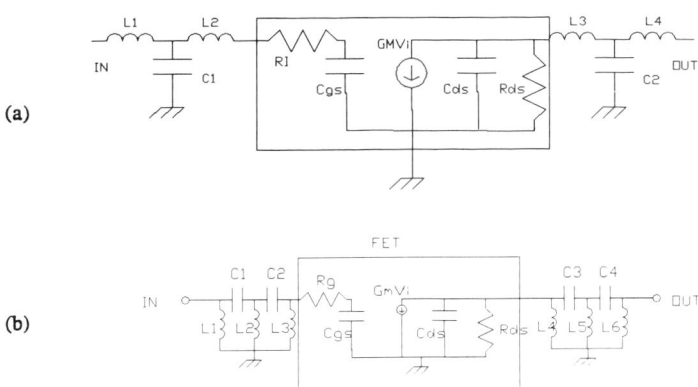

Fig. 5.3 Lumped element (a) low-pass and (b) high-pass matching of FET amplifiers.

5.1.3 Fano's Limit

Calculations by Fano [3] have shown a bandwidth limit for reactively matched amplifiers. Assuming that the input matching is provided by a high-pass network, and the output matching is provided by a low-pass network, the gain of the transistor amplifier is given by

$$g_a(f) = g_0 \left(\frac{f}{f_h}\right)^{-k} \tag{5.2}$$

where g_0 is the gain at f_h and $k = 2$ for 6 db/octave gain rolloff which is typical for a single stage. The mismatch loss for flat gain from f_l to f_h must be

$$L_M(f) = K \left(\frac{f}{f_h}\right)^k \tag{5.3}$$

where K is a gain reduction factor, which predicts how far below MAG at the high end the gain must be to realize flat gain over the bandwidth f_l to f_h.

If the high-pass input matching network is lossless and reciprocal, Fano [4] has shown that

$$\pi \tau_{hP} \geq \int_0^\infty \frac{1}{\Omega^2} \ln \left| \frac{1}{\Gamma_{in}} \right| d\Omega \qquad (5.4)$$

where
- $\Omega = f/f_h$ = the normalized frequency,
- $\tau_{hP} = 2\pi f_h R_{in} C_{in}$,
- R_{in} = the transistor's input resistance (R_g),
- C_{in} = the transistor's input capacitance (C_{gs}),
- Γ_{in} = the input reflection coefficient

Equation (5.4) tells us that Γ_{in} cannot be zero over any finite bandwidth. If Γ_{in} is sloped with frequency such that it is lower at f_h (high end) than at f_l (low end), we can then show that

$$\int_{f_l}^1 \frac{1}{\Omega^2} \ln \left| \frac{1}{1 - K(f/f_h)^K} \right| d\Omega \leq 2\pi \tau_{hP} \qquad (5.5)$$

Figure 5.4 shows how much bandwidth is achievable with no high end-gain reduction for various degrees of input slope.

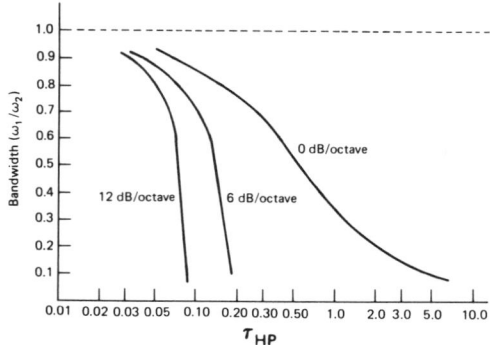

Fig. 5.4 Maximum Fano bandwidth for an FET amplifier limited by its input (high-pass) circuit. No reduction in gain relative to MAG at the high-frequency limit can occur. The most common input slope is 6dB/octave for a single stage. (From [1] pp. 114–118, with permission.)

In a similar way we can see that the output low-pass network bandwidth is governed by

$$\int_{f_l}^{1} \ln \left| \frac{1}{1 - K(f/f_h)^k} \right| d\Omega \leq \frac{2\pi}{\tau_{lP}} \tag{5.6}$$

where

$\tau_{lP} = 2\pi f_h R_{out} C_{out}$,
R_{out} = the output resistance of the transistor (R_{ds}),
C_{out} = the output capacitance of the transistor (C_{ds}).

Again, the achievable bandwidth for a given τ_{LP} is shown graphically in Figure 5.5, assuming various degrees of output slope and no high end-gain reduction.

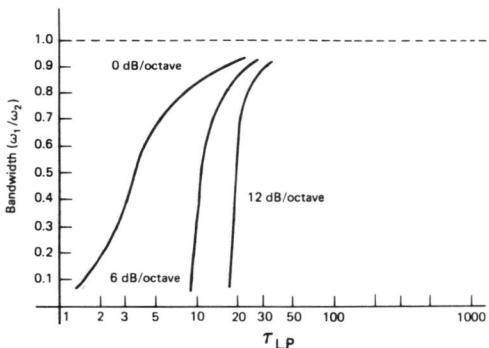

Fig. 5.5 Maximum Fano bandwidth for an FET amplifier limited by its output (low-pass) circuit. No reduction in gain relative to MAG at the high-frequency limit can occur. The most common output slope is 0 dB/octave for a single stage. (From [1] pp. 114–118, with permission.)

5.1.4 Stability

An amplifier is unconditionally stable if the source or load impedances necessary to cause an instability are outside of the $\Gamma = 1.0$ circle (that is, unachievable without a negative resistance). An amplifier is conditionally stable if the impedances that produce instability do not include 50 Ω pure real at either the input or output port.

A stability factor, k, of an amplifier can be defined in terms of its S-parameters as

$$k = \frac{1 - |S_{11}|^2 - |S_{22}|^2 + |D|^2}{2|S_{12}||S_{21}|} \tag{5.7}$$

where S_{11}, S_{21}, S_{12}, and S_{22} are the complex S-parameters of the amplifier and $D = (S_{11} S_{22} - S_{12} S_{21})$. In terms of the k factor, the stability criteria are

For unconditional stability, $k > 1.0$.
For conditional stability, $k < 1.0$.

For the designer to be sure that an amplifier design is unconditionally stable at *all* frequencies, not just in the specified band, is very important.

5.1.5 Noise Match

Matching for the lowest possible noise figure over a band of frequencies requires that a particular source impedance be presented to input of the transistor. This noise optimizing source impedance is called Γ_{opt}, and may be obtained from measurements or from the manufacturer's data sheet. The noise figure of a transistor amplifier with an arbitrary source reflection coefficient, Γ, is given by

$$F = F_{min} + 4R_N \frac{|\Gamma - \Gamma_{opt}|^2}{(1 - |\Gamma|^2)|1 + \Gamma_{opt}|^2} \tag{5.8}$$

where
F_{min} = the transistor's minimum noise figure,
R_N = the transistor's noise resistance, normalized to 50 ohms.

5.1.6 Power Match

Just as low noise matching requires that an optimum input impedance be applied to a transistor, matching for power requires that an optimum output impedance be applied to a transistor. The ideal large-signal "load line" impedance of an FET was calculated in Section 2.1.5. To achieve the highest possible power output at every frequency, it is important that an FET's output be matched to an impedance whose real part is equal to $R_l = (V_{BGD} - V_p - V_S)/I_M$, as given in (2.14).

In practice, R_l is placed in parallel with the small-signal C_{DS} of the FET to create a large-signal FET output model. R_l may then be incorporated into a set of large-signal S-parameters, or the computer program can use the "large-signal" model directly, for optimizing an output matching network. Figure 5.6 shows a comparison between the large-signal and the small-signal output models.

5.1.7 Harmonic Distortions and Intermodulation Products

Because microwave amplifiers are nonlinear devices, they generate harmonic and intermodulation distortion under large-signal conditions. Detailed analyses of dis-

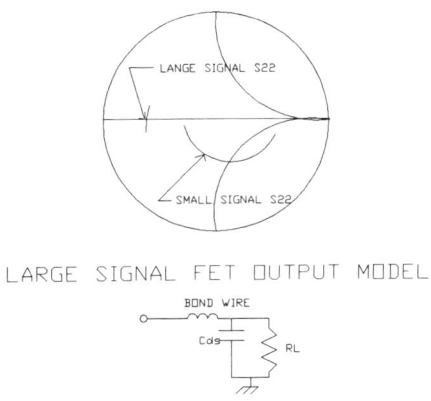

Fig. 5.6 A comparison of the small-signal and large-signal FET output models.

tortion are not readily available at present, although the harmonic balance techniques show promise of providing nonlinear analytic tools in the near future. For now, a fairly good estimate of harmonic and intermodulation distortion can be made using the intercept point concept [5]. For second harmonic distortion, the difference, Δ, in power between the fundamental and the second harmonic levels is

$$\Delta = IP_2 - P_{out}$$

where IP_2 is the second-order intercept point.

For two-tone third-order intermodulation products, the difference, Δ, in power between the main tones and the third-order products is

$$\frac{\Delta}{2} = IP_3 - P_{out}$$

where IP_3 is the third-order intercept point.

Both IP_2 and IP_3 must be determined experimentally. However, a good rule of thumb is that IP_2 is equal to the amplifier's 1 dB compression point plus 20 dB, and IP_3 is equal to the 1 dB compression point plus 10 dB.

5.1.8 Gain Drift with Temperature

All amplifiers experience some gain drift with changing temperature. This change in gain with temperature must be planned into the design of most amplifiers to meet the minimum gain requirements at temperature extremes. Sometimes circuits called *compensators* are employed to counteract the natural gain drift of an amplifier and reduce the effects of temperature on gain. PIN diode attenuators often are used as temperature compensators.

The gain of GaAs FET devices decreases with increasing temperature as a direct result of the saturated drift velocity's change with temperature in GaAs. Temperature drift of gain works out to a very universal number of $-.015$ dB/°C per FET stage. Almost all GaAs FET amplifiers obey this rule.

5.2 AMPLIFIER PERFORMANCE BY BASIC TYPE

5.2.1 Reactively Matched Amplifiers

Reactively matched amplifiers are so called because they make use of lossless matching networks to achieve gain compensation by selectively creating reflections at the interface between the matching network and the device. Figure 5.7 shows several typical topologies for reactively matched amplifiers using microstrip lines and stubs as matching elements. Even the simplest L network is capable of providing flat gain over more than an octave of bandwidth. It is natural for the reflected power from the matching network to be quite high at the low end of the band, with this class of amplifier.

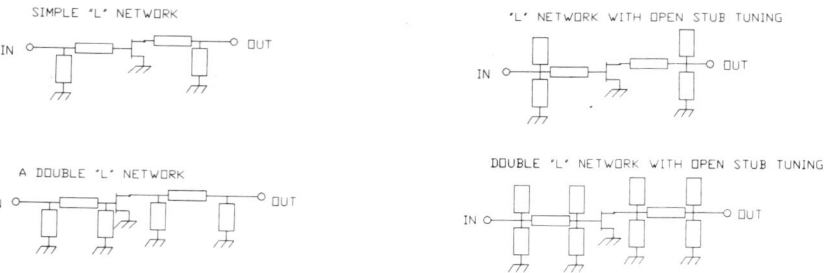

Fig. 5.7 Simple FET amplifier matching networks.

Reactively matched amplifier circuits do not exactly conform to the ideal lumped element topologies discussed in Section 5.1.2 They usually are adaptations of high-pass matching circuits to a microstrip format. Notice that the series capacitors, which are difficult to realize in microstrip have been replaced by microstrip lines. Also, reactively matching networks of proper design are capable of providing low-noise matching or power matching.

Reactively matched amplifiers can be top performers in terms of bandwidth, gain, and power output. Because these matching networks are composed exclusively of lossless reactive elements, the maximum gain, minimum noise figure, and maximum power output can be achieved within the bounds of Fano's limit for a given FET with a properly designed matching network. The only real disadvantage of this class of amplifier is its relatively poor low-frequency match. As we shall see in the next section, the balanced amplifier concept will solve this problem.

5.2.2 Balanced Amplifiers

A way to improve the input and output match of a reactively matched amplifier is to use Lange couplers to create a so-called balanced configuration. Figure 5.8 shows the signal flow paths within a balanced amplifier. The input power is split into two equal signal paths offset in-phase by 90° by the first Lange coupler. The two signal paths, which are each attenuated by 3 dB, are amplified by two *identical* "single-ended" amplifiers. These two signal paths are then recombined in a second Lange coupler. The two paths add in-phase at the 90° output port of the second coupler and cancel at the 0° output port because the two signal paths already have been phase-shifted by 90° in the first Lange coupler. A termination is provided at the cancelling port. Because the two signal paths add in-phase at the 90° output, there is a 3 dB net power gain, which cancels out the 3 dB power loss in the input Lange coupler. The net gain from the input to the output simply is the gain of *one* of the single-ended amplifiers. However, the power output is the combined power of the two amplifiers, which is 3 dB above the power of an individual amplifier.

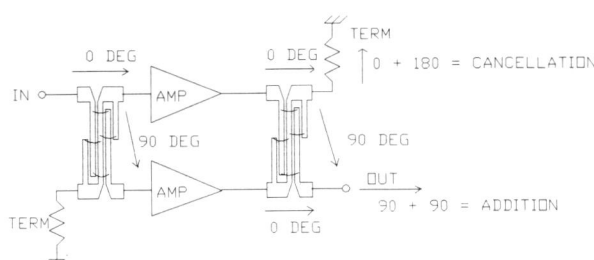

Fig. 5.8 Basic operation of a balanced amplifier.

The input or output return loss of a balanced amplifier is good no matter how poor the match of the basic amplifier because any reflections from the amplifiers always will add in-phase at the terminations (dissipating the reflections). Reflections will be out-of-phase and therefore self-cancelling at the input or the output ports. Thus, a balanced amplifier's input and output match is determined by its Lange couplers and terminations.

Balanced amplifiers have the following *advantages*:
1. They have high gain, approaching MAG at f_h.
2. They have high power output; typically 3 dB higher than that of one amplifier.
3. They have good input-output return loss, independent of the amplifier's matching.
4. They can be cascaded with other amplifiers, lowering the labor costs of a high gain multistage amplifier.

Also, balanced amplifiers have the following *disadvantages*:
1. Their bandwidth is limited to about 3 : 1 by the Lange couplers.
2. Their dc power consumption is two times the dc power of a single-ended amplifier.
3. Their size is large (especially at low frequencies) due to the two Lange couplers.
4. They require twice the number of active devices as a single-ended amplifier.

5.2.3 Lossy Match Amplifiers

Lossy match amplifiers [6] use resistors in their matching networks to provide lossy gain compensation at low frequencies, while maintaining a good input and output match over a broad bandwidth. The gain of a lossy match amplifier always is somewhat lower than an equivalent reactively matched amplifier because of the presence of loss in its matching elements. In some applications, the simplicity, and small size, of a single-ended lossy match amplifier with good input-output matches is a very useful trade-off in spite of reduced gain and power output.

Figure 5.9 shows the basic topology of a lossy match amplifier. The shorted stub-resistor network acts as a gain equalizer, providing high attenuation at low frequencies and little or no attenuation at high frequencies. The series inductors provide conjugate matching into and out of the device at high frequencies. The *advantages* of a lossy match amplifier are

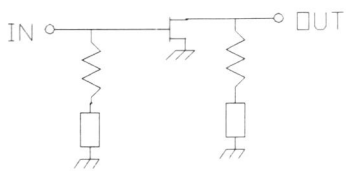

Fig. 5.9 The basic topology of a lossy match amplifier.

1. They provide good gain and reasonable input-output return loss with a single-ended topology, over a broad bandwidth (up to 4 : 1).
2. They have relatively low dc power consumption.
3. They are small in size.
4. They can be cascaded.

The *disadvantages* are
1. They have lower gain than reactively matched amplifiers.
2. They have lower power output than other amplifiers.
3. The have a higher noise figure at low frequencies due to losses in the input matching network.

5.2.4 Feedback Amplifier

Applying negative feedback from the drain to the gate of an FET [7] is an effective way of simultaneously achieving flat gain and good input-output match. Consider the basic topology of a GaAs FET feedback amplifier shown in Figure 5.10. At sufficiently low frequencies where the FET's parasitic model elements do not enter into the calculation, the voltage gain with resistive (R) feedback from the output to the input is

$$g_V = \frac{V_l}{V_i} = \frac{R}{Z_0} \tag{5.9}$$

This assumes that the open-loop gain of the FET is much greater than g_V. The FET's open-loop voltage gain is given by

$$g_V = g_m Z_0 \tag{5.10}$$

Therefore, if feedback is to work, the open-loop gain must be greater than the gain with feedback. This means

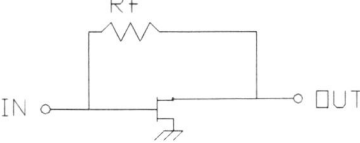

Fig. 5.10 The basic topology of a feedback amplifier.

$$g_m Z_0 > \frac{R}{Z_0}$$

or

$$g_m > \frac{R}{Z_0^2} \tag{5.11}$$

For $Z_0 = 50\ \Omega$ and $g_V = 5.0$ (or 7 dB)

$$\frac{R}{Z_0} = 5.0$$

or

$$R = (5.0)(50.0) = 250\ \Omega$$

and

$$g_m > \frac{250}{(50)^2} = .100\ \text{S or }100\ \text{mS}$$

This condition fixes the lower limit on the FET's transconductance in a feedback amplifier design. The power gain of a feedback amplifier is

$$g_P = \frac{P_{\text{out}}}{P_{\text{in}}}$$

where

$$P_{\text{out}} = \frac{V_l^2}{Z_0}$$

$$= \frac{(R/Z_0)^2 V_i^2}{Z_0}$$

$$= \frac{R^2 V_i^2}{Z_0^3}$$

and

$$P_{\text{in}} = \frac{V_i^2}{4 Z_0}$$

Therefore, the power gain is

$$g_P = \frac{1}{4}\left(\frac{R}{Z_0}\right)^2 \qquad (5.12)$$

or, in terms of decibels:

$$g_P = 20 \log\left(\frac{R}{2Z_0}\right) \qquad (5.13)$$

Of course, parasitic elements within the FET will modify this result. Equation (5.13) should be regarded as a rule of thumb that provides a first estimate of the gain of a feedback amplifier. In practical amplifiers, the feedback resistor itself would become a series L-R-C network due to layout and biasing parasitic elements. Some matching may be necessary to account for the nonideal nature of both the FET and the feedback resistor. In very broadband designs, feedback may be ineffective at the high-frequency end of the band due to parasitic effects. Such amplifiers have a dual nature. They function as feedback amplifiers at low frequencies and as reactively (or lossy) matched amplifiers at high frequencies. Figure 5.11 shows a typical topology for this type of "practical" feedback amplifier.

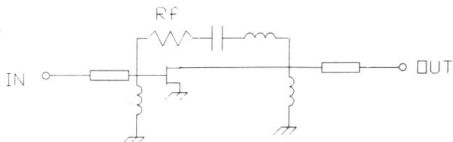

Fig. 5.11 The topology of a practical feedback amplifier including layout and bias parasitic elements.

The *advantages* of feedback amplifiers are
1. Their small size.
2. Their simple topology, with few component parts.
3. Their broadband widths.
4. Their low gain drift with temperature (at least, at those frequencies where feedback dominates).
5. Their ability for cascading (good input-output match).

Their *disadvantages* include
1. Their tricky biasing with the feedback loop in place (they must have a dc block between the gate and the drain).
2. The loss associated with the feedback resistor that erodes the amplifier's output power and noise figure over the lower-frequency end of the band.

3. The need for an FET with a transconductance greater than 100 mS for the amplifier to work properly (this is not a hard and fast rule; in practice FETs with 50 to 100 mS usually work well).

5.2.5 Distributed Amplifiers

Distributed amplifiers were first developed in the 1930s as a way of obtaining very broad band amplification with vacuum tubes triodes. During the last decade, there has been a resurgence of interest in distributed microwave amplifiers using GaAs FET devices [8]. Such amplifiers are a very natural extension of the distributed amplifier concept, because, in most ways, the GaAs FET behaves as a solid-state triode. Distributed GaAs FET amplifier designs can provide very flat gain from dc to over 20 GHz—a very significant result. Distributed amplifiers have come into their own with the availability of GaAs MMIC fabrication techniques, because the small size of the MMIC lends itself to the construction of a distributed amplifier. However, a successful distributed amplifier also may be constructed using hybrid techniques. The reason that distributed amplifiers offer outstanding (gain × bandwidth) products is that they are *not* matched amplifiers and therefore not subject to Fano's bandwidth limit. A distributed amplifier works on a different concept, called *traveling wave amplification*.

Refer to Figure 5.12 for the description and calculations which follow. Two artificial transmission line structures together with a number of FETs form a solid-state distributed amplifier. The first transmission line, which is connected to the input, is called the *gate line*. The second transmission line, which is connected to the output, is called the *drain line*. Each line is terminated in a matched load.

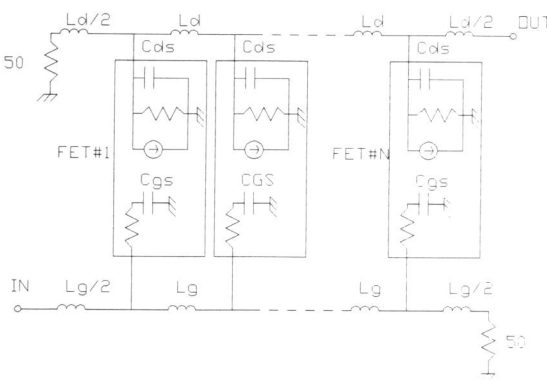

Fig. 5.12 Basic topology of a distributed amplifier.

The shunt capacitors of the gate line are provided by the gate capacitance, C_{GS}, of the FETs. The shunt capacitors of the drain line are provided by the FET's drain capacitance, C_{DS} plus the supplemental value, C_1.

The characteristic impedance of the gate line and the drain line is

$$Z_0 = \sqrt{L_G/C_{GS}} = \sqrt{L_D/(C_{DS} + C_1)} \tag{5.14}$$

where L_G and L_D are the inductance per section of the gate line and the drain line, respectively.

A distributed amplifier provides gain because a voltage that appears at the gate of one of the FETs is amplified via the FET's transconductance, producing a current in the drain line. To produce useful gain, it is very important that these drain currents add in-phase as the signal moves along the drain line toward the amplifier's output. Therefore, the phase shift between FETs along the drain line must be exactly the same as the phase shift between FETs along the gate line. The phase shift of a single section of artificial transmission line is given by

$$\Phi_G = 2\pi f \tau_G \tag{5.15}$$

in the gate line, and

$$\Phi_D = 2\pi f \tau_D \tag{5.16}$$

in the drain line, where τ_G is the gate line time delay per section, and τ_D is the drain line time delay per section.

The condition that will ensure constructive interference of all FET currents, at all frequencies, is

$$\phi_G = \phi_D \tag{5.17}$$

or

$$\tau_G = \tau_D \tag{5.18}$$

Now, for any artificial transmission line, the delay per section is equal to $\sqrt{LC}$; therefore, (5.18) can be rewritten as

$$\sqrt{L_G C_{GS}} = \sqrt{L_D(C_{DS} + C_1)} \tag{5.19}$$

For a realistic amplifier, the impedance conditions (5.14) and the phase conditions (5.19) must be satisfied simultaneously. Inspection shows that the straightforward simultaneous solution of (5.14) and (5.19) is given by the condition that

$$L_G = L_D = L \tag{5.20}$$

and

$$C_{GS} = C_{DS} + C_1$$

or

$$C_1 = C_{GS} - C_{DS} \tag{5.21}$$

A third condition that must be imposed if the amplifier is to work in a 50 Ω system (using 5.14)) is

$$Z_0 = 50 = \sqrt{L/C_{GS}}$$

or

$$L = 50^2 C_{GS} \tag{5.22}$$

We are now in a position to calculate the gain and bandwidth of a distributed amplifier. Ideally, the bandwidth will be from dc to a frequency a little less than the cut-off frequency, f_c, of the gate and drain artificial lines. The cut-off frequency is given by

$$f_C = \frac{1}{\pi Z_0 C_{GS}} \tag{5.23}$$

The current out of each FET adds in-phase along the drain line and is divided equally between the output and the termination.

The gain of a distributed amplifier (neglecting loss) is very simply one-half of the output current per FET times the number of FETs, times the load resistance, divided by the input voltage:

$$G = V_{out}/V_{in} = \frac{n g_m V_i Z_0 / 2}{V_i}$$

or, in decibels:

$$G = 20 \log_{10} \left(\frac{n Z_0 g_m}{2} \right) \tag{5.24}$$

where

n is the number of sections (same as the number of FETs),
Z_0 is the input and output characteristic impedance (50 Ω),
g_m is the transconductance of an individual FET.

Equations (5.23) and (5.24) are elegantly simple expressions for the cut-off frequency and the gain of a distributed amplifier. Multiplying these expressions together to yield the *maximum gain-bandwidth* (GBW) product of a distributed amplifier:

$$\text{GBW} = \left(\frac{1}{\pi Z_0 C_{GS}}\right)\left(\frac{n Z_0 g_m}{2}\right)$$

$$\text{GBW} = \frac{n g_m}{2\pi C_{GS}} \tag{5.25}$$

The cut-off frequency of the individual FETs, f_t is

$$f_t = \frac{g_m}{2\pi C_{GS}}$$

Therefore,

$$\text{GBW} = n f_t \tag{5.26}$$

This important result simply states that the maximum gain-bandwidth product of an ideal distributed amplifier is equal to the cut-off frequency of its individual FETs times the number of FETs.

Gain and bandwidth may be traded off by changing the gate width, w, of the individual FETs. For a high cut-off frequency, use a small C_{GS}, corresponding to a small gate width. Under these conditions, n must be raised to build up the gain, because g_m will be low in the individual FETs. This is why high-frequency distributed amplifiers use *many* small FETs. The value $n = 6$ or 7 is not unheard of in amplifiers operating to 20 GHz or more. At low frequencies only a few FETs are used, but they are proportionally larger.

To predict the correct size FET for an arbitrary cut-off frequency and gain, we introduce the concept of FET equivalent circuit parameters normalized to width. Define these parameters as follows:

$$\begin{aligned} g_m &= w g'_m \\ C_{GS} &= w C'_{GS} \\ I_{DSS} &= w I'_{DSS} \end{aligned} \tag{5.27}$$

where

g'_m is the FET's transconductance per unit width,
C'_{GS} is the FET's gate capacitance per unit width,
I'_{DSS} is the FET's saturated drain current per unit width.

Using the definition in (5.27) and (5.23), we can rewrite the amplifier's cut-off frequency as

$$f_C = \frac{1}{\pi Z_0 w C'_{GS}}$$

or

$$w = \frac{1}{\pi Z_0 C'_{GS} f_C} \tag{5.28}$$

The value of w is plotted as a function of f_C in Figure 5.13, assuming a typical value of $C_{GS} = 1.0$ pF/mm. The individual cut-off frequency is independent of FET width because

$$f_t = \frac{g_m}{2\pi C_{GS}} \equiv \frac{w g'_m}{2\pi w C'_{GS}} = \frac{g'_m}{2\pi C'_{GS}} \tag{5.29}$$

Therefore, the maximum gain-bandwidth product always is given by (5.26) as

$$\text{GBW} = G f_C = n f_t$$

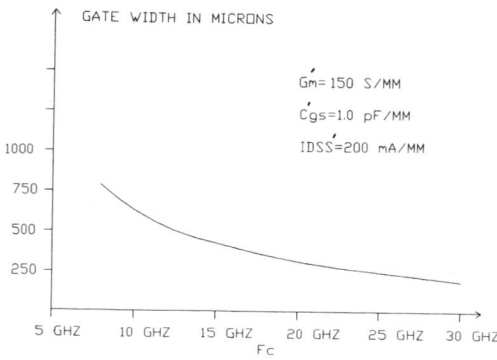

Fig. 5.13 Distributed amplifier design information relating FET gate width to cut-off frequency.

which is independent of gate width. In terms of decibels, gain is equal to

$$G = 20 \log_{10}\left(\frac{nf_t}{f_c}\right) \tag{5.30}$$

Gain is plotted in Figure 5.14 as a function of cut-off frequency for several values of n. We assume $C'_{GS} = 1.0$ pF/mm and $g'_m = .15$ s/mm.

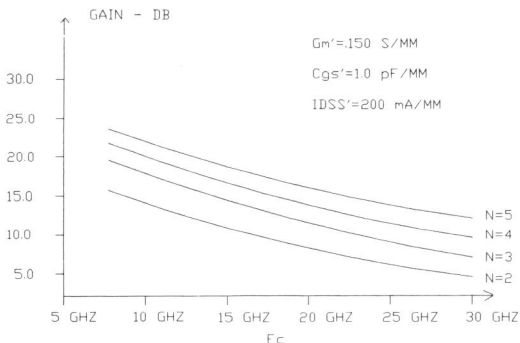

Fig. 5.14 Distributed amplifier design information relating gain to cut-off frequency and the number of sections.

The value of inductance in the artificial distributed lines also may be calculated from the cut-off frequency by going back to the expression for impedance (5.14):

$$Z_0 = \sqrt{L/C_{GS}} = \sqrt{L/wC'_{GS}}$$

or

$$Z_0^2 = \frac{L}{wC'_{GS}}$$

Combining with (5.28)

$$Z_0^2 = \frac{\pi Z_0 C'_{GS} f_c L}{C'_{GS}}$$

After canceling Z_0 and C'_{GS}, and solving for L, we have

$$L = \frac{Z_0}{\pi f_C} \tag{5.31}$$

L is plotted in Figure 5.15 as a function of cut-off frequency.

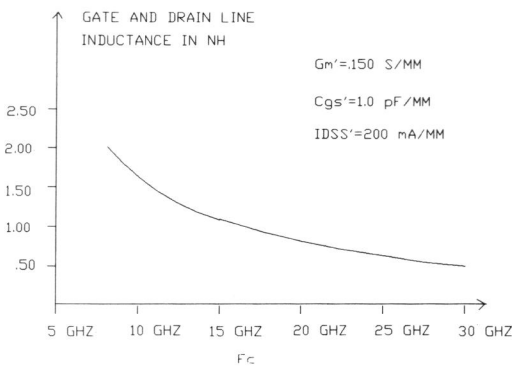

Fig. 5.15 Distributed amplifier design information relating gate and drain line inductance to cut-off frequency.

The dc current of any distributed amplifier may be calculated in a similar way. Assume that the FETs are to be biased at

$$I_{DS} = \frac{I_{DSS}}{2} \tag{5.31a}$$

From (5.27), the drain-to-source current of an individual FET is

$$I_{DS} = \frac{w I'_{DSS}}{2}$$

For the entire amplifier,

$$I_{dc} = n I_{DS} = \frac{n w I'_{DSS}}{2} \tag{5.32}$$

From (5.26),

$$G f_C = n f_t$$

or

$$n = \frac{Gf_C}{f_t}$$

Combining with (5.28) and (5.29) yields

$$I_{dc} = G \frac{I'_{DSS}}{Z_0 g'_m} \qquad (5.33)$$

I_{dc} versus f_C for several values of n is plotted in Figure 5.16. Note that (5.33) acquires its dependence on n and f_C from the G term. For this calculation, g'_m is assumed to be .15 S/mm, and I_{DSS} is assumed to be 20 mA/mm.

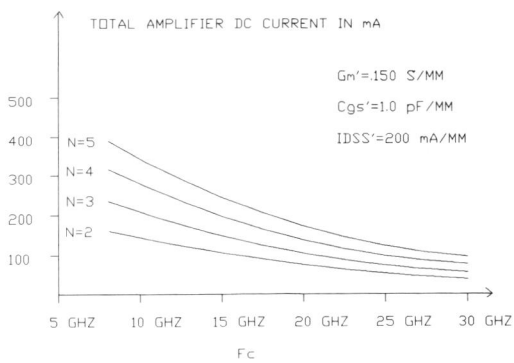

Fig. 5.16 Distributed amplifier design information relating total DC current to the cut-off frequency and the number of sections.

A simple graphic procedure for designing distributed amplifiers involves first determining w based on the choice of cut-off frequency, using Figure 5.13. Next, determine the value of n from the desired value of gain using the graph in Figure 5.14. Then, Figure 5.15 gives L from the choice of cut-off frequency. Finally, the dc current is found as a function of cut-off frequency from Figure 5.16. Remember that a capacitor $C_1 = w(C'_{gs} - C'_{bs})$ must be connected to ground from the drain node of each FET.

This design information may be obtained using the HP-41C program listed in Section 4.6.1. Bear in mind that this design information does not include the effects of losses in either the FETs or the circuit. This simple design process represents a good transition into a Touchstone or SuperCompact file that makes use of exact models for the FETs and the inductors, including losses. In most practical cases, the inductors are constructed using microstrip lines as discussed in Section 3.5.

5.3 EXAMPLE 1: AN MIC-BALANCED REACTIVELY MATCHED AMPLIFIER FOR THE 6 TO 18 GHz BAND

The design of any practical amplifier is a multistep process. In this chapter we carry out the design procedure for synthesizing the ideal circuit topology. The rest of the design, taking the design to final layout will be discussed in Chapter 8. The design procedure leading to the simulation of an ideal circuit is

1. Define a specification;
2. Choose a topology;
3. Choose an active device based on MAG or Fano's bandwidth limit;
4. Address power or noise matching conditions;
5. Choose a bias circuit;
6. Perform computer simulations and optimizations to realize the specifications in Step 1.

Not every design requires all six steps. Only those steps necessary for a given design are used (i.e., power matching could be skipped in a low-noise amplifier design). Let us now follow these design steps for Example 1.

Step 1: Specifications

We choose typical specifications for a 6–18 GHz balanced amplifier:
- Gain = 6 dB ± 1 dB from 6 to 18 GHz;
- Input-output return loss greater than 10 dB, from 6 to 18 GHz;
- Single polarity dc power supply voltage;
- Reverse isolation greater than 15 dB, from 6–18 GHz.

Step 2: Choose a Topology

Let us choose the simple L reactive matching network, shown in Figure 5.7A. However, initial simulations indicate that the simple L network does not provide sufficient flexibility over the 6 to 18 GHz band to simultaneously achieve gain compensation and output match for power. To improve the circuit's matching flexibility, a two-element microstrip step transformer circuit is added to both the input and the output of the basic topology. This circuit becomes the basic single-ended amplifier, shown in Figure 5.17. Two identical single-ended amplifiers are combined with two Lange couplers and two terminations to form the balanced

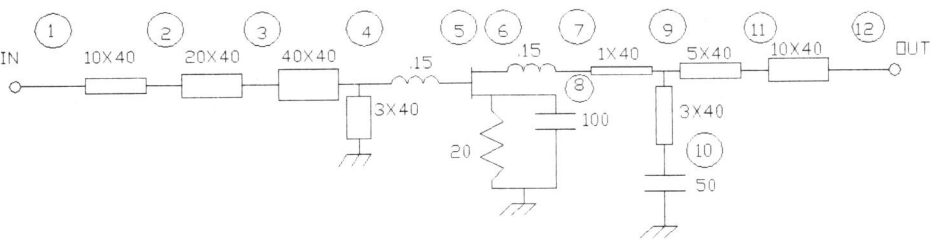

Fig. 5.17 Single-ended 6 to 18 GHz amplifier topology including node numbers and initial element values.

amplifier, which is capable of providing good return loss and high power output over the full bandwidth.

The balanced amplifier topology already has been shown in Figure 5.8, and it will be referred to during the development of a Touchstone file for the total balanced amplifier.

Step 3: Choose an Active Device

The most important criterion in choosing an active device is to be sure that the device has a MAG greater than the specified gain at the highest operating frequency in the band of interest. The specified gain is 6.0 dB, so the device must have MAG greater than 6.0 dB at 18 GHz. Several factors reduce the overall gain to less than MAG. The Lange couplers contribute about .5 dB overall loss to a balanced amplifier. Imperfect matching at the high end also can account for a 1.0 dB loss in gain. Therefore, we are seeking a device with more than 7.5 dB MAG at 18 GHz. The choice is the Harris HMF-0310 GaAs FET device whose data sheet MAG specification is 8.0 dB at 18 GHz. A second reason for choosing the HMF-0310 is its excellent 1 dB gain compression point (+18 dBm). The manufacturer's data sheet supplies all the necessary biasing, S-parameter, and processing information.

Fano's limit predicts the specified bandwidth is achievable with this device. Assuming $R_i = 15\ \Omega$ and $C_{GS} = .30$ pF (typical for this class of FET), $\tau_{HP} = .50$, predicting a bandwidth in excess of 10:1 with no reduction in high-end gain, assuming all the compensation is in the input circuit.

Step 4: Power or Noise Matching Conditions

Because no power or noise figure specifications are in this example, the step will be skipped. However, other examples will demonstrate methods for both noise and power figure matching.

Step 5: Choose a Bias Circuit

Because the specifications call for a single-polarity dc power requirement, we will make use of self-biasing techniques. The basic circuit for self-biasing is shown in Figure 5.18.

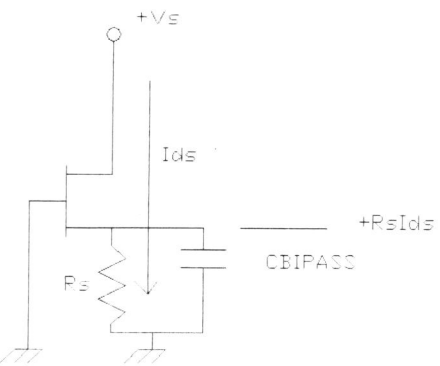

Fig. 5.18 A self-biasing circuit for a GaAs FET amplifier.

A key feature of self-biasing is that a source resistor, R_S, placed between the FET's source and ground, raises the source above ground by a voltage equal to

$$R_S I_{DS}$$

where I_{DS} is the drain-to-source dc current.

A bypass capacitor in shunt with R_S fixes the FET's source at RF ground.

By connecting the FET's gate to dc ground (through an RF choke, or a shorted stub that also may be a part of the gate matching circuit), the voltage between the gate to source is

$$V_{GS} = -R_S I_{DS}$$

V_{GS} has the correct negative polarity to properly bias the FET. The value of V_{GS} may be controlled with the value of R_S. I_{DS} is controlled by R_S, via V_{GS}. Normally, I_{DS} is set to $I_{DSS}/2$, for maximum gain and power output.

Step 6: Basic Circuit Simulation and Optimization

Referring to Figure 5.17, we begin a Touchstone simulation of the single-ended amplifier by first numbering all circuit nodes (including bias circuit, to model its RF effects, if any) and assigning an initial guess to each element value. If a technique is available initially to calculate these parameters by hand, now is the time to use it. Unfortunately, there is no simple way to do this calculation for a reactively matched amplifier. Therefore, as an initial guess, we assume the length of each stub and transmission line is one-eighth wavelength ($\lambda/8$) at mid-band (12 GHz). Because $\epsilon_e \doteq 6.7$ for Al_2O_3, at 12 GHz, the expression for $\lambda/8$ becomes

$$\lambda/8 = \frac{C}{8\sqrt{\epsilon_e}f} \doteq 50 \text{ mils at 12 GHz}$$

The transmission line and stub lengths all are set to 50 mils as a first guess. For a 10 mil thick substrate, the stub widths are initially assumed to be narrow (3 mils) because these stubs act as high-inductance chokes in this circuit. The transmission line elements of the input circuit are stepped progressively wider as the FET's gate is approached because the FET's input impedance is less than 50 Ω at these frequencies. The output circuit transmission line elements are assumed to be very narrow because the FET's drain impedance is very high and are widened progressively as the 50 Ω output terminal is approached.

The Touchstone file for this initial single-ended circuit is written including these initial guesses. Next, the circuit element lengths and widths are optimized by Touchstone by designating each of them to be variables of the optimization process and using the specified gain and bandwidth (not return loss, because return loss will be poor in a single-ended design) as goals of the optimization.

Figure 5.19 shows the performance plots before and after optimization. Note how the performance after optimization closely matches the specified goals.

Next, the Lange couplers and the terminations are combined with the Touchstone single-ended amplifier model to create a total balanced-amplifier circuit model. Be sure to follow the Touchstone syntax for Lange couplers exactly, paying particular attention to proper node numbering. Lange couplers are approximately a quarter wavelength long at mid-band, which is about 90 mils on a 10 mil thick alumina substrate at 12 GHz. Coupler finger widths are approximately 1.4 mils, and finger spacing is about .8 mils. The Touchstone optimizer, or Touchstone's "tune" mode, can be used to fine tune these values.

The Touchstone file for the balanced amplifier circuit (shown in Table 5.1) is created as an extension of the singled-ended amplifier file. The two-port describing the single-ended amplifier, AMP, is treated just as any other circuit element and combined with the Lange couplers and their terminations into a final "top-level" two-port network, called BAL. The circuit diagram used to create the BAL model is shown in Figure 5.20. In Figure 5.21, the plots of calculated gain and return loss for the basic balanced amplifier are given. Notice that gain essentially is unchanged from the single-ended case, but the return loss is considerably improved and

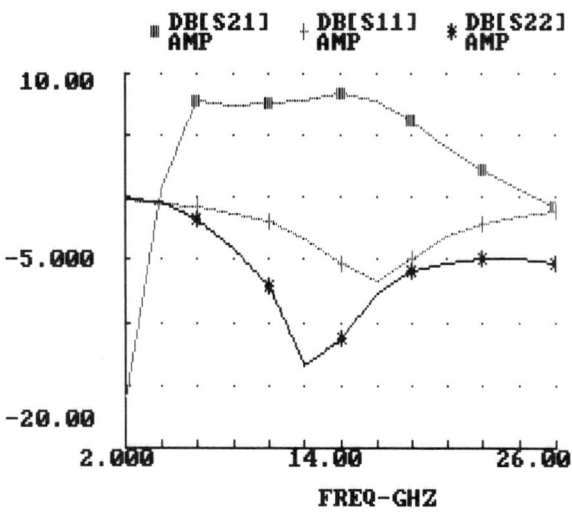

Fig. 5.19 The simulated performance of the single-ended 6–18 GHz reactively matched amplifier.

Table 5.1 Touchstone—Ver(1.30–May-17-85)—Ser(15738-1598-1000)—Con(1CCC22124P) 302.CKT
06-22-89 11:50:05

```
! THE BASIC BALANCED AMPLIFIER TOPOLOGY  FOR 6 TO 18 GHZ.
! THE GaAs FET IS A HARRIS HMF0310.
CKT
      ! THE SINGLE ENDED AMPLIFIER
      MSUB ER=9.9 H=10 T=.2 RHO=1 RGH=0
      MLIN 1 2 W=13.7 L=45.5
      MLIN 2 3 W=31.9 L=46.2
      MLIN 3 4 W=41.5 L=3.6
      MLSC 4 W=2.89 L=52.05
      IND 4 5 L=.10
      S2PA 5 6 8   HMF0310
      IND 6 7 L=.10
      RES 8 0 R=20
      CAP 8 0 C=100
      MLIN 7 9  W=1.0 L=6.6
      MLIN 9 10 W=6.27 L=61.61
      CAP 10 0 C=50
      MLIN 9 11   W=4.69 L=15.20
      MLIN 11 12 W=12.8 L=6.36
      DEF2P 1 12 AMP
! CONNECT TWO AMPS IN A BALANCED CONFIGURATION
      MLANG 1 2 3 4 W=1.4 S=.8 L=80
      MLANG 7 8 5 6 W=1.4 S=.8 L=80
      AMP 4 5
```

Table 5.1 cont'd.

```
    AMP   2  7
    TFR   1  0  W=10  L=10  RS=50  F=0
    TFR   6  0  W=10  L=10  RS=50  F=0
    DEF2P 3  8  BAL

FREQ

    SWEEP  2  26  2

OUT

    AMP  DB[S21]  GR1
    AMP  DB[S11]  GR1
    AMP  DB[S22]  GR1
    BAL  DB[S21]  GR2
    BAL  DB[S11]  GR2
    BAL  DB[S22]  GR2
    BAL  DB[S12]  GR3

GRID

    RANGE  2  26  2
    GR1  -20  10  5
    GR2  -20  10  5
```

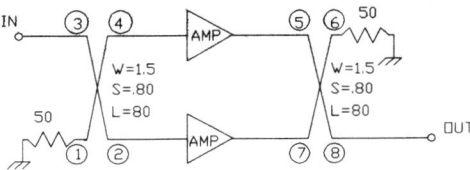

Fig. 5.20 The "high-level" circuit diagram used to create the Touchstone file for the basic balanced amplifier.

greater than 12 dB at all points in the 6 to 18 GHz band. Reverse isolation is also plotted in Figure 5.21.

The next step in the design process is to include any parasitic elements associated with an actual layout, into the circuit model. This work is continued in Section 8.1.

5.4 EXAMPLE 2: AN MIC LOSSY-MATCH AMPLIFIER FOR THE 2 TO 6 GHz BAND

Lossy match amplifiers use resistive gain equalizing networks to "throw away" gain at the low end of the amplifier's band to accomplish gain compensation. The most commonly used equalizer circuit is the simple shunt mounted series *R-L*

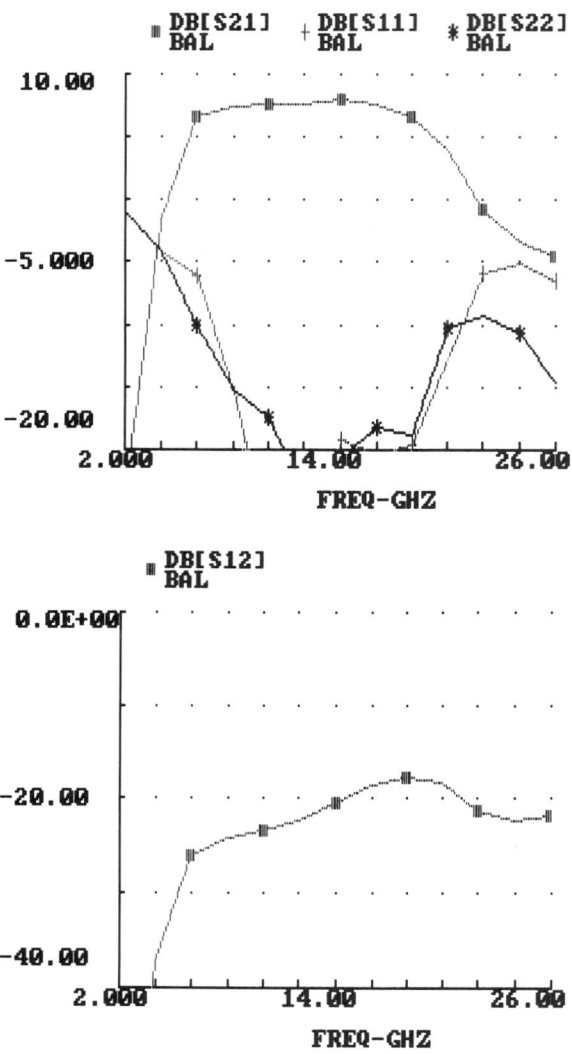

Fig. 5.21 The simulated performance of the balanced 6–18 GHz reactively matched amplifier.

network, which is shown in Figure 5.22. The lossy equalizer flattens gain in the following way. At high frequencies, the inductor is nearly an open circuit, effectively stopping the resistor from passing any current and contributing loss to the circuit. At low frequencies, the inductor is nearly a short circuit, placing the resistor in shunt between the FET's gate and ground, creating maximum loss. By

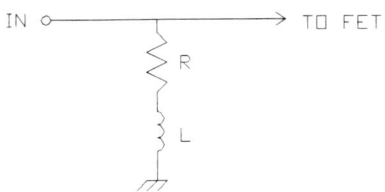

Fig. 5.22 A resistive gain equalizer circuit used in the lossy-match amplifier.

judiciously choosing the values of R and L, gain compensation and good input-output match can be achieved simultaneously. The price paid for the simplicity of this approach is the degradation of noise figure at low frequencies, if the lossy-match circuit is at the input, and the degradation of output power at low frequencies if the lossy-match circuit is in the output. For applications where neither the noise figure nor the power output are of critical importance (such as gain blocks), the lossy-match amplifier provides a simple, inexpensive alternative to some of the more complicated circuit topologies. The design procedure for a lossy matched amplifier follows.

Step 1: Specifications

- Gain = 10 ± 1 dB from 2 GHz to 6 GHz.
- Input-output return loss greater than 5 dB.
- Single supply dc bias.
- Reverse isolation in excess of 20 dB.

Step 2: Choose a Topology

The preceding specification is a natural choice for a lossy-match amplifier topology because it contains no power or noise figure requirements. The gain level is difficult for a single stage, but is relatively easy to achieve with two stages in cascade. The two-stage topology also makes the reverse isolation requirement relatively easy to achieve. To create good input *and* output return loss, resistive gain equalizer circuits will be used in both the input and the output circuits of both stages. These shunt-mounted networks can be used to apply dc bias to the FETs provided that the R-L network's grounded end is RF bypassed by a large value capacitor. Self-bias will be used on both stages to provide single bias polarity operation. The basic topology will use distributed equalizer elements as practical replacements for the ideal lumped elements. The basic topology for the lossy match 2–6 GHz amplifier is given in Figure 5.23.

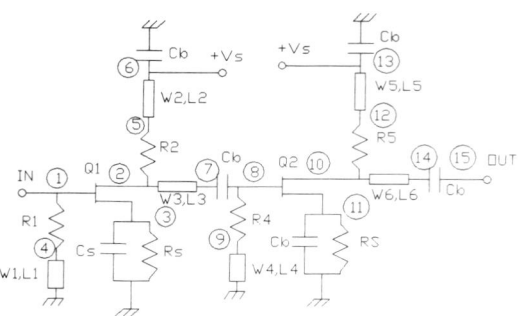

Fig. 5.23 The topology of a two-stage lossy match amplifier using microstrip line matching elements instead of lumped inductors.

Step 3: Choose an Active Device

As a minimum, the MAG of the active device must be equal to 11/2 = 5.5 dB at 6 GHz. Because the lossy-match technique is not very efficient in terms of using the FET's maximum available gain, it is wise to add a safety factor of at least 3 dB. Therefore, the FET we choose must have at least MAG = 8.5 dB at 6 GHz. A good choice for this application is the Litton-Dexcel DXL 3501, which is a low-cost 1.0 micron gate length device that provides MAG = 10.0 dB at 8 GHz.

To check Fano's limit for this device, first estimate that C_{GS} = .45 pF and R_i = 10 Ω. Therefore, τ_{HP} = .169, which leads to a 3 : 1 bandwidth with no degradation in MAG at the high end, assuming 6 dB/octave gain compensation. A 3 : 1 bandwidth is just sufficient for the 2 to 6 GHz band, so we may expect some degradation in high-end gain due to this limit.

Step 5: Choose a Bias Circuit

Because of the specification requirement for single polarity bias supply, a self-bias circuit will be employed. The two gain stages must be dc isolated from each other by an interstage blocking capacitor to prevent dc interactions. Each FET has its own source resistor and its own source capacitor. The shunt drain inductors serve as bias chokes. Because the equalizer resistors are connected between the inductor and the FET's drain, dc current flowing through these resistors will cause some voltage drop. Therefore, the supply voltage must be higher than the FET's drain-to-ground voltage by the amount $I_{DS} \times R_{eq}$. If the bias points are connected in parallel, the total supply current will be the sum of the two FET currents, or $2I_{DS}$.

Step 6: Basic Circuit Simulation and Optimization

The simulation and optimization of the lossy-match circuit begins with writing a Touchstone file for the circuit topology shown in Figure 5.23. Although a lumped element topology theoretically is feasible, the lumped inductors are difficult to fabricate as an MIC circuit. A more easily manufactured design uses microstrip transmission lines and stubs as direct replacements for the lumped inductors. Such a distributed topology will be considered from the start.

The microstrip topology shown in Figure 5.23 is used to write a Touchstone file, which is listed in Table 5.2. All transmission lines and stubs are initially set to $\lambda/8$ at mid-band (about 150 mils on an Al_2O_3 substrate). The resistors initially are set to 50 Ω.

Table 5.2 Touchstone—Ver(1.30–May-17-85)—Ser(15738-1598-1000)—Con(1CCC22124P) 310.CKT 06-22-89 13:02:22

```
! THE BASIC TOPOLOGY OF A TWO STAGE 2 TO 6 GHZ LOSSEY MATCH AMPLIFIER.
! THE GaAs FETS ARE LITTON/DEXCEL DXL3501'S
CKT
      MSUB ER=9.9 H=10 T=.2 RHO=5 RGH=0
      RES  1  4  R\27.10060                    ! R1
      MLSC 4  W\1.63410   L\84.25263           ! W1,L1
      S2PA 1  2  3  DX3501
      RES  3  0  R=20
      CAP  3  0  C=100
      RES  2  5  R\0.33339                     ! R2
      MLIN 5  6  W=1.0    L\340.95210          ! W2,L2
      CAP  6  0  C=50
      MLIN 2  7  W\1.52958 L\79.09347          ! W3,L3
      CAP  7  8  C=50
      RES  8  9  R\48.90438                    ! R4
      MLSC 9  W=1.0       L\193.93790          ! W4,L4
      S2PA 8  10 11 DX3501
      RES  11 0  R=20
      CAP  11 0  C=100
      RES  10 12 R\211.54190                   ! R5
      MLIN 12 13 W\0.08999 L\55.05793          ! W5,L5
      CAP  13 0  C=50
      MLIN 10 14 W=1.0    L\144.68100          ! W6,L6
      CAP  14 15 C=50
      DEF2P 1   15 AMP

FREQ
      SWEEP 1 8 .5
OUT
      AMP DB[S21] GR1
      AMP DB[S11] GR1
      AMP DB[S22] GR1
```

Table 5.2 cont'd.

```
GRID
    RANGE 1 8  .5
    GR1 -20 20  5
OPT
    RANGE 2 6.5
    AMP DB[S21] =10    40
    AMP DB[S11] <-10   10
    AMP DB[S22] <-10   10
```

The optimizer goals are as follows.

RANGE 2.0 6.5
AMP dB[S_{21}] = 10.0
AMP dB[S_{21}] < −12.0
AMP dB[S_{22}] < −10.0
AMP dB[S_{11}] < −10

Allow all microstrip widths and lengths to be variables of the optimization process. The blocking and bypass capacitors are fixed at 50 pF (which is enough to ensure less than 2.0 Ω reactance at 2.0 GHz).

The postoptimization gain, reverse isolation, and return loss calculations are shown in Figure 5.24. The gain is in excess of 10 dB across the 2 to 6 GHz band, and

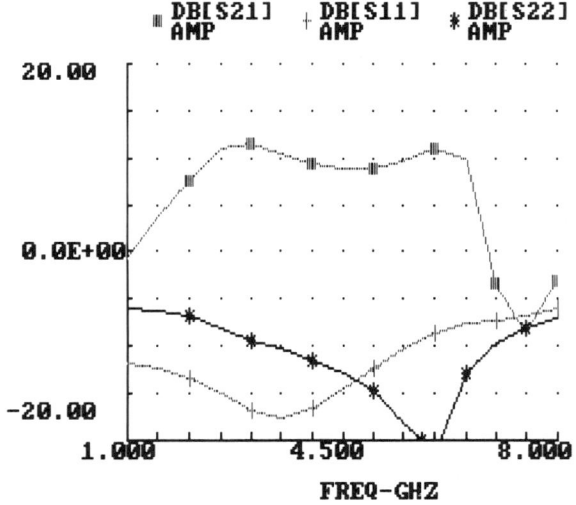

Fig. 5.24 Simulated performance of the 2 to 6 GHz lossy-match amplifier.

the return loss is in excess of 7 dB across the band. Table 5.2 gives the optimized Touchstone file.

In Section 8.2, the layout of this circuit will be described along with a final analysis that includes all layout parasitic elements.

5.5 EXAMPLE 3: AN MMIC FEEDBACK AMPLIFIER FOR THE 2 TO 8 GHz BAND

Feedback amplifiers have become very popular among MMIC amplifier designers because they offer an almost unbeatable combination of performance and simplicity. We will design the amplifier described in this section following three principles that specifically relate to MMIC designs. These principles are

1. FETs are treated as continuous variables of the design. That is, in MMIC designs, the FET is a variable of the design process, right along with the passive circuit elements.
2. All elements of the topology, including *all* parasitic elements, must be included in the model because MMIC chips *cannot* be experimentally modified by tuning.
3. To use a minimum amount of active GaAs "real estate" for an MMIC design is very important economically. This requirement often is in direct opposition to performance goals, and the modeling requirement of item 2, because a small, cramped chip will have more parasitic elements due to coupling paths and tight bends, which defy modeling and erode performance in unpredictable ways. Good engineering judgment must be exercised in each case.

With these special MMIC design criteria in mind, let us proceed with the design of the feedback amplifier.

Step 1: Specifications

- Bandwidth = 2 to 8 GHz.
- Gain = 10 dB minimum, ripple less than ± 1.5 dB.
- Input-output Return Loss = 10 dB minimum.
- Power Output = + 15 dBm at 1 dB compression.
- dc Current = 200 mA maximum

Step 2: Choose a Topology

The first step in the design is to determine the parameters of the FETs. Referring to (5.13), we calculate the feedback resistance needed to produce 5 dB gain per stage.

$$5.0 = 20 \log \left(\frac{R}{2 \times 50} \right)$$

or

$$\frac{R}{100} = 10^{5/20} = 1.778$$

or

$$R = 177.8 \,\Omega$$

The minimum FET transconductance may be calculated from (5.11) as

$$g_M > \frac{R}{50^2} = \frac{177.8}{50^2} = 71 \text{ mS}$$

Assuming that the MMIC amplifier will be fabricated at a foundry whose process design rule for FET transconductance is 120 mS/mm, the minimum FET width will be

$$w = \frac{71 \text{mS}}{120 \text{ (mS/mm)}} = .59 \text{ mm or } 590 \text{ } \mu\text{m}$$

Let us also assume the foundry design rule for I_{DSS} is 300 mA/mm. Two 600 μm FETs, biased at $I_{DSS}/2$ would have a total current of

$$I_{dc} = \frac{300 \text{ mA/mm} \times 2 \times .6}{2} = 180 \text{ mA}$$

Note that 180 mA is within our current specification, so the FETs may be biased in parallel.

Basic lumped-element two-stage feedback amplifier topology is shown in Figure 5.25. Matching elements are provided at the input and output to ensure good return loss at high frequencies, because a pure feedback amplifier circuit, once optimized for flat gain across the band may be somewhat mismatched at the high end. Self-bias circuits are included in the sources of both FETs, and drain bias is connected to the circuit via two high-inductance chokes.

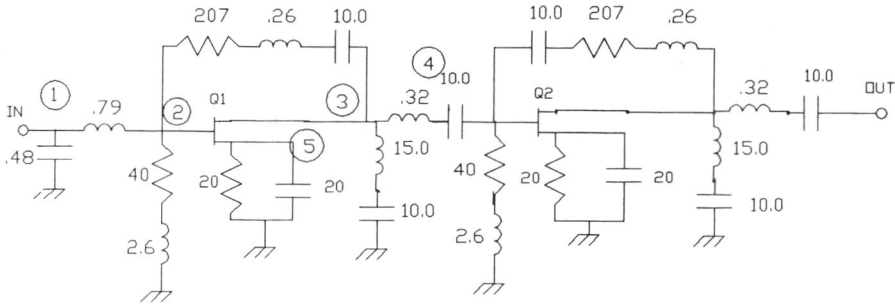

Fig. 5.25 The basic topology of a two-stage feedback amplifier for the 2 to 8 GHz band.

Step 3: Choose an Active Device

In MMIC circuit design, the circuit designer also is the device designer. Most foundries include basic FET building blocks as a part of their design rules. These building blocks, or standard cells, consist of a gate finger of some length and width, plus source and drain contacts. Specific FETs for a given application are designed by either changing the width of the gate finger, or adding additional gate fingers. Good practice always is to have an even number of gate fingers, so that the outside edge of the FET will be a source contact on both sides. This practice greatly simplifies source grounding.

A typical multifinger FET structure is shown in Figure 5.26. This particular FET has four gate fingers, each with a gate width, w, and a total gate width of $4w$. The source contacts are interconnected by a metal layer called an *air bridge*, which runs "in the air" above the gate fingers.

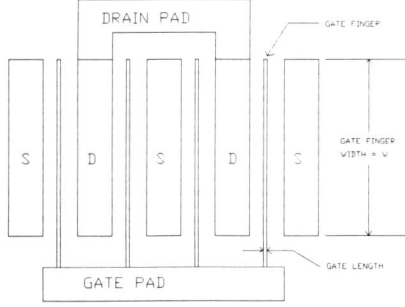

Fig. 5.26 A four gate "finger" interdigitated FET structure with a gate length of l and a total gate width of $4W$.

For a 600 μm FET with four gate fingers, $w = 600/4 = 150$ μm. In most foundry processes, FETs with gate lengths of .5 μm are used exclusively for operation above 6 GHz.

The equivalent circuit model of these FETs is shown in Figure 5.27. The element values are given by the foundry's design rules; usually normalized to 1 mm (1000 μm). For a given gate width, all capacitors and the transconductance in the model are calculated by multiplying their normalized value times ($w_{total}/1000$ μm).

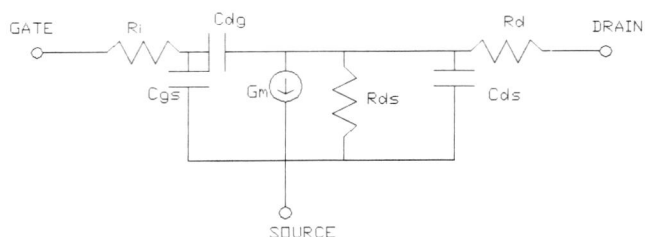

Fig. 5.27 The equivalent circuit of a foundry FET. Element values are normalized to 1000 μm gate width.

All resistors in the model are calculated by multiplying their normalized values times (1000 μm/w_{total}).

For the 600 μm FET used in this circuit, the equivalent circuit parameters are calculated from a typical set of normalized parameters:

$$C_{GS} = \left(\frac{600}{1000}\right) \times 1.00 \text{pF/mm} = .60\text{pF}$$

$$C_{DG} = \left(\frac{600}{1000}\right) \times .20 \text{pF/mm} = .12\text{pF}$$

$$g_m = \left(\frac{600}{1000}\right) \times 120 \text{ mS/mm} = 72\text{mS}$$

$$C_{DS} = \left(\frac{600}{1000}\right) \times .30 \text{pF/mm} = .18\text{pF}$$

$$R_i = \left(\frac{1000}{600}\right) \times 5.0 \text{ } \Omega - \text{mm} = 8.2 \text{ }\Omega$$

$$R_{DS} = \left(\frac{1000}{600}\right) \times 100 \text{ } \Omega - \text{mm} = 165 \text{ }\Omega$$

This equivalent circuit can be included element by element in a Touchstone model for the total amplifier, or the element values may be inserted into one of Touchstone's standard FET models.

Step 4: Power and Noise Matching Conditions

There is no noise figure specification for this amplifier, so we need not be concerned with noise. However, there is a power specification of +15 dBm at the 1 dB gain compression point. Because some power is consumed in the feedback resistor, the chosen FETs must have the capability of generating +15 dBm power with a comfortable margin. Referring to Section 2.1.5, we calculate the maximum power output from equation (2.13) as

$$P_{out} = I_M \frac{(V_{BGD} - V_P - V_S)}{8}$$

Assume

$$I_M = 1.20\, I_{DSS}$$
$$V_{BGD} = +15\text{ V}$$
$$V_P = 3\text{ V}$$
$$V_S = 1\text{ V}$$

For our choice of FETs,

$$I_{DSS} = \left(\frac{600}{1000}\right) 300 \text{ MA/mm} = 180 \text{ mA}$$

So,

$$P_{out} = \frac{(1.2)(.18)(15 - 3 - 1)}{8} = .297 \text{ W or } +24.7 \text{ dBm}$$

which exceeds +15.0 dBm by a very comfortable margin.

Step 5: Choose a Bias Circuit

Self-biasing is used on both FET gain stages. Each FET has a resistor and a capacitor connected between its source and ground. Drain bias is applied via choke inductors. The two drain bias lines may be connected to a common bias supply (about +6.0 V). The total supply current will be 2 × ($I_{DSS}/2$). Each FET gate is grounded through a choke inductor thus completing the self-biasing circuit.

Step 6: Basic Circuit Simulation and Optimization

At this point, important decisions must be made concerning the nature of the inductors. The options are microstrip line *inductors, spiral* inductors, or, in the case of chokes, external wire coils mounted off-chip. We choose on-chip spiral inductors. However, these inductors will remain ideal inductors through the initial circuit optimization and will be replaced by models of the corresponding spiral inductor during the layout phase. Some of these small inductors, like the inductor in the feedback loop will be realized as microstrip transmission lines. The blocking capacitor, C_B, will be fixed at the highest practical value for the particular foundry process. (Assume C_B = 10 pF for this design.) The topology (including these changes) to be optimized by Touchstone is shown in Figure 5.25.

The optimized Touchstone file for the two-stage feedback amplifier is given in Table 5.3. All the inductors and capacitors (except blocking and bypass capacitors) are designated as goals of the optimization process.

Table 5.3 Touchstone—Ver(1.30–May-17-85)—Ser(15738-1598-1000)—Con(1CCC22124P) 379.CKT
07-14-89 11:12:19

```
! BASIC TOPOLOGY FOR THE TWO STAGE 2 TO 8 GHZ. FEEDBACK AMPLIFIER.
! THE GaAs FETS ARE .5X600 MICRON FOUR FINGER FOUNDRY FETS
! ALL CIRCUIT ELEMENTS ARE LUMPED
CKT
    ! SINGLE STAGE
    IND 1 2 L\0.79231
    CAP 1 0 C\0.48068
    SRL 2 0 R\39.99951   L\2.62993
    FET 2 3 0 G=.07 T=3 F=0 CGS=.60 GGS=0 RI=8.2 CDG=.12 CDC=0 CDS=.18 RDS=160
    SRLC 2 3 R\207.70050   L\0.26975   C=10
    IND 3 4 L\0.32788
    IND 3 0 L\15.14249
    DEF2P 1 4 AMP

    ! CASCADE TWO STAGES
    AMP 1 2
    AMP 2 3
    DEF2P 1 3 CAMP

FREQ
    SWEEP 1 10 1

OUT
    CAMP DB[S21] GR1
    CAMP DB[S11] GR1
    CAMP DB[S22] GR1
    CAMP DB[S12] GR2

GRID
    RANGE 1 10 1
```

Table 5.3 cont'd.

```
    GR1 -15 15 5
    GR2 -30 0 5

OPT

    RANGE 2 8
    CAMP DB[S21]=10.0    50
    CAMP DB[S11]<-10     20
    CAMP DB[S22]<-10     20
    GR3 -40 0 10

OPT

    RANGE 17 19
    BAL DB[S21]=7.0 40
    RANGE 6 17
    BAL DB[S21]=7.0 20
    AMP DB[S22]<-15  5
```

The goals for this optimization are

Range 2.0 6.0
AMP dB[S_{21}] = 10.0 dB
AMP dB[S_{11}] < -10 dB
AMP dB[S_{22}] < -10 dB

The graphed performance after optimization is shown in Figure 5.28. The gain exceeds 10 dB across the 2 to 8 GHz band, and the return loss exceeds 9 dB across the band.

See Section 8.3 for the layout of this circuit together with the final Touchstone optimization including all layout parasitic elements and the spiral inductor models.

5.6 EXAMPLE 4: AN MIC TWO-STAGE LOW-NOISE AMPLIFIER FOR THE 4.5 TO 5.0 GHz BAND

Low-noise GaAs FET amplifiers often are constructed using package FETs and soft board (Duroid™, which is copper-clad Teflon-Fiberglas, $E_R = 2.2$) microstrip boards. This is in contrast with most microwave amplifier hybrids, which use gold metalized alumina boards and chip FETs. However, the packaged FET/Duroid construction technique is ideally suited to many low-noise amplifiers, because packaged FETs may be conveniently presorted for noise figure, so that the lowest noise devices can be put in the amplifier's front end where they will do the most good. Second, because of the low conductor loss of its microstrip lines, Duroid is an excellent transmission medium in low-noise amplifiers because it offers lower

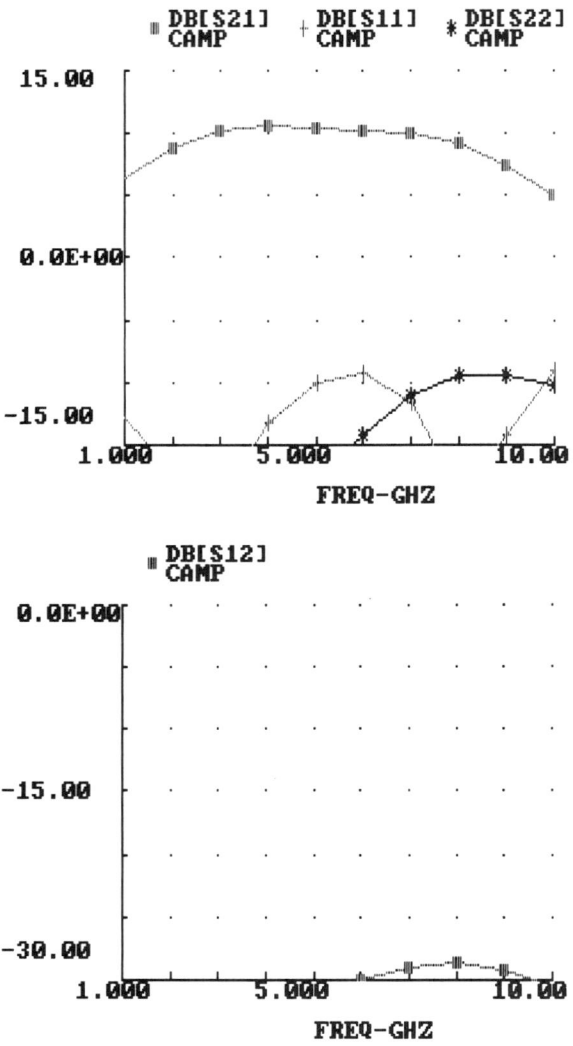

Fig. 5.28 The simulated performance of the 2 to 8 GHz two-stage feedback amplifier.

loss than an equivalent alumina line. Low transmission line loss is very important as any losses within the matching circuits will add directly to the overall noise figure of the amplifier.

A number of very low noise GaAs FETs are available from a number of suppliers. NEC, Toshiba, and Litton-Dexcel all sell extremely low-noise FETs.

These devices have gate widths under .5 μm, and special structural features such as recessed gates and N^+ source and drain contact layers to keep internal losses (which raise noise figure) to an absolute minimum. An alternate device structure, called the *high electron mobility transistor* (HEMT) offers even lower noise figures (and higher associated gain) than conventional FETs. HEMT devices are expected to replace FETs in most low-noise amplifier applications. However, at this time, FETs offer very low noise figures (under 2.0 dB at 18 GHz) at a reasonable price. HEMTs are currently too expensive for all but a few specialized applications.

Most low-noise amplifiers are cascaded, multistage designs. It is critically important that the first stage be extremely low noise; however, subsequent stages must also be reasonably low noise if the overall noise figure is to be low. Usually, low-noise narrowband amplifiers are all single-ended designs with interstage matching FETs. We now proceed with the design process.

Step 1: Specifications

- Bandwidth = 4.5 to 5.0 GHz.
- Gain = 20 dB minimum ± 1.5 dB maximum ripple.
- Noise Figure = Less than 2.0 dB across the band.
- Bias Conditions = + 3V at 30 mA maximum.
  − 3V at 1 mA maximum.

Step 2: Choose a Topology

At 5 GHz, the 20 dB gain requirement dictates a two-stage design. The input matching circuit of a low-noise amplifier must be capable of synthesizing Γ_{opt} of the GaAs FET over the specified bandwidth, while maintaining very low insertion loss. In fact, relatively conventional matching networks such as the simple L network perform quite well in low-noise amplifiers. Modifying the simple *L* topology slightly by adding an extra series line section at the input will help synthesize Γ_{opt}. Adding an open stub to the interstage region between FETs helps facilitate matching between the stages. The sources of the FETs are dc grounded, which means negative bias must be applied to their gates through a shorted stub. The reason for using a gate bias supply, instead of self-bias, is to make sure the FET's bias current easily can be adjusted to its minimum noise figure point. A basic two-stage low noise amplifier topology is shown in Figure 5.29.

Step 3: Choose an Active Device, and Noise Matching

The FET chosen for both stages of this amplifier is Nippon Electric Company's NEC4583. This FET is a packaged .3 μm gate length device with a data sheet

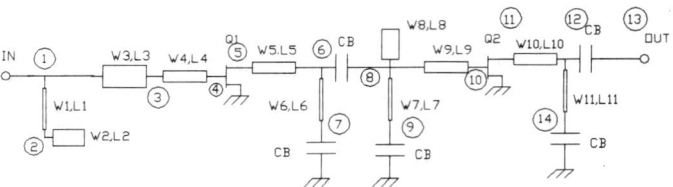

Fig. 5.29 The basic topology of a two-stage 4.5 to 5.0 GHz low-noise amplifier.

minimum noise specification of .8 dB at 8 GHz. Similar devices are available from other suppliers, so the NEC4583 represents typical performance of a class of extremely low-noise devices.

To set up a Touchstone file that analyzes and optimizes the noise figure of this amplifier, the noise parameters of the FET must be entered into a device S-parameter file. These noise parameters are frequency, minimum noise figure in dB, Γ_{opt} (both magnitude and angle), and noise resistance (normalized to 50 Ω).

To construct a device's noise data file, first enter the standard S-parameters at each frequency, and then *at the end of the S-parameter table*, begin again at the lowest frequency to enter the noise parameters. The data file for the NEC 4583 is shown in Table 5.4.

Table 5.4 S-Parameters and Noise Parameters for the NEC 4583

Freq.	S_{11}		S_{21}		S_{12}		S_{22}	
3.0	.94	−54	2.7	128	.05	49	.70	−40
4.0	.92	−71	2.6	111	.06	37	.68	−53
5.0	.85	−104	2.3	80	.08	11	.66	−76

Freq.	F_{min}	$\vert \Gamma_{opt} \vert$	$/\Gamma_{opt}$	$R_n/50$
2.0	.45	.78	29	.67
4.0	.50	.75	58	.65
6.0	.60	.73	90	.40

Step 4: Power and Noise Matching Conditions

Because noise figure is of paramount importance in this example, noise matching becomes an integral part of the device selection process and therefore is included in step 3.

Step 5: Choose a Bias Circuit

Because the bias circuitry is an integral, inseparable part of the matching circuitry in this example, this step is a part of step 2.

Step 6: Basic Circuit Simulation and Optimization

The basic topology will be optimized by Touchstone for a compromise between noise figure, gain, and input-output match. Because each of these parameters is counterbalanced by the others, the performance of a single-ended low-noise amplifier is reached through compromise. A minimum noise figure requires input mismatch, which reduces gain. With low-noise amplifiers, choices are to be made about the relative importance of gain, noise figure, and match in a particular application. This compromise becomes particularly difficult for wide bandwidths. Fortunately, most low-noise amplifiers are narrowband (specific communication and radar bands). A balanced configuration will relieve the input match problem; however, the loss of the input coupler adds directly to the device's noise figure. At low frequencies, the coupler's large size may be a problem. Quite often, input match is ignored up to the point of potential instability, as indicated by reflection gain (i.e., $RL > 1.0$). Sometimes a low-loss ferrite isolator is placed in front of the amplifier to take care of the input match. Output match is less of a problem because the FET's drain circuit interacts less strongly with noise figure than its gate circuit.

To model a multistage low-noise amplifier with Touchstone, the CKT block has to be organized in a special way. A series of two-port networks are defined that contain the following:

Input Matching	Stage 1
First FET	
Output Matching	Stage 1
Input Matching	Stage 2
Second FET	
Output Matching	Stage 2

This process can continue through as many as three stages. The optimized Touchstone file for the basic topology shown in Figure 5.29 is given in Table 5.5. Initially, all microstrip lines are set to $\lambda/8$ and all microstrip stubs are set to $\lambda/4$.

The circuit is optimized for the following goals:

Range   4.4   5.1   (Weight)
       AMP   dB[NF] < 1.5   50

AMP	dB[S_{21}] = 20.0	10	
AMP	dB[S_{12}] < 22.0	10!	This condition ensures a reasonable isolation
AMP	dB[S_{11}] < −5	10	
AMP	dB[S_{22}] < −15	5	

The simulated performance of the LNA after optimization is given in Figure 5.30, and the optimized value of all elements is given in Table 5.5. Section 8.4 discusses the layout and final simulation of this low noise amplifier.

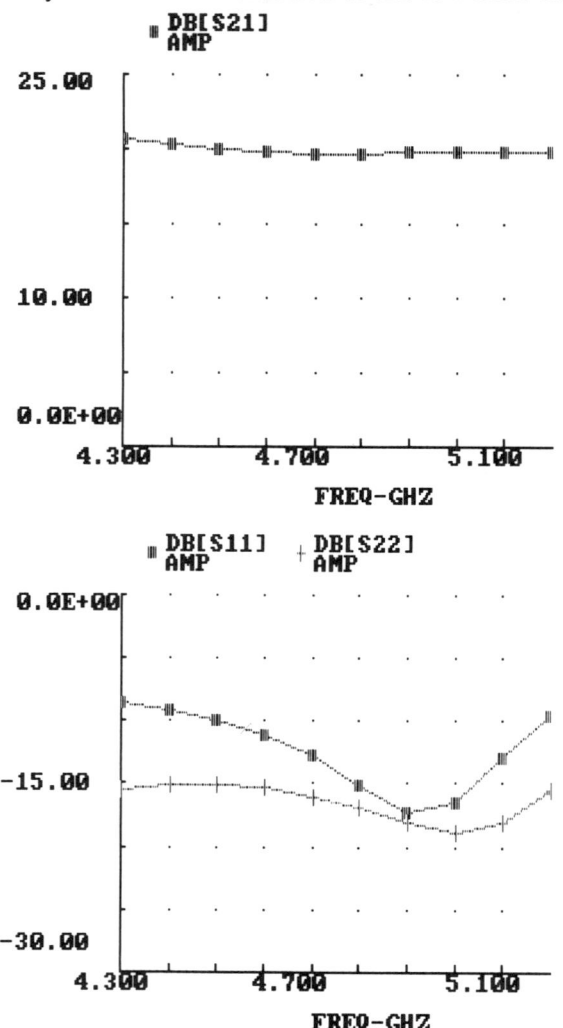

Fig. 5.30 The simulated performance of a two-stage, 4.5 to 5.0 GHz, low-noise amplifier.

Table 5.5 Touchstone—Ver(1.30–May-17-85)—Ser(15738-1598-1000)—Con(1CCC22124P) 114.CKT
06-23-89 18:25:00

```
! THE BASIC TOPOLOGY OF A THE TWO STAGE LOW NOISE AMPLIFIER FOR 4.5 TO 5.0 GHZ
! THE GaAs FETS ARE NE04583.

CKT

    MSUB ER=2.2 H=25 T=1 RHO=1 RGH=0
    MLIN 1 3 W\5.13924 L\655.16740
    MLEF 3   W\74.07652 L\61.86726
    DEF1P 1   BIAS

! FIRST STAGE

    BIAS 1
    MLIN 1 3 W\235.57830 L\197.99930
    MLIN 3 4 W\5.26721 L\133.66300
    DEF2P 1 4 NAIN        ! INPUT NETWORK

    S2PA 4 5 0   NE04583
    DEF2P 4 5 NA2P        ! FET #1

    BIAS 7
    MLIN 5 7 W\34.99665 L\487.12970
    CAP 7 8   C=50
    DEF2P 5 8 NAOUT       ! OUTPUT NETWORK

! SECOND STAGE

    BIAS 14
    MLEF 14 W\183.06430 L\46.12932
    MLIN 14 16 W\32.88865 L\215.31910
    DEF2P 14 16 NBIN      ! INPUT NETWORK

    S2PA 16 17 0   NE04583
    DEF2P 16 17 NB2P      ! FET #2

    BIAS 21
    MLIN 17 19 W\30.37581 L\121.34780
    MLEF 19 W\9.56070 L\824.71880
    MLIN 19 20 W\3.65715 L\90.56828
    MLEF 20 W\111.25990 L\141.98380
    MLIN 20 21 W\70.40086 L\85.54895
    CAP 21 22 C=50
    DEF2P 17 22 NBOUT     ! OUTPUT NETWORK

! CASCADE THREE STAGES

    NAIN 1 2
    NA2P 2 3
    NAOUT 3 4
    NBIN 4 5
    NB2P 5 6
    NBOUT 6 7
```

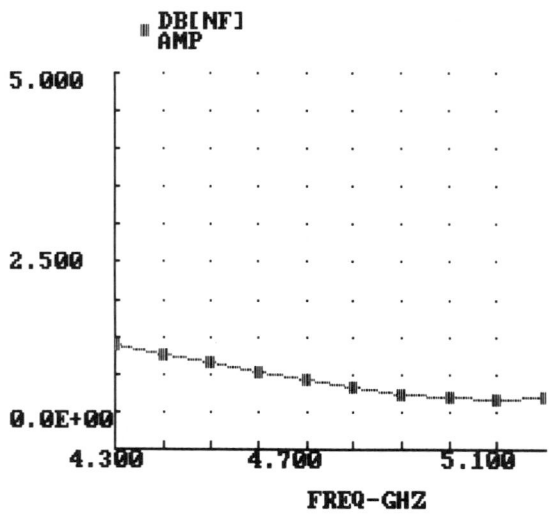

Fig. 5.30 (continued)

5.7 EXAMPLE 5: AN MMIC DISTRIBUTED AMPLIFIER FOR THE 2 TO 20 GHz BAND

When properly designed, MMIC distributed amplifiers have the ability to provide nearly constant gain over a decade of bandwidth with excellent input and output matching. They are often fabricated on a single .100 × .100 inch (or smaller) chip of GaAs. Unfortunately, achieving either low noise or high power over a decade bandwidth along with high gain and good match is very difficult. Therefore, MMIC distributed amplifiers often are used as the "gain blocks" in a cascaded chain, preceded by a special low-noise stage, or followed by a special power stage.

By sticking to a gain block type of distributed amplifier, the design procedure becomes a straightforward matter of strategically using the relationships developed in Section 5.2.5. Once the basic circuit elements are calculated, a Touchstone model is used to optimize for flat gain and good match over the required bandwidth.

Step 1: Specifications

- Bandwidth = 2 to 20 GHz;
- Gain = 7.0 dB ± .5 dB;
- Input-output Return Loss = −15 dB;
- Bias = +5V at 150 mA maximum,
  −5V at 10 mA maximum.

Step 2: Choose a Topology

As in the case of the MMIC feedback amplifier, the FET's design is an important variable of the overall amplifier's design process. Let us begin our topological selection process by choosing the FETs. (This will combine steps 2 and 3, which often is the case with MMIC design.) Assume that both the drain line and the gate line characteristic impedances are 50 Ω. Note that in some designs, gain is traded off with match by simply raising the impedance of the artificial transmission lines (see (5.24)).

Begin by calculating the FET's C_{GS} from the amplifier cut-off frequency, using (5.23). Assume that the highest operating frequency is no more than 70% of the cut-off frequency.

$$C_{GS} = \frac{1}{\pi Z_0 f_C}$$

where $Z_0 = 50\ \Omega$,

$$f_C = \frac{f_{high}}{.70} = 28.5 \text{ GHz}$$

Therefore,

$$C_{GS} = \frac{1}{\pi (50)(28.5 \cdot 10^9)} = .22 \text{ pF}$$

Assuming this circuit will be fabricated at a foundry whose FETs have a normalized C_{GS} of 1.0 pF/mm. We can calculate the width of the individual FETs as

$$.22 \text{ pF} = \left(\frac{w}{1000}\right)(1.0 \text{ pF/mm})$$

$$= .22 \text{ mm or } 220\mu\text{m}$$

Next, we calculate the number of sections from the gain specification and the equation for gain (5.24). First let us assume for the chosen foundry process, the FET's transconductance is 120 mS/mm. Therefore, the transconductance of the individual 220 μm FETs is calculated as

$$g_m = 120 \text{ mS/mm} \left(\frac{220}{1000}\right) = 26.4 \text{ mS}$$

From the equation for gain,

$$G = 7.0 \text{ dB} = 20 \log_{10}\left(\frac{nZ_0 g_m}{2}\right)$$

Taking the antilog and completing the algebra, we have

$$n = 3.39$$

Always round off n to the next higher integer to build in some safety margin. Therefore, the amplifier's topology will be a four-section distributed amplifier structure using 50 Ω gate and drain lines, and 220 μm FETs. The inductance per section of the artificial lines can be calculated from (5.22)

$$\begin{aligned} L &= 50^2 C_{GS} \\ &= 50^2(.22 \cdot 10^{-12}) \\ &= .55 \text{ nH} \end{aligned}$$

Because the chip will be constructed using narrow microstrip transmission lines playing the role of inductors, such a substitution now can be accomplished by using (3.20) for the equivalent microstrip transmission line length for a given inductance:

$$l = \frac{\lambda_g}{2\pi} \sin^{-1}\left(\frac{\omega L}{Z_0}\right)$$

Assume $f_H = 20$ GHz, $\epsilon_e = 8.5$, and $Z_0 = 100$ Ω. (We assume the foundry can fabricate sufficiently narrow microstrip lines on GaAs so that their characteristic impedance is 100 Ω.)

$$\begin{aligned} l &= \frac{[3 \cdot 10^{10}/(\sqrt{8.5} \cdot 20 \cdot 10^9)]}{2\pi} \sin^{-1}\left[\frac{2\pi(20 \cdot 10^9)(.55 \cdot 10^{-9})}{100}\right] \\ &= .0819 \sin^{-1}(.69) \\ &= (.0819)(.7615) = .062 \text{ cm or 25 mils} \end{aligned}$$

The drain line loading capacitances are calculated from (5.21) as

$$C_1 = C_{GS} - C_{DS}$$

Assume

$C'_{DS} = .30$ pF/mm

So

$$C_{DS} = .30 \left(\frac{220}{1000}\right) = .066 \text{ pF}$$

Therefore

$$C_1 = .22 - .066 = .15 \text{ pF}$$

This basic distributed amplifier topology is shown in Figure 5.31.

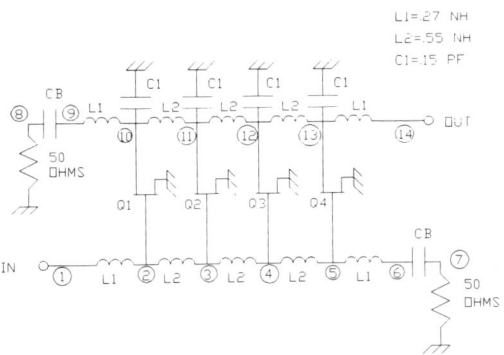

Fig. 5.31 The basic topology of the 2 to 20 GHz MMIC distributed amplifier.

Step 3: Choose an Active Device

The FETs chosen for the distributed amplifier will appear very similar to those chosen for the feedback amplifier discussed in Section 5.5. Before we consider the actual FET design, let us check the overall amplifier bias current consumption for our choice of a 220 µm FET, to be sure it is within the specified current limit. Assume the bias current per FET is $.5 I_{DSS}$. The total current will be

$$I_{\text{total}} = 4 \times (.5)(300 \text{ mA/mm}) \left(\frac{220}{1000}\right) = 132 \text{ mA}$$

which is safely under the 150 mA specification.
Following the FET design in Section 5.5, we choose a four-gate finger device.

The width of each gate finger is

$$w = \frac{220}{4} = 55 \mu m$$

The FET's layout will be identical to the FET layout shown in Figure 5.26.

Using the FET model shown in Figure 5.27 and typical foundry normalized FET parameters, we may calculate the element values as

$$C_{GS} = \left(\frac{220}{1000}\right) \times 1.0 \text{ pF/mm} = .22 \text{ pF}$$

$$C_{DG} = \left(\frac{220}{1000}\right) \times .20 \text{ pF/mm} = .044 \text{ pF}$$

$$g_m = \left(\frac{220}{1000}\right) \times 120 \text{ mS/mm} = 26.4 \text{ mS}$$

$$C_{DS} = \left(\frac{220}{1000}\right) \times .30 \text{ pF/mm} = .066 \text{ pF}$$

$$R_i = \left(\frac{1000}{220}\right) \times 5.0 \text{ }\Omega - \text{mm} = 22.7 \text{ }\Omega$$

$$R_{DS} = \left(\frac{1000}{220}\right) \times 100 \text{ }\Omega - \text{mm} = 454 \text{ }\Omega$$

These element values may be inserted directly into Touchstone's FET model, or a separated FET model may be created within the amplifier's file.

Step 4: Power and Noise Figure Matching Conditions

There are no specified power or noise requirements for this amplifier because it is a gain block; however, let us calculate the maximum expected power output using (2.13).

$$P_{\text{out}} = \frac{I_M(V_{BGD} - V_P - V_S)}{8}$$

assuming

$$I_M = 1.20 I_{DSS}$$

$V_{BGD} = +15$ V
$V_P = 3$ V
$V_S = 1$ V

Assuming that for our choice of foundries $I_{DSS} = 300$ mA/mm, the total saturated drain current drawn by all four 220 μm FETs is

$$I_{DSS} = 4 \times (300)\left(\frac{200}{1000}\right) = 264 \text{ mA}$$

The calculated maximum power output is

$$P_{out, max} = \frac{1.20(.264)(15 - 3 - 1)}{8}$$
$$= .435 \text{ W or } + 26.4 \text{ dBm}$$

Such high power may never be realized by this distributed amplifier because of the difficulty in practice of synthesizing a drain circuit that places the optimum load resistance, R_L, simultaneously across *each* FET at all frequencies (see equation (2.14)). Of course, one half of $P_{out, max}$ is dissipated in the drain line termination, so the most output power that could be expected to be delivered to a load would be +23.4 dBm.

Distributed amplifiers can be expected to generate somewhat less than their calculated maximum power output. A good rule of thumb is to calculate the dc power input of the amplifier, and then calculate the output power, assuming a power-added efficiency of 10%. In this particular case, the maximum dc power is

$$P_{DC} = (+8 \text{ V})(132 \text{ mA}) = 1056 \text{ mW}$$

Assuming 10% power-added efficiency, the power output is calculated to be

$$P_{out} = (.10)(1056 \text{ mW}) = 105.6 \text{ mW or } + 20.2 \text{ dBm}$$

which is less than the maximum power calculated earlier, but more consistent with experience with this class of amplifier.

Step 5: Choose a Bias Circuit

Both the gate line and the drain line circuits are dc isolated from their terminations by a blocking capacitor, C_B. A good way to bias a distribute amplifier is to use two dc supplies (i.e., a positive supply for the drain circuit and a negative bias for the gate circuit). A technique must be developed for applying these voltages to their respective lines without disturbing the RF performance. A very straightforward technique for doing this is to provide off-chip inductive bias chokes connected directly to the amplifier's input and the output terminals, as shown in Figure 5.32.

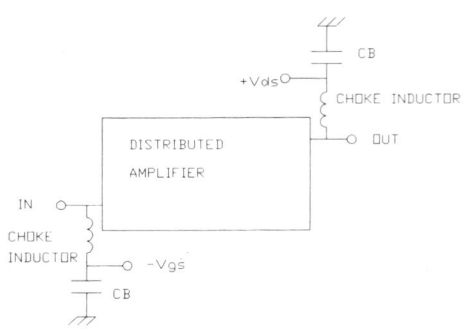

Fig. 5.32 Simple off-chip bias network for the MMIC distributed amplifier.

An alternative on-chip bias scheme would involve placing both a gate bias choke and a drain bias choke on the chip. These chokes would be spiral inductors. The reactance of such devices must be at least 100 Ω at 2 GHz. Their inductance is calculated as

$$L_C = \frac{100}{2\pi(2 \cdot 10^9)} = 7.95 \text{ nH}$$

To design an 8 nH spiral inductor that is resonance-free over the entire 2 to 20 GHz band is very difficult. (See Section 4.6.)

There are several alternative biasing approaches, all of which are application specific. They are listed so that you may be aware that these options exist.

- On-chip drain inductor, and a 100 Ω on-chip resistor from the gate line to a gate bias pad. A resistor can be used to provide gate bias, because little or no dc gate current is drawn by this device.
- Self-bias (single positive supply voltage) can be achieved by mounting the entire MMIC chip on top of an MIS bypass capacitor. The gate line must be grounded through a resistor; and an appropriate source resistor is connected from the top of the bypass capacitor to ground. The source resistor will be determined by the relationship

$$R_S = \frac{V_\text{gate}}{I_{DSS}/2}$$

where V_gate is the gate-to-source voltage at $I_{DSS}/2$ drain current.

With this scheme, drain bias must be provided either with an on-chip spiral inductor, or with an off-chip choke.
- An on-chip resistor is used to provide gate bias. Drain bias is delivered through an on-chip active load, which is an FET whose gate and source are connected together and whose width is chosen so that its I_{DSS} is equal to the total current drawn by the entire amplifier. Because an active load is biased well into current saturation, its RF resistance is quite high (several hundred ohms), making it act as an RF choke.

The price paid for using an active load is a higher than normal supply voltage. This is because the drop (three to four volts) across the active load is added to the normal V_{DS} of the FETs. Losses associated with the active load can reduce gain, and any parasitic elements associated with the active load can cause gain roll off at high frequencies. However, active loads are *much* smaller than spiral inductors and free of resonance. Sometimes an active load is a good biasing alternative if real estate is limited.

Step 6: Basic Circuit Simulation and Optimization

The basic topology, shown in Figure 5.31, may be analyzed and optimized using Touchstone. We will choose to use microstrip lines as inductors from the start, although optimizing the topology is possible using lumped inductors and then converting them to microstrip lines. The advantage of starting with microstrip line inductors is that the parasitic shunt capacitance of these lines is included in the model from the beginning, allowing the optimizer to adjust other circuit elements to compensate for these inherent parasitic capacitances.

The microstrip line widths should all be set to whatever value is the minimum line width for the particular foundry process to be used. Minimum width assures that the inductance per unit length will be maximized, and the capacitance per unit length will be minimized. The exception to this statement is if a particular foundry process offers extremely narrow lines (air bridge or first metal) that prove to be very lossy. A judgment may be necessary concerning the trade-off of loss or size. Very narrow lines make small size possible, but contribute additional loss. Assume in this example a minimum line width of .50 mils and a corresponding metal thickness of .10 mils. The optimized Touchstone file for basic distributed amplifiers topology is given in Table 5.6.

The on-chip blocking capacitors are to be 10 pF which is the maximum practical value for most foundry processes. The bias circuit is modeled as two

Table 5.6 Touchstone—Ver(1.30–May-17-85)—Ser(15738-1598-1000)—Con(1CCC22124P) 304.CKT
06-26-89 10:19:55

```
! THE BASIC TOPOLOGY OF THE FOUR SECTION 2 TO 20 GHZ DISTRIBUTED MMIC AMPLIFIE
! THE GaAs FETS ARE .5X230 MICRON FOUNDRY FETS
! DISTRIBUTED ELEMENT MODEL
VAR

    L1\16.83917
    L2\37.65157
    W1=0.50
    L3=5
    CB=10
    C1\0.06361
    R1\30.89591
    R2\81.69948

CKT

    ! GaAs FET MODEL
    FET 1 2 3 G=.035 T=1.7 F=0 CGS=.23 GGS=0 RI=8 CDG=.005 CDC=0 CDS=.08 RDS=14
    DEF3P 1 2 3 TRAN

    ! DISTRIBUTED AMP MODEL
    MSUB ER=12.5 H=5 T=.2 RHO=5 RGH=0
    MLIN 1  2     W^W1  L^L1
    MLIN 2  3     W^W1  L^L2
    MLIN 3  4     W^W1  L^L2
    MLIN 4  5     W^W1  L^L2
    MLIN 5  6     W^W1  L^L1
    MLIN 8  10    W^W1  L^L1
    MLIN 10 11    W^W1  L^L2
    MLIN 11 12    W^W1  L^L2
    MLIN 12 13    W^W1  L^L2
    MLIN 13 14    W^W1  L^L1
    CAP 8 9   C^CB
    CAP 10 0  C^C1
    CAP 11 0  C^C1
    CAP 12 0  C^C1
    CAP 13 0  C^C1
    CAP 6 7   C^CB
    RES 7 0   R^R1
    RES 8 0   R^R2
    TRAN 2 10 0
    TRAN 3 11 0
    TRAN 4 12 0
    TRAN 5 13 0
    DEF2P 1 14   AMP

FREQ

    SWEEP 1 26 1

OUT

    AMP DB[S21] GR1
    DEF2P 1 7    AMP

FREQ

    SWEEP 4.0  5.5  .1
```

Table 5.6 cont'd.

```
OUT
    AMP DB[S21] GR1
    AMP DB[NF]  GR2
    AMP DB[S11] GR3
    AMP DB[S22] GR3
GRID
    RANGE 4.3  5.2  .1
    GR1 0 25 5
    GR2 0 5 .5
    GR3 -30 0 5
OPT
    RANGE 4.3  5.2
    AMP DB[NF]<1.5     50
    AMP DB[S21]=20.0   40
    AMP DB[S11]<-10    10
    AMP DB[S22]<-15    10
```

off-chip choke inductors (16 nH), which are placed from the input to ground and from the output to ground.

Several hundred iterations of the Touchstone optimizer produces the performance shown in Figure 5.33. The optimizer has flattened the gain response to about

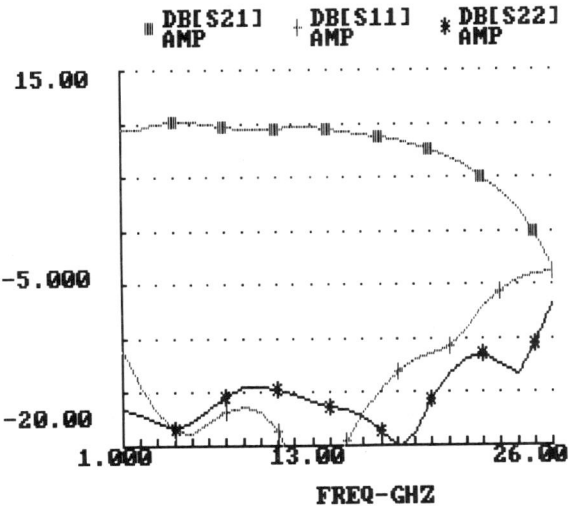

Fig. 5.33 The simulated performance of the four-section 2 to 20 GHz distributed amplifier, with distributed gate and drain line inductors.

8 dB across the band and improved the return losses. The next step, which is discussed in Section 10.2, is to create the MMIC layout for this circuit and then reanalyze its performance, including all layout parasitic elements.

5.8 EXAMPLE 6: A ONE-WATT MIC BALANCE POWER AMPLIFIER FOR THE 4 TO 10 GHz BAND

Many applications, including radar transmitters and test instruments, require relatively high power outputs (1 W or more). In some applications, high power is required over bandwidths in excess of an octave. Special design techniques are necessary to ensure the delivery of full power output from power GaAs FETs over such wide bandwidths. This example will demonstrate these techniques by designing a 1 W, 4 to 10 GHz amplifier. Such techniques easily are extended to different bands and different devices. In this example, the output circuit will be specially tailored to provide the optimum load impedance for power output (see Section 2.1.5) across the entire bandwidth of operation. The Touchstone optimizer is used to optimize power match.

Step 1: Specifications

- Bandwidth = 4 to 10 GHz.
- Small-Signal Gain = 5 dB minimum.
- Input-output Return Loss = 10 dB minimum.
- Power output (at saturation) = +30 dBm (1 watt).
- dc Bias = 6 V at 450 mA maximum.

Step 2: Choose a Topology

We will choose a balanced reactively matched amplifier topology (similar to example 1) because of its flexibility in terms of tailored output impedance for broadband power matching and its ability to provide good input-output return loss regardless of the match into and out of the single-ended amplifier. The basic topology is shown in Figure 5.34. This circuit is a variation on the single L matching network. The difference lies in relocating the shorted stubs to the FET's gate and drain, rather than at the input and the output terminals. This topology allows matching lower impedance levels, which is an important advantage with power FETs, whose large gate widths mean low impedance levels.

The starting values for all the circuit elements are obtained by using a rule of thumb that at mid-band (7 GHz, in this case), the shorted stubs are $\lambda/4$ long, and the line sections are $\lambda/8$ long. Therefore, the calculated stub and line lengths are

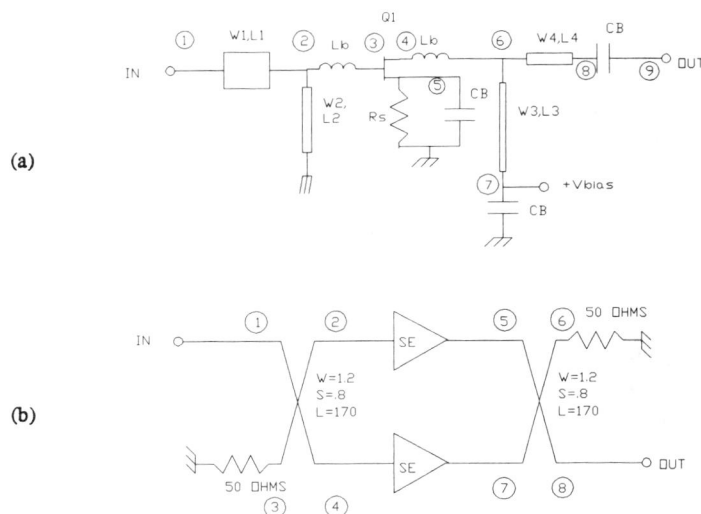

Fig. 5.34 Basic (a) single-ended and (b) balanced topology of the 4 to 10 GHz power amplifier.

$$\lambda/4 = \frac{3 \cdot 10^{10}}{(4)\sqrt{6.3}(7 \cdot 10^9)} = 4.26 \text{ cm or } 168 \text{ mils}$$

$$\lambda/8 = 84 \text{ mils}$$

The stub widths initially are chosen to be the minimum width allowed by the Al_2O_3 fabrication process (about 2 mils). The line widths are chosen to be midway between 50 Ω and the FET's input or output impedance (depending on whether it is in the input circuit or the output circuit).

Step 3: Choose an Active Device and Large-Signal Considerations

The specifications require that the active device be capable of in excess of 5 dB small-signal gain at 10 GHz and in excess of +27 dB/m saturated power. Furthermore, the circuit must be configured in such a way that self-biasing is easy from a layout point of view. The operating current per device is required to be less than 225 mA. The GaAs FET chosen for this application is the Mitsubishi MGF2116 chip. The MGF2116 has a MAG of 9 dB at 10 GHz; and a saturated output power of +28 dBm at a typical operating current of 200 mA. Although the MGF2116 is available from the manufacturer in a number of different package styles for narrow band application, this application will use the chip form of the device because broadband matching requires low device parasitics.

Let us calculate the device's maximum power output, and its large-signal load impedance using relationships (2.13) and (2.14):

$$P_{out} = \frac{I_M(V_{BGD} - V_P - V_S)}{8}$$

and

$$R_L = \frac{(V_{BGD} - V_P - V_S)}{I_M}$$

For the MGF2116, assume that $I_M = 500$ mA, $V_{BGD} = 14$ V, $V_P = 3$ V, $V_S = 1.5$ V; therefore

$$P_{out} = \frac{(.5)(14 - 3 - 1.5)}{8} = .59 \text{ W or } +28 \text{ dBm}$$

and

$$R_L = \frac{(14 - 3 - 1.5)}{.5} = 19 \Omega$$

As discussed in Section 2.1.5, a large-signal FET model is created by substituting the large-signal load resistance R_L, for R_{DS} in the FET's equivalent circuit. Only S_{22} changes for the large-signal case when converting this model to a set of S-parameters. The new values of S_{22} may be calculated from the output model alone by paralleling C_{DS} and R_L. C_{DS} may be calculated from the small-signal S-parameters, or using a Smith chart to convert S_{22} to $Y = G + jB$ and then calculating C_{DS} from B using the equation

$$B = 2\pi f C_{DS}$$

For the MGF2116 chip, $C_{DS} = 1.2$ pF. Once C_{DS} has been determined, a Touchstone model for the FET's large signal output circuit (R_L in parallel with C_{DS}) is used to calculate the large signal values of S22. Figure 5.35 compares small signal and large signal S_{22}s for the MGF2116. Table 5.7 gives the complete large-signal S-parameters of the MGF2116. Bond wire inductances ($L_B = .10$ nH) have been included in the final S-parameter model.

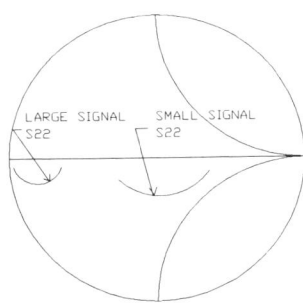

Fig. 5.35 A comparison between the small-signal output impedance and the large-signal output impedance of a MGF2116 FET. The respective impedances are shown on a Smith chart plot.

Table 5.7 Large-Signal S Parameters of the Mitsubishi MGF2116 Chip Operating at +7 V at 200 mA

Freq.	S_{11}		S_{21}		S_{12}		S_{22}	
4.0	.74	−172	2.12	43	.06	−12	.57	−158
6.0	.78	161	1.73	10	.06	−30	.67	−157
8.0	.82	122	1.48	−30	.07	−43	.75	158
10.0	.79	80	1.26	−63	.08	−52	.81	159

Step 4: Choose a Bias Circuit

As discussed in step 1, we intend to self-bias the FETs in this amplifier. Therefore, the source contact must be RF grounded via a capacitor and connected to ground through a resistor, R_S. The value of R_S is determined by the proper operating current (50 percent I_{DSS}) and the gate-source voltage necessary to produce this operating current.

$$R_S = \frac{V_{GS}}{I_{DS}}$$

For the MGF2116, $V_{GS} = 1.2$ V, and $I_{DS} = 200$ mA. Therefore,

$$R_S = \frac{1.2}{.20} = 6\,\Omega$$

Step 5: Power and Noise Figure Matching Conditions

This example has no noise matching requirements. Power matching calculations are worked out in step 3 as a part of choosing the active device.

Step 6: Basic Circuit Simulation and Optimization

Using the circuit diagram 5.35, a Touchstone file is created and listed in Table 5.8. The blocking and bypass capacitors, C_B, are set to 50 pF. All microstrip line lengths and widths are made variables of the optimization. A primary goal for the optimizer is to make the *single-ended* output return loss exceed 7 dB to provide a broadband match to ensure high power output. The optimized performance of the circuit is shown in Figure 5.36. All optimized circuit parameters are given in the Touchstone file in Table 5.8.

Table 5.8 Touchstone—Ver(1.30–May-17-85)—Ser(15738-1598-1000)—Con(1CCC22124P) 307.CKT
06-23-89 20:25:27

```
! THE BASIC TOPOLOGY OF THE 4 TO 10 GHZ BALANCED POWER AMPLIFIER.
! THE GaAs FETS ARE MITSUBISHI MGF2116'S.
! LARGE SIGNAL S PARMETERS ARE CALCULATED USING THE LOAD LINE METHOD.
CKT
    ! SINGLE ENDED
    MSUB ER=9.9 H=25 T=.2 RHO=5 RGH=0
    MLIN 1 2 W\10.83756 L\11.06689
    MLSC 2 W=2.0         L\164.88640
    IND 2 3 L=.10
    S2PA 3 4 7   MG2116L
    IND 4 5 L=.10
    CAP 7 0 C=200
    RES 7 0 R=10
    MLSC 5 W\2.04447 L\23.91598
    MLIN 5 6 W\84.52541 L\111.78160
    DEF2P 1 6  AMP

    ! BALANCED
    MLANG 1 2 3 4  W=1.2  S=.8  L=170
    MLANG 5 6 7 8      W=1.2  S=.8  L=170
    AMP 2 5
    AMP 4 7
    RES 3 0 R=50
    RES 6 0 R=50
    DEF2P 1 8    BAL

FREQ
    SWEEP 3 11 1

OUT
    AMP DB[S21] GR1
    AMP DB[S11] GR1
```

Table 5.8 cont'd.

```
    AMP  DB[S22]  GR1
    BAL  DB[S21]  GR2
    BAL  DB[S11]  GR2
    BAL  DB[S22]  GR2

GRID

    RANGE 3 11 1
    GR1  -15  15  5
    GR2  -30  15  5

OPT

    RANGE 4 10
    BAL  DB[S21]=6    50
    BAL  DB[S11]<-15  10
    BAL  DB[S22]<-15  10
    AMP  DB[S22]<-7   30
```

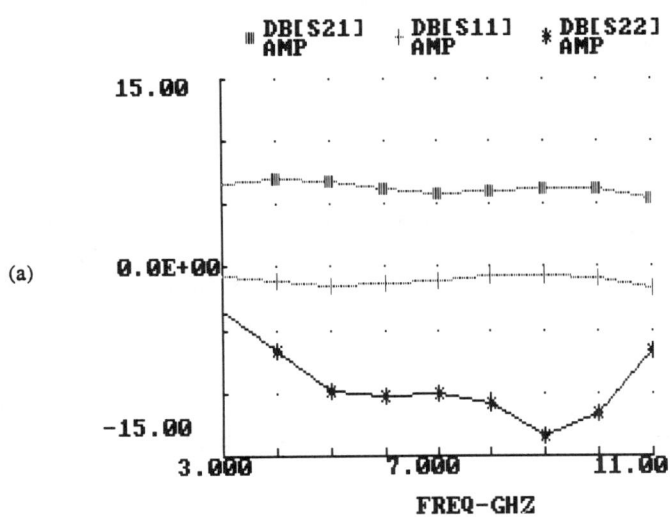

(a)

Fig. 5.36 Simulated performance of the (a) high-power, single-ended amplifier and (b) 1-W balanced 4 to 10 GHz amplifier.

(b)

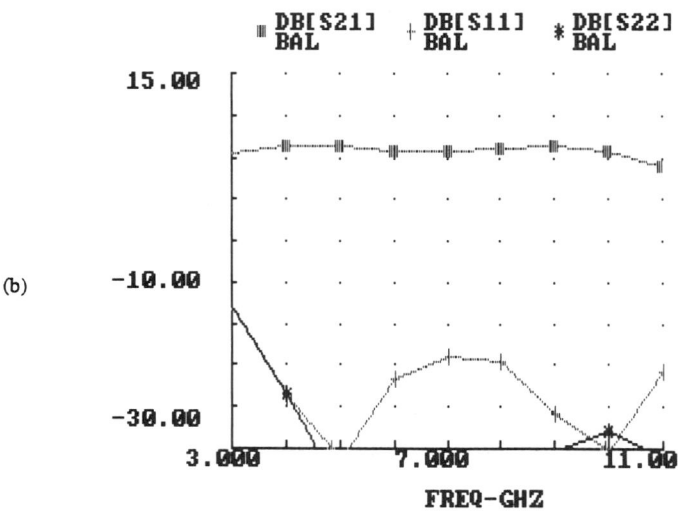

Fig. 5.36 (continued)

5.9 EXAMPLE 7: A BALANCED MIC PIN ATTENUATOR FOR THE 6 TO 18 GHz BAND

Often a microwave amplifier's gain must be held constant over a wide range of temperatures. Because the gain of a GaAs FET amplifier varies with temperature by approximately $-.015$ dB/°C per stage (see Section 5.1.8), a technique is needed to *compensate* for this gain drift. A PIN diode attenuator has been used successfully to control the gain of the entire cascaded amplifier chain by varying the PIN attenuator's attenuation automatically so that it cancels out the amplifier's natural gain drift. A thermistor or a temperature sensing diode can provide a temperature reference, which is then processed into a control signal for the PIN attenuator, using analog or digital techniques.

We will discuss the design only of the PIN attenuator in this section. The control circuits can be designed using a number of standard digital and analog techniques. As discussed in Section 2.4, PIN diodes are dc current controlled microwave resistors, obeying a relationship

$$R = \frac{K}{I_0} \qquad (5.34)$$

where

K is a constant (approximately 10 Ω/mA),
I_0 is the diode's dc current.

We neglect the effects of series resistance and shunt capacitance for this initial calculation.

By connecting the PIN diode in shunt with a 50 Ω transmission line (as shown in Figure 5.37), the amount of transmission mismatch loss is directly controlled by the PIN diode's dc current.

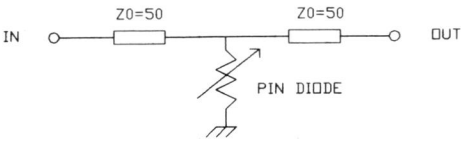

Fig. 5.37 Basic single-ended PIN diode attenuator topology.

The mismatch created by the PIN diode is given in terms of a reflection coefficient as

$$\Gamma = \frac{50 - R}{50 + R} \tag{5.35}$$

The mismatch transmission loss, in dB, is

$$L = -10 \log_{10}(1 - |\Gamma|^2) \tag{5.36}$$

Combining (5.34), (5.35), and (5.36), we obtain an expression for the attenuation of the basic single-ended PIN diode attenuator in terms of the dc control current flowing through the PIN diode:

$$L = -10 \log_{10}\left[1 - \left(\frac{50 - K/I_0}{50 + K/I_0}\right)^2\right] \tag{5.37}$$

Next, we calculate the maximum attenuation of the PIN attenuator, assuming that 20 mA is the maximum available control current:

$$L = -10 \log_{10}\left[1 - \left(\frac{50 - 10/20}{50 + 10/20}\right)^2\right]$$
$$= -10 \log_{10}(.039)$$
$$= -14 \text{ dB}$$

Experience indicates that 14 dB (down) is a realistic attenuation control range for this type of PIN attenuator.

A problem with the basic PIN attenuator is that, because of its reflective nature, the input and output return losses are very poor. This is a serious problem in many applications. This problem is solved easily by constructing a *balanced* attenuator, which is topologically similar to a balanced amplifier. Like the balanced amplifier, the balanced attenuator uses a pair of Lange couplers to cancel out the reflections from the PIN diodes, ensuring good input and output return loss over the bandwidth of the couplers. We proceed step by step through the design of the balanced PIN attenuator.

Step 1: Specifications

- Bandwidth = 6 to 18 GHz.
- Attenuation Range = Up to −14 dB.
- Input-output Return Loss = 10 dB minimum.
- Control Current = Less than 40 mA total (two diodes).
- Attenuation Flatness with Frequency = ±1 dB maximum.

Step 2: Choose a Topology

The basic shunt-mounted diode, single-ended configuration discussed earlier is chosen as the basic topology. Lange couplers are used to connect the attenuators into a balanced configuration. Series capacitors between each diode and the Lange couplers serve as dc blocks. Bias current is fed to each PIN diode through a 10 nH choke inductor. This basic balanced topology is shown in Figure 5.38. The Touchstone file for the basic balanced topology is given in Table 5.9.

The Touchstone file is divided into three sections: the diode model, and the single-ended circuit model, and the balanced configuration. The Lange couplers are the same as in the balanced amplifier discussed in Example 1.

Step 3: Choose an Active Device

We choose a chip PIN diode instead of a beam lead PIN diode because the chip is more easily grounded in the shunt circuit configuration used here. The model of the PIN diode is

$R_S = .2 \, \Omega$
$C_0 = .08 \, \text{pF}$
$R_d = 10/I_{dc} \, (\text{mA}) \, \Omega$

Table 5.9 Touchstone—Ver(1.30–May-17-85)—Ser(15738-1598-1000)—Con(1CCC22124P) 373.CKT
06-26-89 10:53:30

```
! THE BASED TOPOLOGY OF A 6 TO 18 GHZ BALANCED PIN ATTENUATOR
VAR
    RS=.50
    RD=1000
    CD=.08
    L1=50
    WC=1.4
    SC=.8
    LC=80
    CB=50

CKT
    MSUB ER=9.9 H=10 T=.15 RHO=5 RGH=0

    ! PIN DIODE MODEL
    RES 1 2 R^RS
    PRC 2 3 R^RD C^CD
    DEF2P 1 3 PIND

    ! SINGLE ENDED ATTENUATOR

    CAP 2 5 C^CB
    MLIN 5 6 W=10 L^L1
    IND 6 7 L=.15
    PIND 7 0
    IND 7 19 L=16
    CAP 19 0 C^CB
    IND 7 8 L=.15
    MLIN 8 9 W=10 L^L1
    CAP 9 10 C^CB
    DEF2P 2 10 SE

    ! BALANCED ATTENUATOR

    MLANG 1 2 3 4 W^WC S^SC L^LC
    RES 3 0 R=50
    SE 2 10
    SE 4 16
    MLANG 10 17 16 18 W^WC S^SC L^LC
    RES 17 0 R=50
    DEF2P 1 18 BAL

FREQ
    SWEEP 5 20 1

OUT
    BAL DB[S21] GR1
    BAL DB[S11] GR2
    BAL DB[S22] GR2
    AMP DB[S11] GR1
    AMP DB[S22] GR1

GRID
    RANGE 1 26 1
    GR1 -20 15 5
```

Table 5.9 cont'd.

```
OPT
    RANGE 2 20
    AMP DB[S21]=10 50
    AMP DB[S11]<-15 10
    AMP DB[S22]<-15 10
GRID
    RANGE 5 20 1
    GR1 -30 0 5
    GR2 -30 0 10
```

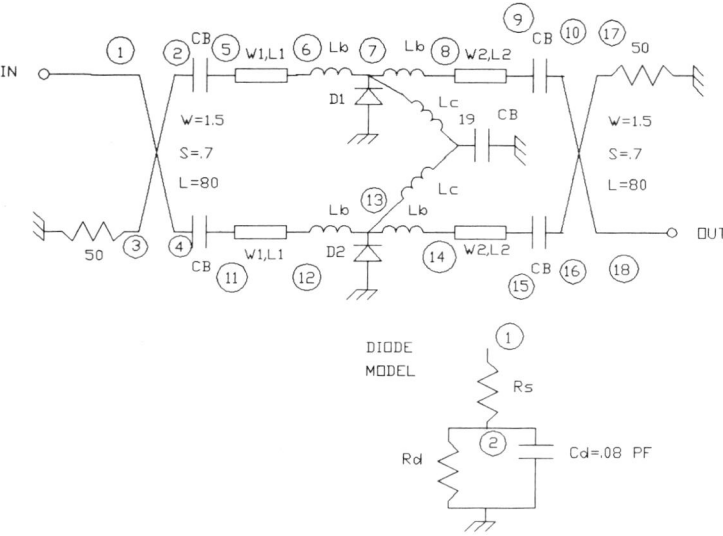

Fig. 5.38 The topology of a balanced PIN attenuator for the 6 to 18 GHz band. The topology includes a diode model and bond wire inductances.

Step 4: *Choose a Bias Circuit*

As mentioned earlier, bias current is delivered to the PIN diodes via the 10 nH choke inductors. Because each inductor is sufficiently broadband to cover the entire 6 to 18 GHz range without causing any degradation in performance, this simple arrangement is all that is needed to provide bias. Each choke is RF grounded at its supply end with a 50 pF capacitor.

Step 5: Power and Noise Figure Matching Conditions

This step is not applicable in the case of PIN attenuators.

Step 6: Basic Circuit Simulation and Optimization

The Touchstone file for the basic circuit given in Table 5.9. The attenuator is optimized for the following performance.

First optimization:

    Range   6.0   18.0
    dB $[S_{21}] > -1.0$! for zero current
    dB $[S_{11}] < -15$ (i.e., $R = 200,000\ \Omega$)
    dB $[S_{22}] < -15$

Second optimization:

    Range   6.0   18.0
    dB $[S_{21}] = -14$! for $I_{dc} = 20$ mA
    dB $[S_{11}] < -15$ (i.e., $R_D = .5\ \Omega$)
    dB $[S_{22}] < -15$

During the first optimization, Touchstone simulates the case of zero current, and the circuit elements are optimized for flatness and lowest insertion loss. During the second optimization, the circuit elements are optimized for a flat 14 dB attenuation at $I_{dc} = 20$ mA. The final circuit must be a compromise between these two requirements. Actually, very little performance is sacrificed with this compromise. The optimized circuit elements given in Table 5.9 reflect this compromise. In Figure 5.39, the attenuator's simulated performance in its final configuration is

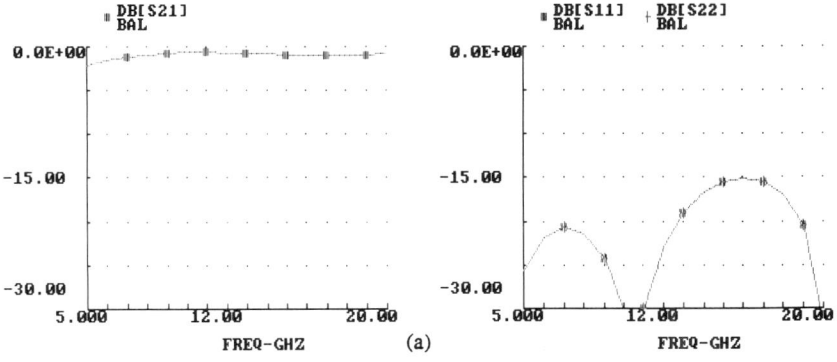

Fig. 5.39 The simulated performance of the balanced PIN diode attenuator, for (a) $I_{control} = 0.0$ mA$_2$ and (b) $I_{control} = 20$ mA.

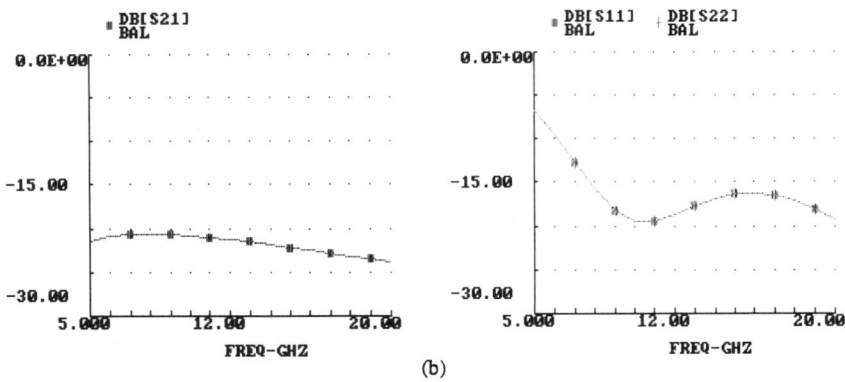

Fig. 5.39 (continued)

shown for two values of current. Note how the attenuation stays quite flat with frequency, indicating the flatness need not be sacrificed to operate the attenuator over a wide range of currents.

REFERENCES

1. G. Vendelin, *Design of Amplifiers and Oscillators by The S-Parameter Method*, John Wiley and Sons, New York 1982, p. 15.
2. *Ibid.*, p. 16.
3. R. M. Fano, "Theoretical Limitations on the Broadband Matching of Arbitrary Impedances," *J. Franklin Inst.*, Vol. 249, January 1960, pp. 57–83; and February 1960, pp. 139–155.
4. Vendelin, *op. cit.*, pp. 109–113.
5. "Application Note on Intercept Points," *Microwave and Millimeter Wave Amplifiers Data Book,* 2d Edition, Avantek, Inc., 1989.
6. K. B. Niclas, "On the Design and Performance of Lossy Match GaAs MESFET Amplifiers," *IEEE Trans. Microwave Theory Tech.*, Vol. MTT-30, Nov. 1982, pp. 1900–1907.
7. K. B. Niclas, "Reflective Match, Lossy Match, Feedback and Distributed Amplifiers: A Comparison of Multi-Octave Performance Characteristics, *IEEE MTT-S Symp. Digest*, 1984, pp. 215–217.
8. J. B. Beyer, S. N. Prasad, R. C. Becker, J. E. Nordman, and G. K. Hohenwarter, "MESFET Distributed Amplifier Design Guide Lines," *IEEE Trans. Microwave Theory Tech.*, Vol. MTT-32, March 1984.

Chapter 6
Oscillator Circuits

6.1 NEGATIVE RESISTANCE CONCEPTS

An oscillator is an energy conversion device that transforms dc power into ac power. Like all energy conversion devices, oscillators operate at less than 100 percent conversion efficiency due to thermodynamic considerations. To model an oscillator from a circuit point of view, techniques have been developed to represent the energy conversion process as a circuit element. The best known of these techniques is called the *negative resistance concept.* To understand a negative resistor, let us first review how a positive (normal) resistor works. Referring to Figure 6.1(a), the generator develops a voltage, V, that it impresses across a positive resistor, R. A current, $I = V/R$ flows *into* the resistor, *dissipating* a power, I^2R.

Now consider the negative resistor shown in Figure 6.1(b). The generator develops a voltage, V, that it impresses across the negative resistor, $-R$. A current, $I = -V/R$ flows *out* of the negative resistor (and into the generator), *generating* a power I^2R.

Consider an application of a negative resistor to the so-called *reflection amplifier,* shown in Figure 6.2. The circulator is a device that separates the incident wave from the reflected wave. The gain, in dBs, of the reflection amplifier is equal to

$$G = 20 \log_{10} \Gamma \tag{6.1}$$

where

$\Gamma = \dfrac{Z - Z_0}{Z + Z_0}$ is the reflection coefficient of an impedance, Z, terminating a transmission line of impedance Z_0.

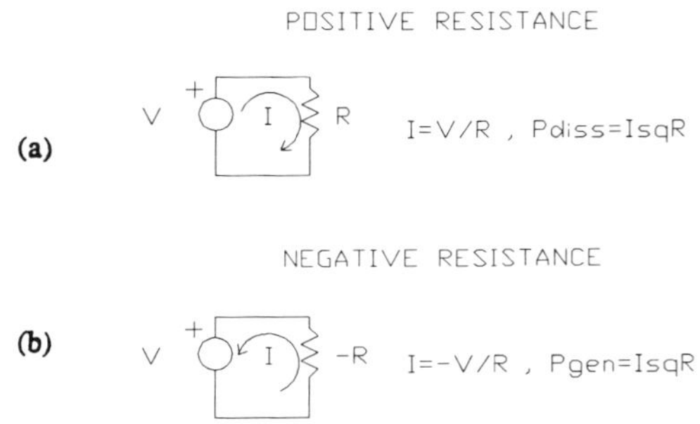

(a) POSITIVE RESISTANCE

V, I, R   I=V/R , Pdiss=IsqR

(b) NEGATIVE RESISTANCE

V, I, -R   I=-V/R , Pgen=IsqR

Fig. 6.1 Positive and negative resistance concepts.

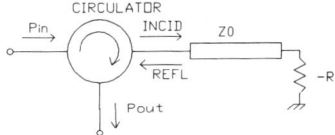

Fig. 6.2 A reflection amplifier using a negative resistor.

If the line is terminated in the negative resistor, $-R$,

$$|\Gamma| = \left|\frac{R + Z_0}{Z_0 - R}\right| > 1.0 \tag{6.2}$$

for all negative resistance, so G is greater than zero for any value of negative resistance.

Because a negative resistance can produce gain, it also can cause oscillation. Oscillators consist of three basic parts: a negative resistance, a resonator, and a load. This basic oscillator topology is shown in Figure 6.3.

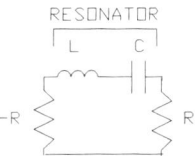

Fig. 6.3 The basic oscillator topology consists of a negative resistor, a resonator, and a load.

Under the condition that $|-R|$ is greater than R_{load}, this circuit has a net gain that will amplify any noise present (some noise always is present in *any* real circuit). The noise will be amplified most strongly at the series resonance f_R of L and C; in fact, the noisy signal will grow in time with an exponential characteristic until it becomes a sine wave at f_R. The sine wave is the signal developed by the oscillator, and it will continue to grow without limit until the negative resistor is reduced in magnitude, by nonlinear effects, to equal R_{load}. When $|-R|$ reaches the value of R_{load}, the sine wave's amplitude reaches a fixed level, and the oscillator is said to have reached *steady state*. The steady state condition means that the ac power *generated* in the negative resistance is exactly equal to the ac power *dissipated* in the load resistance. In real circuits, $|-R|$ changes value because the semiconductor device that produces $-R$ saturates as the signal levels increase, thereby decreasing $|-R|$. From this simple negative resistance oscillator model, we can deduce two important rules for oscillator operation.

1. Start Oscillation Condition

For oscillations to begin, the magnitude of the negative resistance must exceed the load resistance,

$$|-R| > R_{\text{load}}$$

and the oscillations will occur at the circuit's reactive resonance, which is

$$\sum_{i=1}^{N} X_i = 0 \implies f_0 = \frac{1}{2\pi}\sqrt{\frac{1}{LC}} \tag{6.3}$$

2. Steady-State Oscillation Condition

Oscillations cease growing, and reach steady state, at the signal level where the negative resistance just equals the load resistance:

$$|-R(P_{SS})| = R_{\text{load}}$$

where P_{SS} = the steady-state power output. The frequency of oscillation continues to be determined by the circuit's reactive resonance:

$$f_0 = \frac{1}{2\pi}\sqrt{\frac{1}{LC}} \tag{6.4}$$

With the commonly available computer-aided design software tools (such as Touchstone and SuperCompact), only start oscillation conditions can be analytically evaluated. The steady-state conditions are inherently large signal and cannot be simulated by linear analysis programs. However, start oscillation conditions are sufficiently useful to form the basis for most practical oscillator designs.

6.2 TRANSISTOR OSCILLATOR BASICS

Today most microwave oscillators use either GaAs FET or silicon bipolar active devices. Certain two-terminal devices such as Gunn diodes and Impatt diodes also are useful for constructing oscillators, but they will not be considered here. Our attention will now focus on the techniques used to make a GaAs FET or a bipolar transistor into a negative resistance oscillator and how these devices function in oscillator circuits.

A transistor is transformed into a negative resistance when feedback is applied in some form. From an energy point of view, dc power is applied to the transistor in the form of bias voltage and bias current. The biased transistor develops gain over some useful frequency range. When feedback is applied to the transistor, its impedance is transformed into negative resistance at one or more of its terminals. A resonator is connected to the terminal with negative resistance, and a load resistor is connected to the transistor's output terminal (usually its drain or its collector). If the conditions for oscillation are met, oscillation occurs at the resonator's resonant frequency and ac power is delivered to the load. A conversion efficiency describes the ratio of the ac power delivered to the load to the dc bias power. Figure 6.4 diagrams the functional parts of a transistor oscillator.

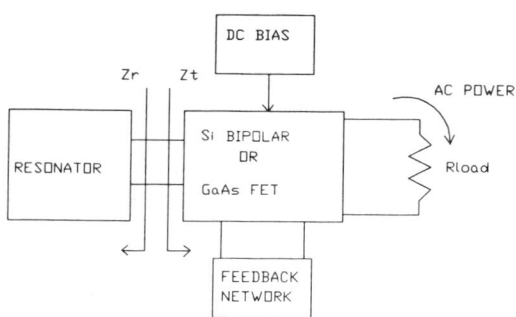

Fig. 6.4 A functional diagram of a microwave transistor oscillator.

Using the diagram in Figure 6.4, we may generalize the oscillation equations to include both the start oscillation conditions and the steady state oscillation conditions. Call Z_R the impedance of the resonator. Call Z_T the impedance of the transistor, *with feedback* and bias connected, and the load resistor connected to its output port. Under these conditions, the start oscillation conditions may be rewritten as follows.

1. Start Oscillation Conditions

$$\text{Re}[Z_R] \leq |\text{Re}[Z_T]|$$

where

$$\text{Re}[Z_T] < 0.0$$

and

$$\text{Im}[Z_R] = -\text{Im}[Z_T] \tag{6.5}$$

2. Steady State Oscillation Conditions

$$\text{Re}[Z_R] = |\text{Re}[Z_T(P_{out})]|$$

where

$$\text{Re}[Z_T] < 0.0$$

and

$$\text{Im}[Z_R] = -\text{Im}[Z_T] \tag{6.6}$$

In other words, oscillations begin whenever the magnitude of the negative resistance of the transistor, with feedback, is greater than the positive resistance of the resonator *and* when the reactance of the resonator is equal in magnitude and opposite in sign from the reactance of the transistor with feedback. The oscillator's steady state power output is reached when the negative resistance of the transistor with feedback has been reduced by saturation to a value *exactly* equal in magnitude, and opposite in sign, to the resistance of the resonator. The negative resistance of a transistor is reduced as the power level increases by a variety of saturation mechanisms, such as current saturation, and voltage breakdown. Figure 6.5 diagrams a typical path by which saturation leads to steady state oscillation.

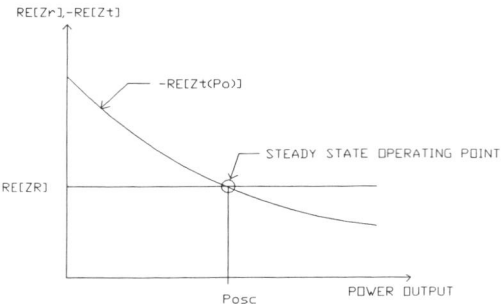

Fig. 6.5 An oscillator reaches steady state operation by virtue of large-signal device saturation.

6.3 FEEDBACK TECHNIQUES FOR OSCILLATOR TRANSISTORS

Transistors form a three-terminal network, and there are many ways to apply feedback between any two of the terminals to cause negative resistance at the third terminal. In this section, we address only two of the most popular feedback techniques. However, one of these techniques is used in the vast majority of practical microwave transistor oscillators. The two circuit techniques are bipolar common-base inductive feedback and GaAs FET common source capacitive feedback. The schematic diagram of these two basic circuits is shown in Figure 6.6.

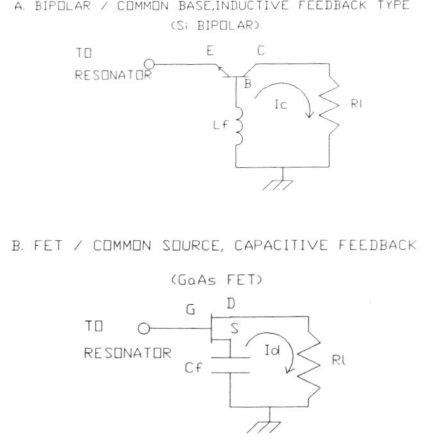

Fig. 6.6 Basic microwave transistor oscillator topology: (a) bipolar or common-base inductive feedback; (b) FET or common-source capacitive feedback.

Both circuits provide feedback by circulating the output RF current, I_0, through the feedback element in such a way that a voltage is developed across the feedback element, which becomes part of the input (resonator) signal. Therefore, by this process, the output signal is fed back to the transistor's input, via the feedback element (either L_f or C_f).

The usual way to select the value of L_f, or C_f for a given transistor type is to write a Touchstone or a SuperCompact file for the basic circuit under consideration and optimized L_f or C_f for a goal of maximum negative resistance at the input terminal over the band of interest.

Very often increased negative resistance and increased bandwidth can be obtained simultaneously by adding a reactive output network between the transistor's output and the load. This output matching network may be simply a shunt L and a shunt C. This output network also may be optimized for maximum negative resistance. We will see how this process proceeds when we get to the specific design examples. The goal of the optimizer always will be to achieve the start oscillation conditions over the desired bandwidth. The feedback element and the output matching network elements are the optimization variables. Both the resistance and the reactance of the resonator must be accounted for in this process, if possible and if the results are to be meaningful. Some nominal forward gain or loss should also be a goal of this optimization process, because excessive loss from the resonator to the load will mean that power generated at the resonator cannot be transferred to the load.

6.4 OSCILLATOR-TRANSISTOR COMPARISON

The "right" transistor for a given application should be chosen from the start of the design process. To help make this choice, a performance comparison has been prepared (see Table 6.1). The table compares various parameters and conditions for the two types of active device, the Si bipolar transistor, and the GaAs FET.

Table 6.1 A Comparison between a Si Bipolar Transistor and a GaAs FET for Microwave Oscillator Operation

Item	Si Bipolar	GaAs FET
Type of Feedback	Common-base inductive	Common-Source capacitive
Typical Bias Conditions	+10 V at 20 mA, −1V at 20 mA	+5V at 50 mA, −1V at 1 mA
Upper Useful Frequency	12 GHz	40 GHz

Table 6.1 (Cont'd)

Item	Si Bipolar	GaAs FET
Lower Useful Frequency	dc	3 GHz (with common-source capacitive feedback)
Magnitude Of Negative Resistance	20 to 100 Ω	10 to 170 Ω
Character Of Active Reactance	Inductive (with common-base inductive feedback)	Capacitive (with common-source capacitive feedback)
Power Output	+5 to +15 dBm	+5 to +15 dBm
Type of Parasitic Element that Could Cause Uncontrolled Spurious Oscillation	Series capacitance between the transistor and the resonator	Series inductance between the FET and the resonator

6.5 FREQUENCY-TEMPERATURE STABILITY AND TUNING LINEARITY

The stability of an oscillator's frequency over a range of temperatures is an important issue in many applications. Temperature stability of the resonator is controlled by factors such as the differential thermal expansion of its materials or the temperature slope of its junction capacitance, in the case of the varactor. Dielectric resonators are naturally very stable owing to their basic material properties; however, varactor and YIG resonators drift substantially with temperature. For this reason, YIG and varactor-tuned oscillators often are operated in temperature-controlled environments, usually making use of devices called *heaters*. Heaters dissipate electrical energy at low temperature to heat up the resonating element (and the transistor in most practical cases) and are turned off by a controller circuit at high temperatures. Typical heaters can maintain the temperature of a YIG or a varactor resonator within 5°C over a 100°C ambient range.

Excellent tuning linearity occurs naturally with YIG resonance phenomena (see Chapter 2), so little has to be done to improve it. YIG-tuned oscillators are capable of tuning linearity within 1 percent with no attempt at linearizing. Varactor-tuned oscillators are a different story. Because of the nonlinear relationship between voltage and a varactor's capacitance ($C = K/V^N$), the relationship between a varactor-tuned oscillator's frequency and its tuning voltage is inherently nonlinear. If a particular application demands linear tuning, a predistortion network, called a *linearizer,* must be connected between the tuning signal and the varactor terminals of the varactor-tuned resonator.

6.6 OSCILLATOR RESONATOR COMPARISON

Three basic types of resonators find application in transistor microwave oscillators: *YIG-tuned resonators* (YTO), *varactor-tuned resonators* (VCO), and mechanically tuned resonators including dielectric resonators. The examples in this chapter will cover YIG resonators, varactor-tuned resonators, and *dielectric resonators* (DRO).

Resonators are characterized in terms of their (1) *tuning range*, (2) *quality factor, Q*, (3) *temperature stability*, and (4) *potential higher-order resonances*. Table 6.2 summarizes this comparison.

Table 6.2 Resonator Comparison

Type of Resonator	Tuning Range	Tuning Linearity	Q	Temperature Stability	Potential for Higher-Order Resonances
Varactor	One octave	2:1	10–50	Poor	Low
YIG	Over a decade	1%	500–3000	≈ -200 kHz/°C	Medium
DRO	10%	—	5,000–30,000	≈ 5 ppm/°C	Medium
Mechanically Tuned Cavity	Under an octave	—	200–1000	≈ -100 kHz/°C	High

The oscillator's FM noise is inversely proportional to the resonator's Q, so high Q is very desirable from a noise standpoint. In a varactor-tuned resonator, Q also is inversely proportional to its tuning range. Therefore, tuning range and noise usually occur in opposition to each other and may be traded off with resonator Q.

A varactor-tuned oscillator's frequency and tuning voltage have a naturally nonlinear relationship. The amount of tuning nonlinearity can be reduced significantly by any one of the three following methods:

1. Reducing the tuning range.
2. Using hyperabrupt junction diodes.
3. Employing external linearizer circuits.

6.7 OSCILLATOR NOISE

Noise sources within the oscillator's active device (bipolar or GaAs FET) will *modulate* the output signal of a microwave oscillator so that noise sidebands are produced about the main "carrier" signal. These noisy sidebands are called *AM noise* and *FM (or PM) noise*. AM noise sidebands result from noisy amplitude modulation of the oscillator by the noise sources; FM noise sidebands result from noisy frequency modulation of the oscillator by the noise sources. The spectra of these sidebands are shown in Figure 6.7.

Two kinds of noise sources are at work in bipolar and GaAs FET devices: low frequency $1/f$ noise sources which originate in trapping states within the semiconductor material, and microwave frequency thermal (white) noise sources. These noise sources modulate the oscillator's carrier signal to form both AM and FM noise sidebands. To understand the modulation process for each kind of source consider the noise equivalent circuit of a microwave transistor oscillator shown in Figure 6.8.

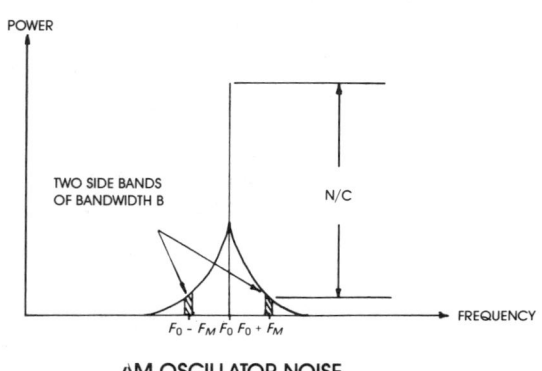

AM OSCILLATOR NOISE

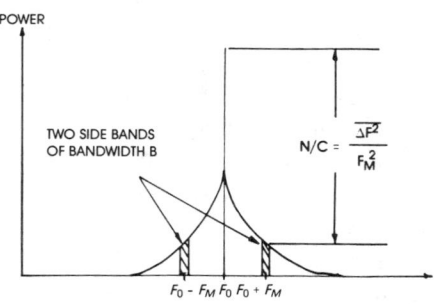

FM NOISE-TO-CARRIER RATIO

Fig. 6.7 Oscillator noise spectra.

The thermal noise source $\overline{i_R^2}$ acts as a randomly varying injected microwave signal to which the oscillator attempts to "lock in frequency[!]." The attempted "locking" phenomenon produces both amplitude and frequency modulation. The low frequency $1/f$ noise sources cause random variations in the transistor's equiva-

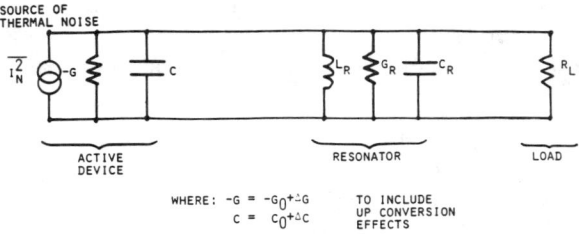

CLASSICAL DIODE OSCILLATOR NOISE MODEL

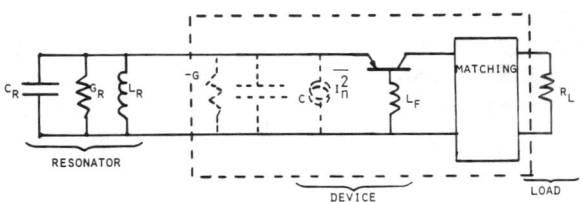

TRANSISTOR OSCILLATOR NOISE MODEL

Fig. 6.8 Noise equivalent circuit of a transistor microwave oscillator.

lent circuit parameters, which, in turn, randomly modulate the oscillator's amplitude and frequency.

Next, we list the four noise equations for a microwave transistor oscillator along with typical (empirically derived) noise parameters for both bipolar transistors and GaAs FETs. We can see that, in analogy to diode negative resistance oscillators[2], the FM thermal noise of a transistor oscillator is

$$\overline{\Delta\omega^2} = \frac{\omega_0^2 k T_n B}{4 Q_r^2 P_{\text{out}}} \tag{6.7}$$

where
- ω_o is the oscillator's angular center frequency,
- k is Boltzmann's constant $(1.38 \cdot 10^{-23} \text{J/K})$,
- T_n is the transistor equivalent noise temperature,
- B is the bandwidth of measurement,
- Q_r is the resonator's unloaded Q,
- P_{out} is the oscillator's output power,
- $\overline{\Delta\omega^2}$ is the oscillator's mean square FM frequency deviation noise.

Thermal noise is dominant for frequencies far from the carrier frequency. However, up-converted ($1/f$) noise is dominant at frequencies close to the carrier. The up-converted FM noise of a transistor oscillator can be shown to be

$$\overline{\Delta\omega^2} = \left[\frac{\omega_0^2\left(\frac{\partial C_d}{\partial V_0}\right)^2}{4Q_rG_r}\right] S_{\Delta V_0}(\omega_m)B \tag{6.8}$$

where

$\left(\dfrac{\partial C_d}{\partial V_0}\right)$ is the active device's capacitance change with dc voltage (i.e., the emitter base junction capacitance in a bipolar transistor),

$S_{\Delta V_0}$ is the spectrum of the low-frequency noise fluctuations,

G_r is the resonator's equivalent parallel conductance (losses),

ω_m is the low angular frequency (modulation frequency), where the noise fluctuations are originating.

The AM noise of a transistor oscillator can be treated in a similar manner. The equation for the AM thermal noise of a transistor oscillator is

$$\text{N/C} = \frac{kT_nBG_r^3/P_{\text{out}}\left(\frac{\partial G_d}{\partial V_1}\right)}{1 + (\omega/\omega')^2} \tag{6.9}$$

where

N/C is the AM noise-to-carrier ratio in the bandwidth, B;

$\left(\dfrac{\partial G_d}{\partial V_1}\right)$ is the device's terminal conductance sensitivity to RF signal amplitude, called the *oscillator's nonlinearity factor*,

$$(\omega')^2 = \frac{\omega_0^2 P_{\text{out}}\left(\frac{\partial G_d}{\partial V_1}\right)}{8Q_r^2G_r^3}$$

and ω' is called the *AM corner frequency*. The AM up-converted noise is given by the expression

$$\text{N/C} = \left(\frac{1}{4}P_{\text{out}}^2\right)\left(\frac{\partial P_{\text{out}}}{\partial V_0}\right)^2 S_{\Delta V_0}(\omega)\frac{B}{1 + (\omega/\omega')^2} \tag{6.10}$$

where

$\left(\dfrac{\partial P_{\text{out}}}{\partial V_0}\right)$ is the oscillator's power sensitivity to changes in dc voltage.

The FM noise mean square frequency deviation may be converted to a noise to carrier ratio by using a relationship from small-signal FM theory:

$$\text{N/C} = \dfrac{\overline{\Delta\omega^2}}{\omega_m^2} \qquad (6.11)$$

For most active devices, the spectrum of the low-frequency noise fluctuations obey a $1/f$ behavior pattern with frequency. This spectrum can be expressed by the simple equation

$$S_{\Delta V_0}(f)B = \dfrac{N}{f^\alpha} \qquad (6.12)$$

where
    N is a strength parameter,
    α is a spectrum shape coefficient (close to 1.00).

These noise relationships may be applied to a particular oscillator by first carefully measuring all of the active device's noise parameters. However, approximated noise calculations may be made using the following typical noise parameters.

The experimentally derived noise parameters for a typical microwave bipolar transistor are

$$\dfrac{\partial C_d}{\partial V_0} = 2 \cdot 10^{-3} \text{ pF/V}$$

$$S_{\Delta V_0}(f) = 10^{-8}/f, \text{ V}^2/\text{Hz}$$

$$T_n = 1000 \text{ K}$$

The experimentally derived noise parameters for a typical microwave GaAs FET are

$$\dfrac{\partial C_d}{\partial V_0} = 2 \cdot 10^{-3} \text{ pF/V}$$

$$S_{\Delta V_0}(f) = 10^{-7}/f, \text{ V}^2/\text{Hz}$$

$T_n = 500$ K

Figure 6.9 shows the typical shape of the AM and FM components of oscillator noise. Figure 6.10 compares the FM noise produced by a bipolar transistor with that produced by a GaAs FET. At very low modulating frequencies (close to the carrier), the bipolar transistor has considerable noise advantage over the GaAs FET due to its significantly lower $1/f$ noise, whereas at frequencies far from the carrier, GaAs FET oscillators may be quieter. Whenever possible, use bipolar transistors in applications where close to the carrier oscillator noise is a critical specification.

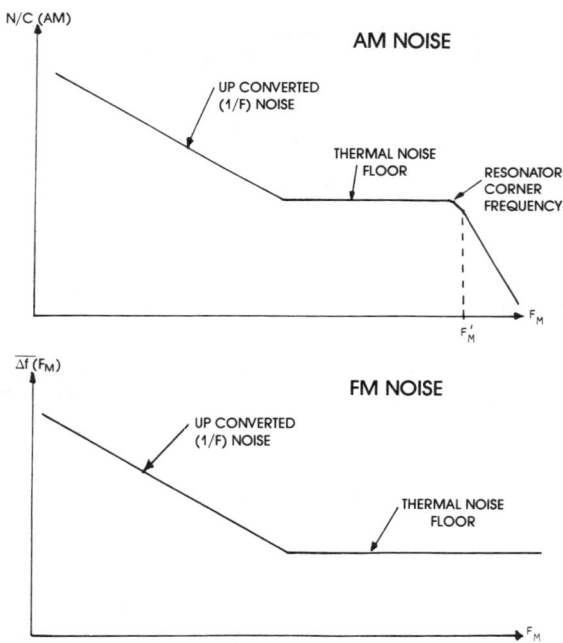

Fig. 6.9 AM and FM oscillator noise spectra.

Other than properly selecting the transistor, the most important thing a designer can do to reduce oscillator noise is raise the resonator's Q. Unfortunately, in some applications, increasing Q may directly affect another critical parameter, such as tuning range or tuning linearity. As usual, considerable judgment must be exercised by the designer.

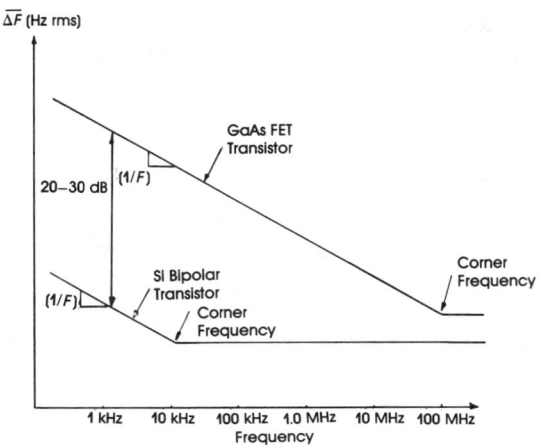

Fig. 6.10 A comparison of ($1/f$) FM noise spectra in Si bipolar transistors and GaAs FETs.

6.8 EXAMPLE 1: A 12 TO 18 GHz FET VARACTOR-TUNED VCO

Because few bipolar transistors oscillate effectively at frequencies above 12 GHz, a designer is forced to use a GaAs FET device in designing fundamental oscillators for these frequencies. The basic topology shown in Figure 6.6(b) will serve as the starting point of this design. The construction media will be an alumina substrate with gold metalized traces. Hybrid assembly techniques will be employed (see Chapters 7 and 8).

Let us follow a step-by-step process for designing a microwave oscillator. Let us begin the 12–18 GHz VCO example.

Step 1: Specifications

Tuning Range = 12 to 18 GHz.
Power Output = +5 dBm minimum.
Tuning Voltage = 0 to −20 V.

Step 2: Choose a Topology

The basic FET oscillator topology shown in Figure 6.6(b) is the starting point of the topology. The additions are a series inductor, between the varactor diode and the FET's gate, to provide an inductive element for resonating the varactor and a second inductor connected between the FET's gate and ground for bias purposes. On the drain side of the FET, a shorted stub and an open stub are connected to the drain terminal to provide matching between the drain and the load. The initial topology is shown in Figure 6.11.

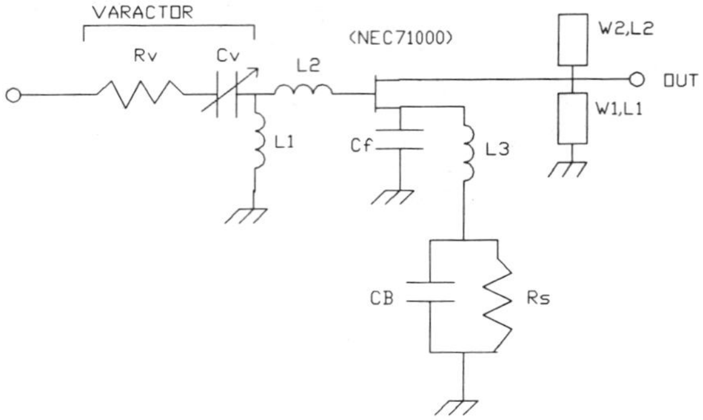

Fig. 6.11 The basic topology of the 12 to 18 GHz varactor-tuned VCO.

Step 3: Choose an Active Device

Because negative resistance is created by a feedback process involving gain, it is very important that the RF transistors have sufficient gain at the high end of the band to create the negative resistance. It also is important that the saturated power of the transistor be well above the oscillator's power output specification. The NEC 710 00 GaAs FET is an excellent choice for this application because it provides over 7 dB gain at 18 GHz and over +15 dBm saturated power. The FET must be mounted in such a way that stray capacitance between its source and ground is minimized, because the optimum source feedback capacitance is very small and easily can be "swamped" by such parasitics. Of course, this rules out packaged FETs, as only chips can be mounted directly on an alumina circuit substrate. The varactor diode is chosen to be a low-loss hyperabrupt device that covers the range of capacitance required by the optimized circuit.

Step 4: Choose a Bias Circuit

This particular circuit is very easy to self-bias with a source resistor. The dc source current is conducted through a choke connected to the FET's source, to a bypass capacitor, and a source resistor connected to ground. The value of the source resistor will be

$$R_S = \frac{V_{GS_1}}{(I_{DSS/2})}$$

where

V_{GS_1} is the value of gate-to-source voltage necessary for a drain current of $IDSS/2$.

The gate itself is connected to ground by L_1 and L_2.

$IDS = (IDSS/2)$

is chosen as a bias point because it provides the best compromise among gain, saturated power, and low harmonic output.

Step 5: Power and Noise Matching Conditions

Because this analysis will involve calculating the start oscillation conditions using linear analysis CAD tools, it is not straightforward to calculate output power or determine the proper matching circuit to maximize the output power. Noise is not a specification; however, if it were a requirement, we would calculate the resonator's Q from the expression

$$Q = \frac{2\pi f_0 C}{G}$$

and apply the noise equations in Section 6.7 to calculate the AM and FM noise sideband levels.

Step 6: Basic Circuit Simulation and Optimization

We begin by creating a Touchstone file for the basic circuit topology shown in Figure 6.11. This file is given in Table 6.3. We can use the optimization capability of Touchstone to extend the "start oscillation conditions" of this circuit over the entire band of interest. Notice that the equivalent circuit of the varactor diode is included as a part of the oscillator circuit. This is done so that all varactor losses are accounted for directly, and so that the calculated resonant frequency will be the operating frequency.

The optimizer block will be as follows:

Opt
    Range   12   18   ! (The band of interest)
    VCO RE [Z_1] <-20 ! The net negative resistance, including losses in the varactor diode will be more than $-20 \, \Omega$ at the start of oscillation.

Table 6.3 Touchstone—Ver(1.30–May-17-85)—Ser(15738-1598-1000)—Con(1CCC22124P)
343.CKT 06-09-89 11:25:19

```
! THE BASIC 12 TO 18 GHZ FET \ VARACTOR  VCO CIRCUIT TOPOLOGY

VAR

   R1=2
   CV=2
   L1=2
   C1=.12
   L2=10
   W1=2
   P1=100
   W2=20
   P2=40

CKT

   MSUB ER=9.9 H=10 T=.2 RHO=5 RGH=0
   RES 1 2 R^R1
   CAP 2 3 C^CV
   IND 3 0 L^L1
   IND 3 4 L=1.5
   S2PA 4 5 6   NEC71000
   CAP 6 0 C^C1
   IND 6 0 L^L2
   MLSC 5 W^W1 L^P1
   MLEF 5 W^W2 L^P2
   DEF2P 1 5  VCO

FREQ

   SWEEP 10 20 1

OUT

   VCO DB[S21] GR1
   VCO RE[Z1] GR2
   VCO IM[Z1] GR2
   VCO  S11

GRID

   RANGE 10 20 1
   GR1 -10 10 5
   GR2 -100 100 10

OPT

   RANGE 12 18
   VCO RE[Z1] < -20   50
   VCO IM[Z1] > -5    10
```

VCO dB $[S_{21}] > -5$!   The forward gain from the resonator to the load will be greater than -5 dB. This condition ensures that the power generated at the resonator will be transferred to the load.

The optimized element values are given in the Touchstone file in Table 6.3. It is very important to choose a varactor diode with sufficient capacitance tunability to cover the desired frequency range. Hyperabrupt varactors are available with tunability ratios exceeding 20:1, which should be sufficient for this and for most VCO applications. By placing the varactor diode in series with the FET's "input," and connecting a resonating inductance, $L2$, between them, the reactive start oscillation conditions as calculated by Touchstone will occur at the frequency where $IM[Z_1] = 0$. This means that the FET-varactor combination will be in resonance at the operating frequency. First, the optimizer is run to maximize negative resistance, $(RE[Z_1])$, and forward gain $dB[S_{21}]$. Next, the varactor's capacitance, C_v, is varied through its operating range to ascertain that the entire frequency range will be covered (i.e., $Im[Z_1] = 0$). The varactor's capacitance must be guessed at first, keeping in mind that 20:1 capacitance ratio is the best that realistic varactors can achieve. The range of values for C_v found during optimizing this VCO were .12 pF to 2.0 pF. Notice that the negative resistance including the varactor's losses exceeds $-15 \ \Omega$ across the band of interest, which is a very healthy start oscillation condition.

Simulated start oscillation performance of the 12 to 18 GHz FET varactor VCO is shown in Figure 6.12.

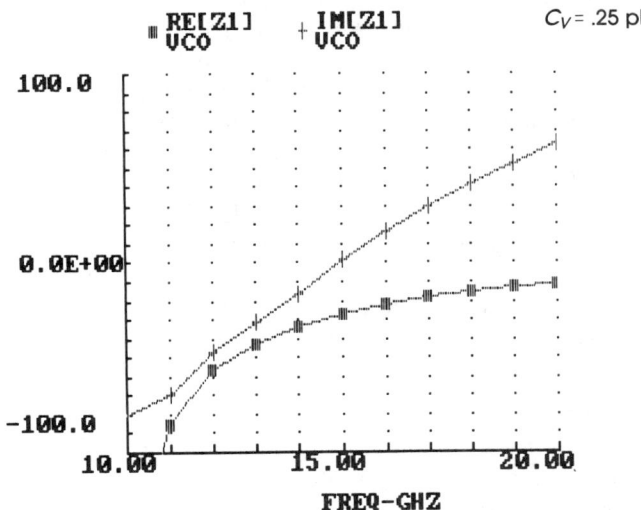

Fig. 6.12 Simulated performance of the 12 to 18 GHz FET varactor VCO.

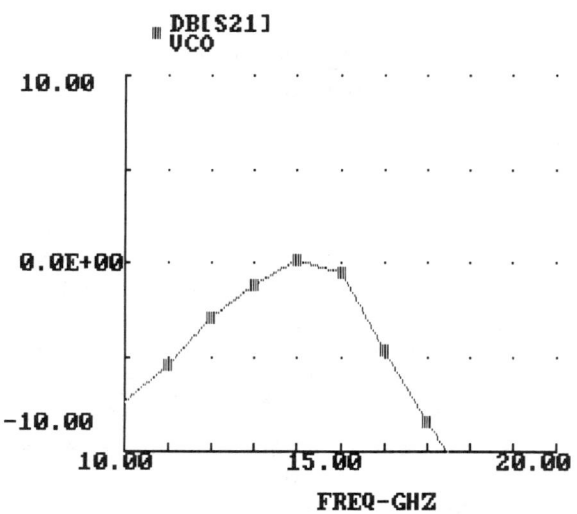

Fig. 6.12 cont'd.

6.9 EXAMPLE 2: A 4 TO 8 GHz BIPOLAR VARACTOR VCO

Noise considerations make bipolar transistors the preferable active devices for VCOs if the frequency range is below 10 GHz. The feedback method most often used with bipolar transistor oscillators is the common-base inductive configuration shown in Figure 6.6(a). As in the case of the FET oscillator, some matching elements are needed between the collector of the transistor and the load to guarantee strong negative resistance, and maximum bandwidth.

This type of oscillator is very predictable and has a very simple equivalent circuit. To first order, the equivalent circuit of a common-base bipolar oscillator has a negative resistance of -10 to -50 Ω, connected in series with its feedback inductor. This means, as a first guess, the feedback inductor and the varactor capacitance should be chosen so that their series resonance frequencies will cover the band of interest. Because of the simplicity of this bipolar oscillator's equivalent circuit, only a 4:1 varactor capacitance variation is needed to tune an octave bandwidth, although in practice more may be required.

Step 1: Specifications

    Tuning Range = 4 to 8 GHz.
    Power Output = +8 dBm.

Tuning Voltage = 0 to −20 V.
Bias Supply = +12 V, −5 V.

Step 2: Choose a Topology

The basic common-base inductive feedback topology was given previously in Figure 6.6(a). To become a practical circuit, it is necessary (1) to add a varactor diode's equivalent circuit at the input, (2) to provide a choke from the emitter terminal of the bipolar to a bypass capacitor to apply emitter bias (negative .7 V), and (3) to provide an output matching network between the collector terminal at the load.

All these items have been included in the basic topology shown in Figure 6.13. Unlike the FET VCO, we need not add inductance in series with the varactor, because the feedback inductance L_f is effectively placed in series with the varactor as a result of the feedback process. The dc bypass and blocking capacitors are used for providing bias. The value of C_V and L_f may be calculated from the resonance condition, which must occur at all frequencies in the band.

$$f_h = \frac{1}{2\pi} \sqrt{\frac{1}{L_f C_{v_1}}}$$
$$f_l = \frac{1}{2\pi} \sqrt{\frac{1}{L_f C_{v_2}}}$$
(6.13)

where
 f_l is the lowest frequency in the band of interest,
 f_h is the highest frequency in the band of interest,
 C_{v_1} is the lowest value of capacitance the varactor can assume,
 C_{v_2} is the highest value of capacitance the varactor can assume,
 L_f is the feedback inductance.

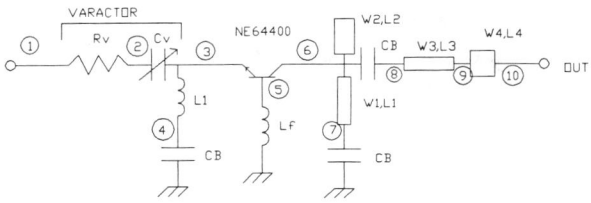

Fig. 6.13 The basic topology of the 4 to 8 GHz bipolar VCO.

Assuming $L_f = .5$ nH, and $f_L = 4$ GHz, and $f_H = 8$ GHz, we calculate C_{v1} and C_{v2} as

$$(2\pi)^2 f_h^2 = \frac{1}{L_f C_{v_1}}$$

or

$$C_{v_1} = \frac{1}{(2\pi)^2 L_f f_h^2} = \frac{1}{(2\pi)^2 (.5 \cdot 10^{-9})(8 \cdot 10^9)^2}$$
$$= .79 \text{ pF}$$

Similarly,

$$(2\pi)^2 f_l^2 = \frac{1}{L_f C_{v_1}}$$

or

$$C_{v_2} = \frac{1}{(2\pi)^2 L_f f_l^2} = \frac{1}{(2\pi)^2 (.5 \cdot 10^{-9})(4 \cdot 10^9)^2}$$

$$C_{v_2} = 3.16 \text{ pF}$$

Step 3: Choose an Active Device

For a bipolar transistor to create substantial negative resistance up to 8 GHz, it must have reasonable gain up to 8 GHz. We choose the NEC 64400, which is widely used in oscillator applications up to 12 GHz.

The initial value of C_v is 3.16 pF. The entire optimization process can be performed at this value. After the optimization has been completed, the exact values of C_v at the band edges may be determined with the help of Touchstone's "tune" mode. In tune mode, any circuit parameter (in this case, C_v) can be changed quickly, creating a family of updated performance curves on the computer's graphics display.

Step 4: Choose a Bias Circuit

Because the specifications tell us that two bias supply polarities are available, let us use the very simple two-polarity bias circuit shown in Figure 6.14. The value of R_E

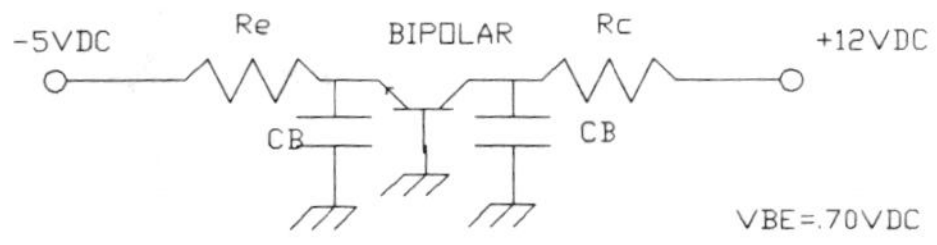

Fig. 6.14 Simple two-supply bias circuit for a bipolar oscillator.

and R_C in Figure 6.14 are determined by the emitter and collector currents plus the desired collector voltage.

$$R_E = \frac{5 - .7}{I_E}$$
$$R_C = \frac{12 - V_{CB}}{I_C} \qquad (6.14)$$

Assuming that $I_e \doteq I_C = 17$ mA, and $V_{CB} = +10$ V,

$$R_E = \frac{5 - .7}{.015} = 286.7 \ \Omega$$
$$R_C = \frac{12 - 10}{.015} = 133.3 \ \Omega$$

R_E and R_C are connected to bypass capacitors within the RF circuit.

Step 5: Power and Noise Matching Considerations

This oscillator has no noise specifications. The power requirement of +8 dBm will be satisfied by the power margin of the NEC 64400 transistor, which is capable of considerably more than +8 dBm.

Step 6: Basic Circuit Simulation and Optimization

A Touchstone file has been written for the basic circuit topology shown in Figure 6.13. This file is listed in Table 6.4.

The optimizer block is

```
Opt
    Range 4   8              ! The band of interest.
    OSC Re[Z₁]<−20           ! Net negative resistance
                               is to exceed 20 Ω in
                               magnitude.
```

Table 6.4 Touchstone—Ver(1.30–May-17-85)—Ser(15738-1598-1000)—Con(1CCC22124P) 383.CKT 07-27-89 20:22:33

```
! THE BASIC TOPOLOGY OF THE 4 TO 8 GHZ VCO
! MODEL USES BOTH DISTRIBUTED AND LUMPED ELEMENTS.

VAR

    RV=2.0
    CV=.25

CKT

    MSUB ER=9.9 H=10 T=.2 RHO=5  RGH=0

    ! OSCILLATOR CIRCUIT

    RES 1 2 R^RV
    CAP 2 3 C^CV
    IND 3 4 L=5
    CAP 4 0 C=50
    S2PA 5 6 3   NE64400
    IND 5 0 L=.70
    MLEF 6   W\2.70472 L\0.30549
    MLIN 6 7   W=2 L\92.31743
    CAP 7 0 C=50
    CAP 6 8 C=50
    MLIN 8 9 W=2.0       L\10.68604
    MLIN 9 10 W\52.94377 L\15.96629
    DEF2P 1 10 OSC

FREQ

    SWEEP 3 9  1

OUT

    OSC RE[Z1] GR1
    OSC IM[Z1] GR1
    OSC DB[S21] GR2
GRID

    RANGE 3 9  1
    GR1 -50 50 10
    GR2 -20 20  5

OPT

    RANGE 4 9
    OSC RE[Z1]<-20   50
    OSC DB[S21]>-5   20
```

OSC dB[S_{21}]>−5   !   The forward gain from the resonator to the load to be greater than −5 dB.

Once the optimizer has adjusted the circuit's elements for maximum negative resistance and forward gain over the band of interest, the actual values of the varactor's capacitance may be determined by the following process. The reactive condition for oscillation is

$$\text{Im}\,[Z_R] = -\text{Im}\,[Z_T]$$

Because our circuit model incorporates the varactor resonator into the transistor's circuit, this oscillation condition becomes, in Touchstone syntax,

OSC Im[Z_I] = 0

Now, using Touchstone's "tune mode," we simply vary the value of C_V until the condition $I_M[Z_I] = 0$ is reached at exactly 4.0 GHz. This is the maximum value of C_v. Next, vary C_v until $I_M[Z_I] = 0$ is reached at exactly 8.0 GHz. This is the minimum value of C_v.

The final optimized circuit element values are shown in Table 6.4. The simulated performance of the 4 to 8 GHz bipolar VCO is given in Figure 6.15.

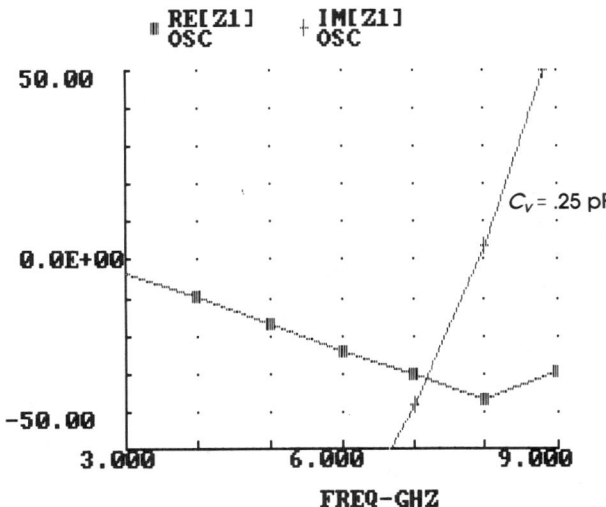

Fig. 6.15 Simulated performance of the 4 to 8 GHz bipolar varactor VCO.

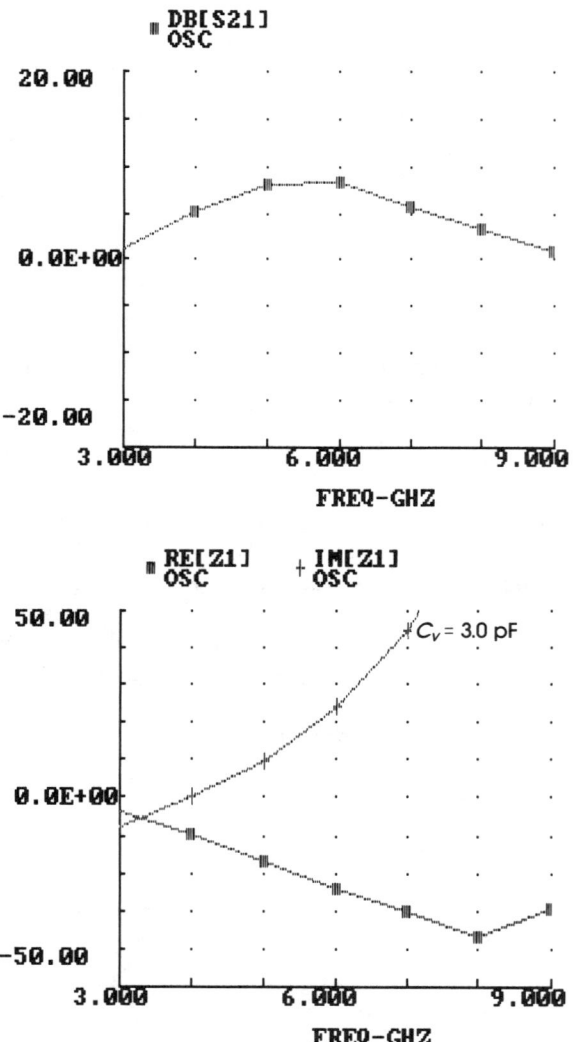

Fig. 6.15 cont'd.

6.10 EXAMPLE 3: A YIG-TUNED 2 TO 8 GHz BIPOLAR OSCILLATOR

The major difference between YIG-tuned oscillators and varactor-tuned oscillators is the YIG resonator itself. The circuit topology, including feedback and output matching elements, is essentially the same in both cases. Similarly, the analysis and optimization techniques for YIG-tuned oscillators is essentially the same as used for varactor-tuned VCOs.

The YIG resonator differs from the varactor resonator in three important ways:
1. A YIG resonator is tuned by a magnetic field at a *very* linear rate.
2. The YIG sphere is a complete resonator and, other than coupling, requires no external element to produce a resonance.
3. The YIG's Q factor is 10 to 100 times higher than that of a varactor diode, leading to very low oscillator noise and a very stable oscillation frequency.

Step 1: Specifications

Tuning Range = 2 to 8 GHz.
Power Output = +10 dBm.
Tuning Linearity = ± 1 percent.
FM Noise = Less than −90 dBC/1 kHz bandwidth at 100 kHz from the carrier.
dc Bias = +10V at 25 mA.

Step 2: Choose a Topology

This requirement is at a low enough frequency so that a bipolar transistor may be used. Because of the low-noise requirement, a bipolar transistor is essential. Therefore, the basic circuit topology will be a common-base bipolar transistor with inductive feedback. The basic topology shown in Figure 6.16 is very similar to that used in Example 2.

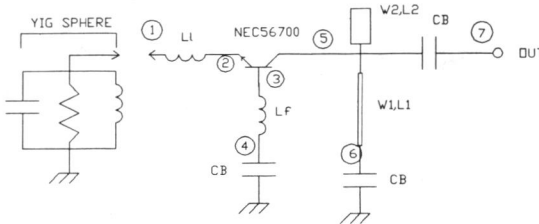

Fig. 6.16 The basic topology of the 2 to 8 GHz YIG-tuned bipolar oscillator.

Unlike a varactor resonator, a YIG resonator requires no external resonating element, although its loop inductance, L_1, must be included in the model. See Section 2.5 for more details.

The reactance slope of the YIG resonator is so high due to its high Q, that the transistor oscillator's circuit reactance has little or no effect in determining the frequency of oscillation. The conditions for oscillation are satisfied on, or very close to, the YIG sphere's resonant frequency. Circuit design is simply a matter of optimizing the circuit's negative resistance and forward gain over the band of interest. The YIG resonator determines the tuning rate, tuning linearity, FM noise (the transistor's noise data enters into this calculation), and temperature stability of the oscillator.

Step 3: Choose an Active Device

The active device must be a microwave bipolar transistor capable of delivering substantial gain up to 8 GHz and capable of generating +10 dBm power output with a comfortable margin. For this design we choose the NEC 56700, which is very similar to the NEC 64400 but with somewhat higher power output at high frequencies.

Step 4: Choose a Bias Circuit

The specification for this oscillator calls for a single-polarity bias supply. Therefore, we must use a self-bias circuit to supply the bipolar transistor's voltages and current. Figure 6.17 shows a bipolar transistor's single-polarity bias circuit.

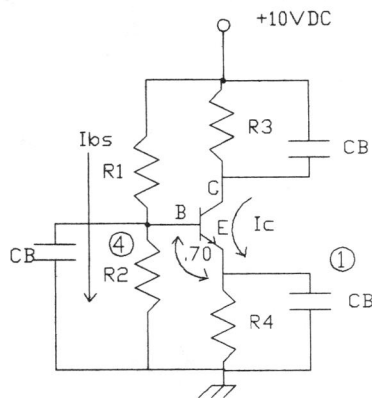

Fig. 6.17 Self-bias circuit for the YIG-tuned bipolar oscillator requires only a single-polarity power supply.

To analyze this bias circuit, make the following assumptions:

$I_{BS} = 5 \text{ mA}$

$I_C = 20 \text{ mA}$

$V_{BE} = .7 \text{ V}$

$V_{CE} = 8 \text{ V}$

The bias circuit is described by the following relationships:

$I_{BS} = 5 \text{ mA} = \dfrac{10.0 \text{ V}}{R_1 + R_2}$

$V_4 = I_{BS} R_2$

$V_1 = I_C R_4$

$10 \text{ V} = I_C R_3 + V + I_C R_4$

$V_4 - V_1 = .7 = I_{BS} R_2 - I_C R_4$

So,

$$R_1 + R_2 = \dfrac{10}{.0005} = 2000 \ \Omega$$

Let

$V_1 = 1.0 \text{ V} = I_C R_4$

$R_4 = \dfrac{1.0}{.020} = 50 \ \Omega$

$V_4 = 1.0 + .7 = I_{BS} R_4$

$R_2 = \dfrac{1.7}{.005} = 340 \ \Omega$

$R_1 = 2000 - R_2 = 1660 \ \Omega$

$10 \text{ V} = I_C R_3 + 8 + 1$

$I_C R_3 = 1 \text{ V}$

$$R_3 = \frac{1}{.020} = 50 \ \Omega$$

Node numbers refer to the points within the RF circuit topology where the bias circuit is connected.

Step 5: Power and Noise Conditions

Because we are able to calculate only the start oscillation condition with linear CAD tools, power must be regarded as determined by the bipolar transistor's intrinsic power generating ability. The NEC 56700 was selected for its ability to generate a power considerably higher than +10 dBm.

We may calculate the FM noise of this oscillator by using (6.7), (6.8), (6.12), and the noise parameters given in Section 6.7. Assume for the purposes of this calculation that the dominant FM noise mechanism is $(1/f)$ up-converted noise, and assume that the YIG resonator's Q is 1000. Then we use (6.8) and (6.12) directly for $f_M = 100$ kHz.

$$\overline{\Delta \omega^2} = \left[\frac{\omega_0^4 \left(\frac{\partial C_d}{\partial V_G} \right)^2}{4 Q_R \ G_R} \right] \left(\frac{N}{f^\alpha} \right)$$

$$= \left[\frac{(2\pi \cdot 8 \cdot 10^9)^4 (2 \cdot 10^{-3} \cdot 10^{-12})^2}{4(1000)(.007)} \right] \frac{10^{-8}}{(100 \cdot 10^3)} = .09 \ \text{Hz}^2/\text{Hz}$$

or

$$\overline{\Delta \omega_{rms}} = .30 \ \text{Hz}_{rms}/\text{Hz}$$

From (6.11),

$$\text{N/C} = \frac{\overline{\Delta \omega^2}}{\omega_M^2}$$

$$= \frac{.09}{(2\pi \cdot 100 \cdot 10^3)^2} = 2.27 \cdot 10^{-13}$$

or

$$10 \ \log[2.27 \cdot 10^{-13}] = -126.4 \ \text{dB}/1 \ \text{Hz BW}$$

In a 1 kHz bandwidth (BW), the noise increases by 30 dB $\left(\text{that is, } 10\log\left(\frac{1000}{1}\right)\right)$:

$$\text{N/C (BW = 1 kHz)} = -96.4 \text{ dB at } f_M = 100 \text{ kHz}$$

which is well within the specification.

Step 6: Basic Circuit Simulation and Optimization

A Touchstone file for the basic circuit topology given in Figure 6.16 is listed in Table 6.5. The optimizer block created for the YIG tuned oscillator requires a

Table 6.5 Touchstone—Ver(1.30–May-17-85)—Ser(15738-1598-1000)—Con(1CCC22124P) 384.CKT 07-27-89 21:06:27

```
! THE BASIC TOPOLOGY OF THE 2 TO 8 GHZ YIG TUNED BIPOLAR OSCILLATOR
! COMMON BASE\INDUCTIVE FEEDBACK.
! MODEL USES A COMBINATION OF DISTRIBUTED AND LUMPED ELEMENTS.
CKT
     MSUB ER=9.9 H=10 T=.2 RHO=1 RGH=0
     IND 1 2 L=1.5
     S2PA 3 5 2 NEC56710
     IND 3 4 L=1.0
     CAP 4 0 C=200
     MLEF 5 W\29.04825 L\102.03430
     MLIN 5 6 W\2.71480 L\334.79480
     CAP 6 0 C=50
     CAP 5 7 C=50
     DEF2P 1 7   OSC
FREQ
     SWEEP 1 10 1
OUT
     OSC RE[Z1]    GR1
     OSC IM[Z1]    GR1
     OSC DB[S21]   GR2
GRID
     RANGE 1 10 1
     GR1 -100 100 10
     GR2 -10 10 5
OPT
     RANGE 2 8
     OSC RE[Z1] <-10    20
     OSC IM[Z1] > 20    10
     OSC DB[S21] >-5     5
```

minimum negative resistance of $-10\ \Omega$ consistent with the low loss associated with the YIG resonator. Everything else in the optimizer block is very similar to the VCO examples.

```
OPT
     RANGE  2     8              ! The band of interest.
     OSC    Re[Z₁]<−10            ! Net negative resistance
                                    is to exceed 10 Ω in
                                    magnitude.
     OSC    dB[S₂₁]>−5            ! The forward gain from
                                    the resonator to the
                                    load is to be greater
                                    than −5 dB.
```

Because the shorted stub in the output matching network acts as an RF choke, it is set initially to the minimum width for thin film processing (1–2 mils). The open stub width is initially set to 15 mils. The final optimized element values are given in Table 6.5. The simulated performance of the optimized 2 to 8 GHz YIG-tuned oscillator is shown in Figure 6.18.

6.11 EXAMPLE 4: A YIG-TUNED 6 TO 18 GHz FET OSCILLATOR

At present, commercially available bipolar transistors are not a practical option for oscillators operating above 10 GHz. This means that for frequencies above

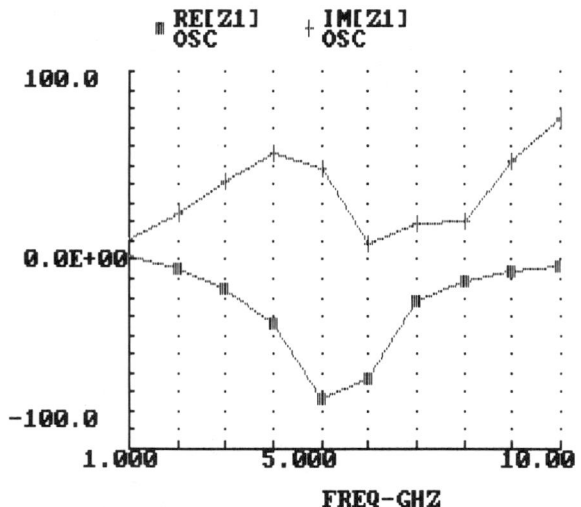

Fig. 6.18 Simulated performance of the 2 to 8 GHz bipolar YIG YTO.

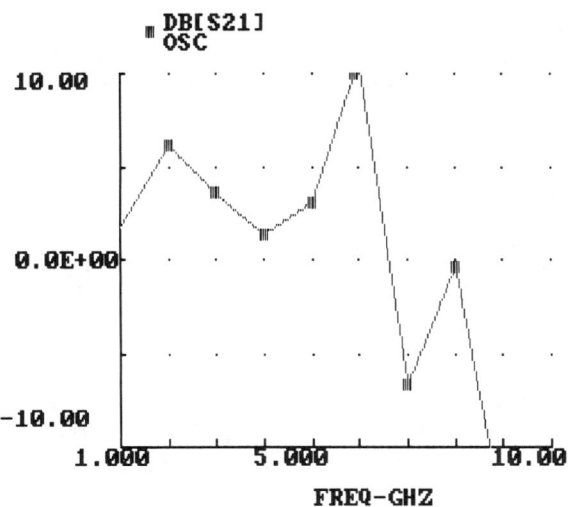

Fig. 6.18 cont'd.

10 GHz, the active device of an oscillator must be a GaAs FET. However, two problems are associated with using GaAs FET devices in YIG oscillators: (1) higher oscillator noise close to the carrier, and (2) the potential for an unwanted fixed frequency parasitic oscillation associated with resonance between the YIG loop inductance and the FET feedback capacitor's capacitive reactance.

Of the two problems, the parasitic oscillation problem is more serious because it can cause the oscillators to "stick" at an uncontrolled fixed frequency. Oscillator noise, on the other hand, may be suppressed sufficiently by the YIG resonator's Q to be acceptable in most applications.

The parasitic oscillation problem is best resolved by moving this spurious resonance to a frequency just below the band of negative resistance, so that the start oscillation conditions are not satisfied at the spurious resonance frequency. This problem never arises with bipolar transistors because their common-base inductive feedback circuit makes them always present an inductive impedance to the resonator, which *cannot* resonate with another inductance, such as the loop's inductance.

Step 1: Specifications

> Tuning Range = 6 to 18 GHz;
> Power Output = +10 dBm;
> Tuning Linearity = ± 1%;
> dc Bias = +10 V at 35 mA.

Step 2: Choose a Topology

The basic topology for a GaAs FET oscillator is shown in Figure 6.6(b). We choose a circuit for this application that is built around this basic topology. The following important modifications have been made to the basic topology.
1. A choke is connected to the source of the FET to allow for self-biasing.
2. A shunt capacitor is added to the FET's gate to lower the YIG-loop parasitic fixed resonant frequency.
3. A drain matching network is added to the circuit to broaden the bandwidth of the negative resistance. One of the best drain networks for an FET oscillator is the gate of a second FET, because an FET's gate is low impedance and highly reflective, and these reflections act as a feedback path that reinforces the negative resistance. Also, a second FET stage acts as a buffer amplifier, thus raising the power output and improving load isolation.

The basic 6 to 18 GHz FET oscillator circuit is shown in Figure 6.19.

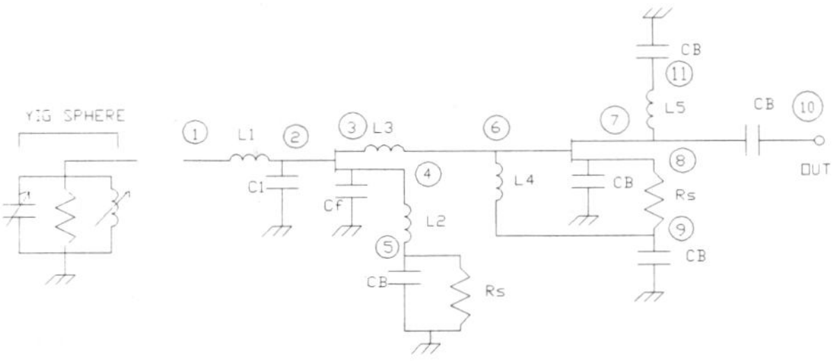

Fig. 6.19 The basic topology of a 6 to 18 GHz YIG-tuned GaAs FET oscillator.

Inductors L_2, L_4, and L_5 are bias chokes and initially will be set to relatively high values (10 nH) in the optimization process. L_1 is the YIG loop inductance and initially set to .5 nH. L_3 is an interstage matching inductance and guessed to be about 1.0 nH. The large capacitors, which all are called C_B, are either bypass or blocking capacitors and fixed at 50 pF. The capacitor C_1 is a shunt capacitance used to lower the YIG loop parasitic resonance frequency to below the band of the negative resistance. Initially, we guess C_1 to be .3 pF. C_f is the feedback capacitance, and we initially guess it to be .5 pF. R_1 and R_2 are biasing resistors that will be discussed in step 4.

Step 3: Choose an Active Device

The GaAs FET for the 6 to 18 GHz YIG oscillator must have strong gain up to 18 GHz and must be capable of generating in excess of +10 dBm output power. A device that fulfills both of these conditions easily is the NEC 71000. Therefore, this device is our choice for both Q_1 and Q_2.

Step 4: Choose a Bias Circuit

If we examine Figure 6.19, we see that the bias circuit already is in place, within the RF circuit. The bias circuit alone is shown in Figure 6.20.

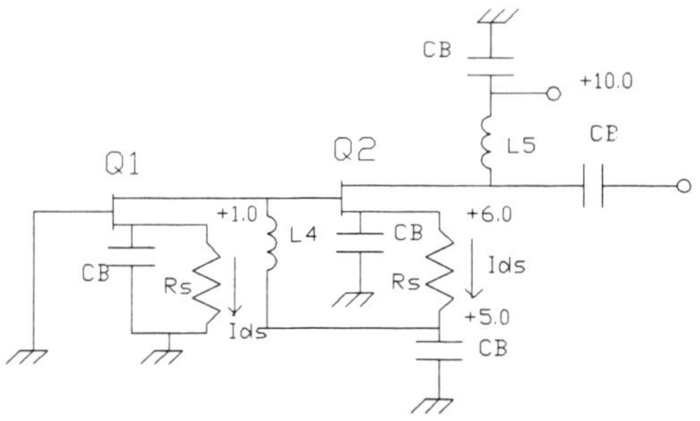

Fig. 6.20 The bias circuit for the 6 to 18 GHz YIG-tuned oscillator.

Bias is applied to the drain of Q_2 through the choke L_5. Drain current, I_{DS}, passes out the source of Q_2, and through resistor, R_S, developing about 1 V across it to provide gate to source bias voltage for Q_2. A second choke L_4 passes I_{DS} to the drain of Q_1. A second source resistor, R_S, provides gate to source bias for Q_1.

Step 5: Power and Noise Conditions

The NEC 71000 GaAs FET is capable of well in excess of +10 dBm power. There are no noise specifications for this oscillator.

Step 6: Basic Circuit Simulation and Optimization

The basic circuit shown in Figure 6.19 is the basis of a Touchstone file given in Table 6.6. A very important aspect of this analysis is to make sure that the parasitic resonance between the loop inductance, L_1, and the rest of the circuit is below the band of negative resistance. Because the tuning bandwidth is 6 to 18 GHz, we will set an optimizer goal to at least $-10\ \Omega$ negative resistance from 5.5 GHz to 18 GHz and let the Touchstone optimizer adjust the circuit elements to place the parasitic resonance below 5.5 GHz. The optimizer block of the Touchstone file for this circuit will become

Table 6.6 Touchstone—Ver(1.30–May-17-85)—Ser(15738-1598-1000)—Con(1CCC22124P) 385.CKT 07-28-89 12:33:10

```
! THE BASIC TOPOLOGY OF THE 6 TO 18 GHZ YIG TUNED FET OSCILLATOR.
! THE TOPOLOGY USES A TWO STAGE FET LINEUP, WITH THE FIRST FET OPERATING
! AS A COMMON SOURCE / CAPACITIVE FEEDBACK OSCILLATOR, AND THE SECOND FET
! OPERATING AS A BUFFER AMPLIFIER WHICH PROVIDES A FAVORABLE LOW LOAD
! IMPEDANCE TO THE FIRST FET.
CKT
    IND  1  2  L=7.0
    CAP  2  0  C=.2
    S2PA 2  3  4   NEC71000
    IND  4  5  L\5.80076
    CAP  4  0  C\0.37086
    CAP  5  0  C=50
    RES  5  0  R=20
    IND  3  6  L\0.65297
    IND  6  9  L\6.22719
    S2PA 6  7  8   NEC71000
    CAP  8  0  C=50
    RES  8  9  R=20
    CAP  9  0  C=50
    IND  7  11 L\0.33256
    CAP  8  10 C=50
    CAP  11 0  C=50
    DEF2P 1 10   OSC

FREQ
    SWEEP 2 25 1

OUT
    OSC RE[Z1]     GR1
    OSC IM[Z1]     GR1
    OSC DB[S21]    GR2
    OSC S11
    OSC S22

GRID
    RANGE 2 25 1
    GR1 -50 50 10
    GR2 -20 20 5

OPT
    RANGE 5.5 18
    OSC RE[Z1] <-10    20
    OSC DB[S21] >-5     5
    RANGE 4.0
    OSC IM[Z1]=0       10
```

Opt
	Range	5.5 18	!	Frequency range of the negative resistance.
	OSC	Re $[Z_1] < -10$	!	Net negative resistance is to exceed 10 Ω in magnitude.
	OSC	dB $[S_{21}] > -5$	!	The forward gain from the resonator to the load is to be greater than -5 dB to ensure that power is transferred to the load.
	Range	4.0	!	Frequency of parasitic resonance.
	OSC	Im $[Z_1] = 0$	!	Resonance conditions.

The optimized simulated performance is shown in Figure 6.21. The optimized circuit parameters are given in Table 6.6.

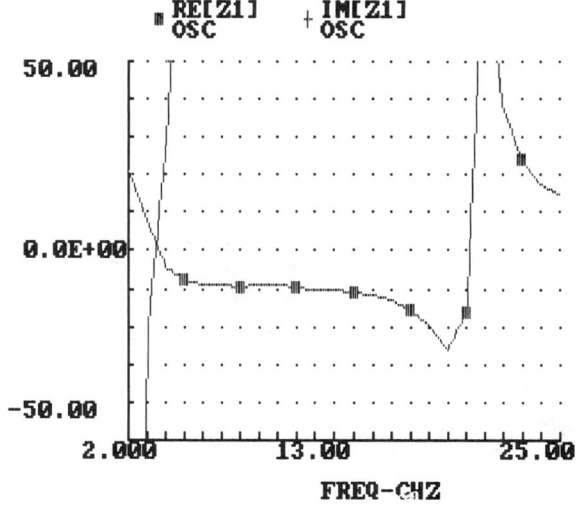

Fig. 6.21 Simulated performance of the GaAs FET-YIG 6 to 18 GHz YTO.

6.12 EXAMPLE 5: AN 8.4 GHz DRO FET OSCILLATOR

Traditionally, highly stable fixed-frequency oscillators have been constructed by one of two techniques. The first technique involves locking the oscillator to a highly stable signal generated by a quartz crystal controlled oscillator operating at a low frequency. The second technique uses a mechanically tuned cavity resonator constructed with low thermal expansion material (such as Kovar) to control the frequency of a free-running microwave oscillator. Both techniques are complicated and bulky and do not lend themselves very well to microelectronic circuit techniques.

A third technique for constructing a highly stable fixed-frequency oscillator involves using a dielectric resonator (see Section 2.6) to control the oscillator's frequency. Practical dielectric resonators can be fabricated to operate from 2 GHz to over 40 GHz with quality factors of several thousand and extremely tight temperature-frequency stability. Oscillators controlled by such resonators will have very low FM noise and be very stable with temperature.

Let us now investigate the design of a typical dielectric resonator oscillator. The DRO will be designed for operation at 8.4 GHz and use a GaAs FET active device.

Step 1: Specifications

Operating Frequency = 8.4 GHz;
Power Output = +10 dBm;
FM Noise = −110 dBC max at 1 MHz from the carrier in a 1 kHz bandwidth;
Temperature Stability = ±5 MHz over the temperature range 0 to +50°C.

Step 2: Choose a Topology

Because we have decided to use a GaAs FET active device, the important question is what circuit techniques to use to apply the feedback that produces negative resistance. Although we have already stated that broadband FET oscillators normally are designed to use common-source capacitance feedback, we violate this rule by operating the oscillator with common-gate inductive feedback.

Common-gate inductive feedback is a very effective circuit technique in fixed-frequency oscillators because the band of negative resistance is relatively narrow and centered about the resonant frequency of the FET's gate capacitance, C_{gs}, and the feedback inductance. For broadband devices, such as VCOs and YTOs, this type of feedback simply is too narrowband; but, for fixed-frequency oscillators, this technique is a simple and effective way of generating strong negative resistance at a desired fixed frequency.

The basic circuit topology of the DRO is shown in Figure 6.22. The dielectric resonator's coupling is modeled as an ideal transformer with an adjustable turns ratio. This resonator is connected to the FET's source. An inductor between the FET's gate and ground provides the feedback, and a parallel inductance and capacitance in the drain circuit provides output matching. The dielectric resonator

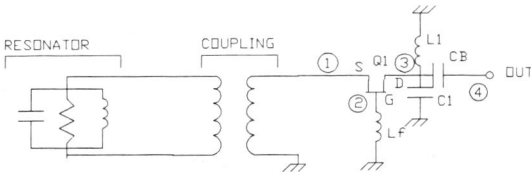

Fig. 6.22 The basic topology of an 8.5 GHz dielectric resonator FET oscillator.

puck must be specially selected for the 8.4 GHz operating frequency. Its approximate dimensions and dielectric constant are $d_r = .303$ inches, $t_r = .134$ inches, and $\varepsilon = 28.6$ (Murata Erie P/N DRD077SC034). The resonator puck's equivalent circuit is calculated from its Q and resonant frequency, and an assumed value for either L or C.

Step 3: Choose an Active Device

The active device is an GaAs FET with substantial gain at 8.4 GHz and a power generating capability well in excess of +10 dBm. We choose an NEC 71000 device because it fulfills all of the oscillator's technical requirements; and it is representative of a whole class of low-power GaAs FETs.

Step 4: Oscillator Noise and Temperature Stability

The FM noise of a dielectric resonator oscillator may be calculated from the equation for FM thermal noise, which is (6.7). We assume that at 1 MHz from the carrier, the up-converted noise has been reduced to a level below that of the thermal noise. Therefore, under these conditions, the oscillator's FM noise is purely thermal.

Using (6.7) and substituting the appropriate power, quality factor, and noise temperature for this oscillator, we calculate that

$$\overline{\Delta\omega^2} = \frac{[(2\pi)8.4 \cdot 10^9]^2(1.38 \cdot 10^{-23})(500)(10^3)}{4(5000)^2(.01)}$$

$$= .019 \text{ Hz}^2/1\text{KHz}$$

The noise-to-carrier ratio is simply the mean square frequency derivation divided by the square of the modulation frequency (6.11):

$$\text{N/C} = 10 \log \left[\frac{\overline{\Delta\omega^2}}{\omega_m^2}\right] = 10 \log \left\{\frac{.019}{[2\pi (100 \cdot 10^3)]^2}\right\}$$

$$N/C = 10 \log [4.81 \cdot 10^{-14}]$$
$$= -133 \text{ dBC/1KHz}$$

The calculated noise-to-carrier ratio is lower than the -100 dBC specification by a considerable margin.

The resonator's temperature-frequency drift may be calculated from the relationships presented in Section 2.6. Normally, the designer "works" with the dielectric resonator's manufacturer to select the proper puck for a specific application in terms of its resonant frequency and temperature stability.

Step 5: Choose a Bias Circuit

The FET dielectric resonator oscillator will have an extremely simple self-bias circuit, using the resonator's termination resistance as a source resistor. Because the dielectric resonator is coupled to the FET's source via a 50 Ω microstrip line, this microstrip line may be terminated in a resistor of approximately 50 Ω, which serves as both a termination and a source-bias resistor. The gate is appropriately grounded by the feedback inductor, and the drain contact is biased to approximately $+5.0$ V through a choke inductor. Such a bias circuit functions exactly like the self-bias networks already discussed in Chapter 5.

Step 6: Basic Circuit Simulation and Optimization

A Touchstone file for the basic circuit topology shown in Figure 6.22 is given in Table 6.7. The Touchstone file is used to calculate reflection gain and impedance at

Table 6.7 Touchstone—Ver(1.30–May-17-85)—Ser(15738-1598-1000)—Con(1CCC22124P) 374.CKT 07-28-89 12:44:53

```
! THE BASIC TOPOLOGY OF AN 8.4 GHZ DIELECTRICLY TUNED FET OSCILLATOR USING
! COMMON GATE / INDUCTIVE FEEDBACK.

CKT
    PRLC 1 0 R=500 L=8.8E-4 C=412
    S2PA 2 3 1   NEC71000
    IND  2 0 L=1.20
    CAP  3 0 C=1.0
    IND  3 0 L=2.0
    CAP  3 4 C=50
    DEF1P 4 DRO

FREQ
    SWEEP 8.300 8.500 .01
```

Table 6.7 cont'd.

```
OUT
    DRO DB[S11] GR1
    DRO RE[Z1] GR2
    DRO IM[Z1] GR2
GRID
    RANGE 8.300 8.500 .01
    GR1 -10 10 1
    GR2 -50 50 5
```

the oscillator's output terminal. The oscillation condition will be fulfilled if the negative real part of this impedance exceeds $-50\ \Omega$ and the imaginary part of the impedance is zero at this frequency. These conditions easily can be met by "hand selecting" the component values using Touchstone's "tune" mode. Although the Touchstone optimizer could do the same job, in such a simple circuit, to select the right component values by a short process of trial and error is often easier. The Touchstone simulation of the optimized dielectric resonator oscillator's performance is shown in Figure 6.23. Final component values are given in the Touchstone file in Table 6.7.

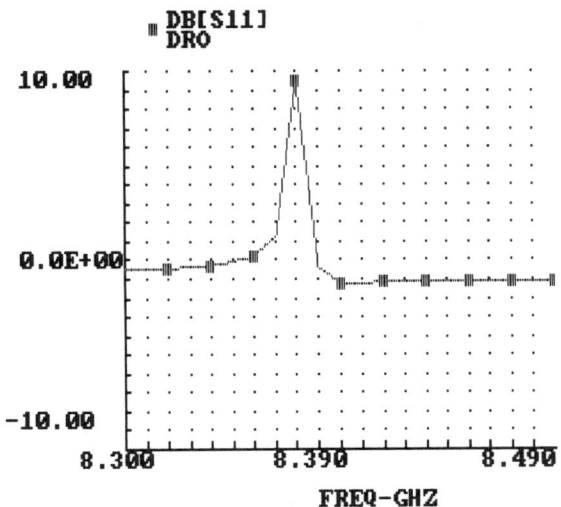

Fig. 6.23 Simulated performance of the FET-dielectric resonator 8.4 GHz Fixed-frequency oscillator.

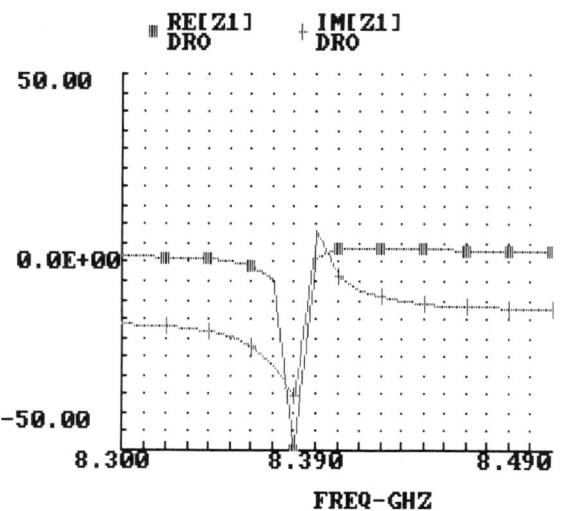

Fig. 6.23 cont'd.

REFERENCES

1. K. Kurokawa, "Noise in Synchronized Oscillators," *IEEE Trans. Microwave Theory Tech.*, Vol. MTT-16, April 1968, pp. 234–240.
2. A. A. Sweet, "A General Analysis of Noise in Gunn Oscillators," *Proc. IEEE,* Vol. 60, August 1972.

Chapter 7
MIC Layout and Fabrication

7.1 LAYOUT STRATEGY

7.1.1 Process Flowchart

Both MIC layout and MIC fabrication are multistep processes. Because layout and fabrication are so interdependent, here we will consider them as part of an overall process. A block diagram of the overall MIC layout and fabrication process is shown in Figure 7.1. The present chapter will explain in detail the purpose and practice of each step. An example of a typical MIC circuit is shown in Figure 7.2.

7.1.2 Design Fabrication Rules

All metalized MIC circuit boards, whether hardboard (alumina, quartz, BeO) or soft board (Teflon-Fiberglas material such as Duroid™) are patterned by photolithographic techniques. The first step in fabrication is to produce high-quality scaled artwork of the circuit to be fabricated. When making this artwork, certain rules must be observed that relate to the particular process used by the photolithographic fabrication laboratory. All fabrication laboratories publish a set of standard design rules to ensure that designers produce artwork compatible with their process. As an example, some typical generic values and specifications often found in a set of MIC design rules are given in Table 7.1.2.

7.1.3 Low Parasitic Design

A design begins with a schematic diagram whose element values came from a Touchstone optimization (as discussed in Chapters 5 and 6). To translate the design into a realistic circuit, the designer must transform the diagram into a physical

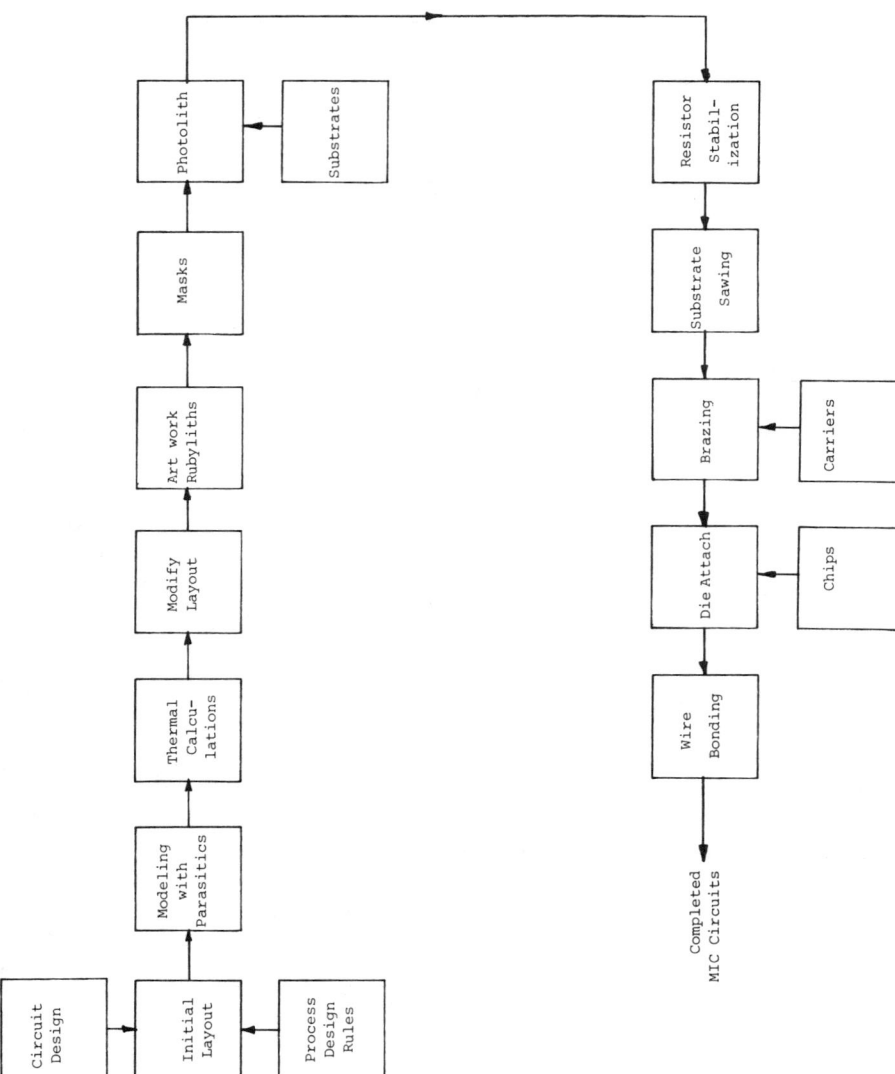

Fig. 7.1 Process flowchart for MIC circuit development.

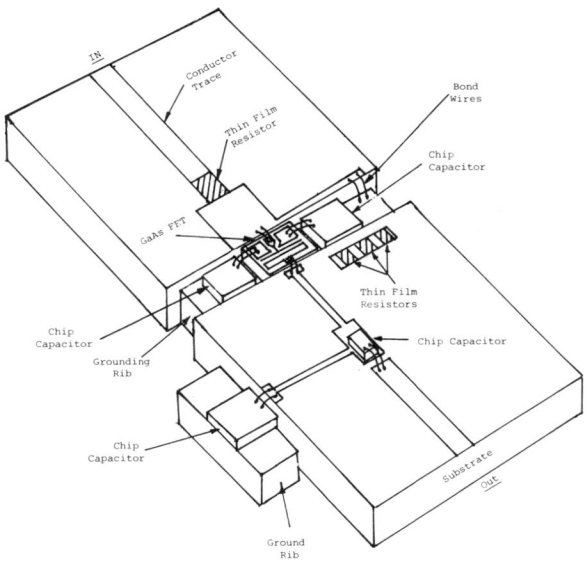

Fig. 7.2 An example of an MIC circuit.

Table 7.1.2 MIC Design Rules

Limit	Typical Value
Minimum Line Width	1.0 mil.
Minimum Line Spacing	.7 mil.
Resistivity of Resistive Layer	50 Ω per square.
Test Resistor	At least one test resistor must be provided that is located in an unused area of the circuit and mid-range in value relative to other resistors on the circuit.
Outer Border	No circuit feature will be located within 2 mils of the outer border.
Design Grid	All design features (with the exception of coupler gaps, which are otherwise noted), will lie upon a 1.0 mil grid.
Etch Factor	Plate up process: none. Etchback process: 0.1 mil per edge of any line. This means artwork must be oversize so that each line is enlarged by .1 mil per edge. During etching this .1 mil will be undercut and removed.
Via Hole Diameter and Spacing	10 mils diameter on 10 mil centers.
Via Hole Inductance	.2 nH.

layout of interconnected components and microstrip line traces. A major challenge of any design is to maintain a circuit's desired electrical performance through this transformation.

Because of discontinuity effects where lines and components connect and strong coupling effects between lines running closely parallel to each other, various nonideal parasitic elements "creep" into designs as an inadvertent result of creating a layout. Such parasitic elements can detrimentally affect the performance of any circuit, especially at high frequencies. It is the job (and the art) of the designer to avoid these parasitics or minimize their effects. Once a layout has been completed, the designer must go back to the Touchstone or SuperCompact file and insert all parasitic elements, for realistically gauging the performance of a given layout. Wherever possible, the following parasitic elements should be avoided. If this is impossible, the number of such elements should be minimized by strategic layout planning:

- Microstrip bends;
- Microstrip steps;
- Microstrip tees;
- Microstrip crosses;
- Close proximity between microstrip lines (resulting in coupling crosstalk);
- Long bond wires.

Figure 7.3 shows the physical appearance of these parasitic elements.

7.1.4 Grounding

Good, low-loss, low-inductance grounds are very important in any realistic circuit layout. Because the basic ground reference in an MIC circuit is the backside

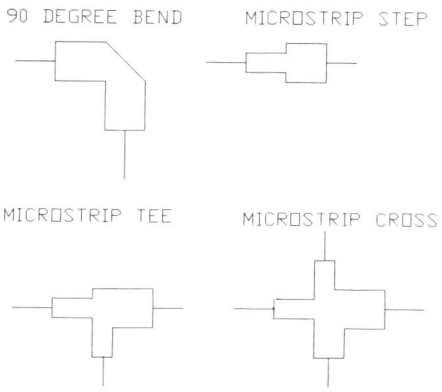

Fig. 7.3 Parasitic elements to be avoided in MIC layouts.

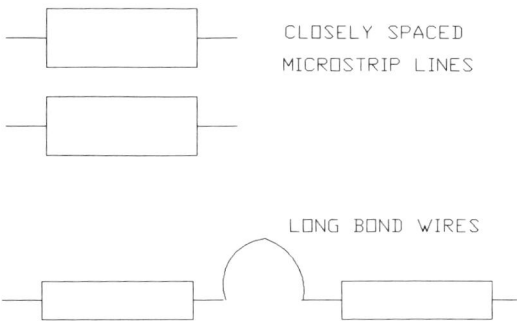

Fig. 7.3 cont'd.

metallization of the circuit's substrate, it becomes necessary to make direct connections between the substrate's bottomside metallization and its topside metallization, to create high-quality grounds.

The following types of grounds are most often used in MIC circuits:

- Sputtered through via holes.
- Ribbon bonds to carrier grounds.
- Bond wires to carrier grounds.
- Substrate "wrap-around" grounds.
- Metal rib-ribbon (or bond wire) ground.

Some of the pros and cons of each type of ground follow.

Type of Ground	Pros	Cons
Via holes.	May be located at any point in a circuit. Small, low inductance.	The designer must observe certain minimum diameter and spacing rules between vias. Expensive to fabricate.
Ribbon bond to carrier.	Low inductance ground. Easy to fabricate. Inexpensive.	Can be located only at the circuit's periphery.
Bond wire to carrier ground.	Easy to fabricate.	Higher inductance than ribbon (use multiwire bonds).
Wrap-around ground.	Convenient.	Higher inductance. Sometimes lossy at microwave frequencies. Precludes the use of step-and-repeat techniques for substrate manufacturing.

Type of Ground	Pros	Cons
Rib-ribbon (or bond wire)	Very low inductance.	Each circuit substrate must be fabricated one at a time, significantly increasing manufacturing cost. Must braze ribs in place while brazing the circuit substrates to the carrier.

Figure 7.4 diagrams the various kinds of MIC Ground Systems.

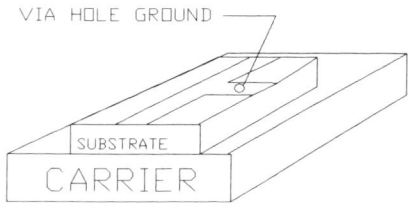

Fig. 7.4 Via hole ground.

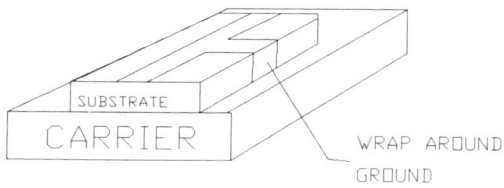

Fig. 7.4 (continued) Wrap around ground.

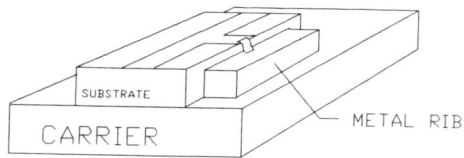

Fig. 7.4 (continued) Metal rib ground.

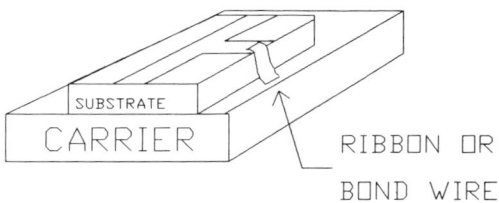

Fig. 7.4 (continued) Ribbon bond to carrier ground.

7.2 PARASITIC MODELS

In a realistic layout, some parasitic elements are unavoidable. Here we explore the models for the various types of parasitic elements, in terms of

- Its appearance in a layout.
- Its equivalent circuit model.
- Mathematical expressions for the elements in the model.
- The Touchstone syntax for each model

All parasitics, plus the equivalent circuits of their models are shown in Figure 7.5. Most parasitic elements do not normally occupy physical area within a circuit, but are best thought of as lumped elements located at a point within a larger distributed network. The major exception is close proximity microstrip lines.

7.2.1 Microstrip End Effects

A fringing capacitance, C_f, at the end of an open circuited microstrip line adds some equivalent extra microstrip line length l_l, which Edwards [1] has shown is equal to

$$l_l = \frac{cZ_0 C_f}{\sqrt{\epsilon_e}} \qquad (7.1)$$

where
 c is the speed of light in free space,
 Z_0 is the line's characteristic impedance
 ϵ_e is the effective dielectric constant

$$C_f/w = \exp\left[2.2 \sum_{\tau=1}^{5} K_\epsilon \left(\log \frac{w}{h}\right)^{i-1}\right] \text{ pF/m} \qquad (7.2)$$

Element	Geometry	Equivalent Circuit	TouchStone Syntax
Bond Wire	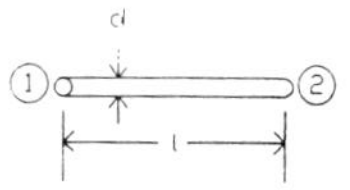		WIRE 1 2 D= L= RHO=
Ribbon	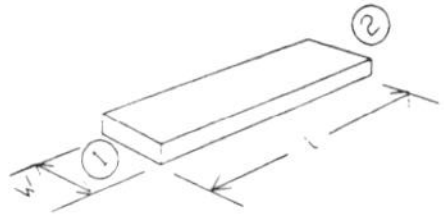		RIBBON 1 2 W= L= RHO=
Thin Film Resistor	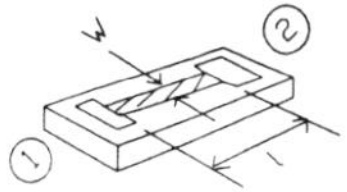	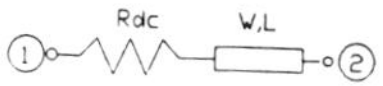	TFR 1 2 W= L= RS= F=
Via Hole Ground	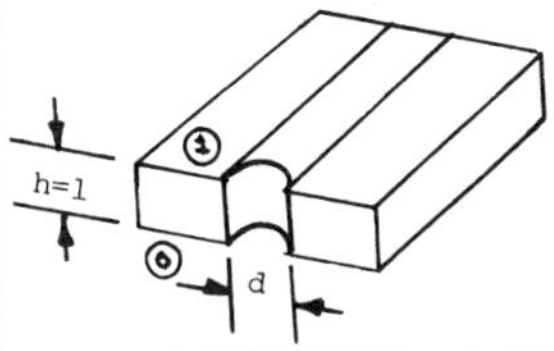		Model as an inductor, using equation (7.14) IND 1 0 L=
Coupling Between Two Parallel Microstrip Lines	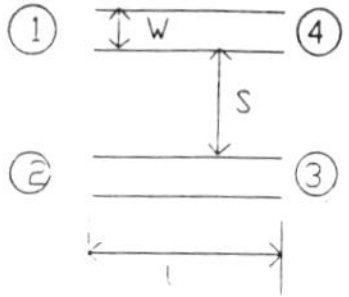	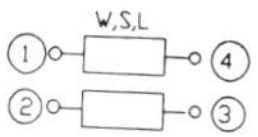	MCLIN 1 2 3 4 W= S= L=

Fig. 7.5 Microstrip parasitic elements.

Element	Geometry	Equivalent Circuit	TouchStone Syntax
Microstrip End Effects	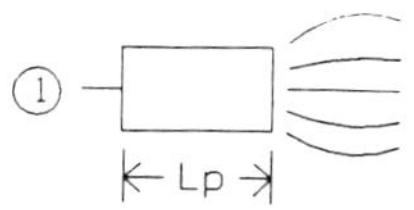	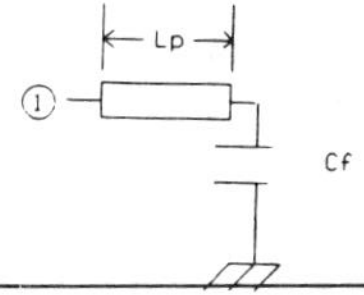	MLEF 1 W= L=
Microstrip Bend 90° Mitered	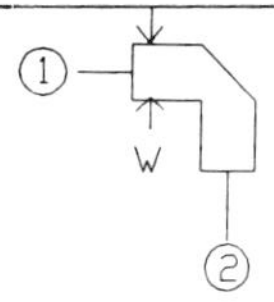	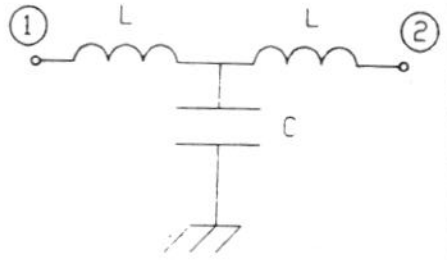	MBEND 1 2 W= ANG= M=
Symmetric Step in Microstrip Width	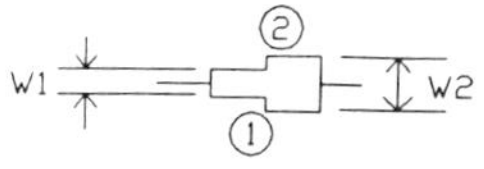	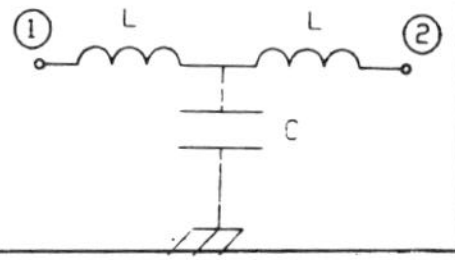	MSTEP 1 2 W1= W2=
Microstrip Tee Junction	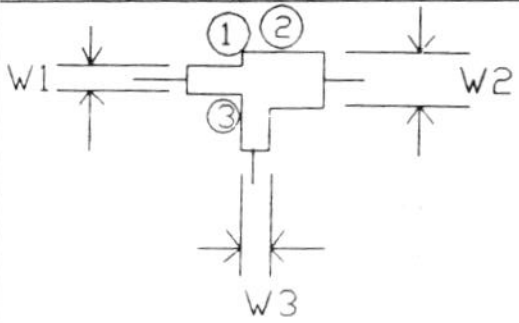	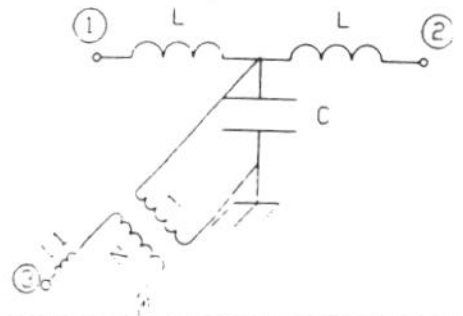	MTEE 1 2 3 W1= W2= W3=
Microstrip Cross Junction	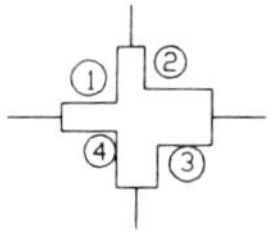	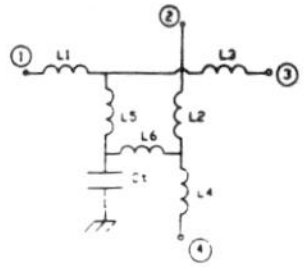	MCROS 1 2 3 4 W1= W2= W3= W4=

Fig. 7.5 (continued).

is the fringing capacitance derived by Silvester and Benedek [2], where
 K_ϵ = a tabulate set of coefficients, see Silvester and Benedek,
 h = the substrate thickness,
 w = the width of the line.

The Touchstone syntax for an open circuited stub that takes into account fringing capacitance at its open end is

 MLEF   1   W=   L=

7.2.2 Microstrip Bends

Microstrip bends occur whenever a microstrip line changes direction. They can account for a wide variety of angles, from just a few degrees to 90°. At angles greater than 90° the line is bending "back on itself," a situation to be avoided. Bends may be radiused from a point or have full or mitered corners. Bends with mitered corners are called *compensated bends* because mitering reduces the bend's electrical parasitic effects. In general, bends appear inductive at microwave frequencies because of the magnetic field "bending" associated with forcing the currents to change direction. A small amount of shunt capacitance is also associated with a bend.

For an uncompensated 90° bend, Gupta, Garg, and Bahl [3] have calculated a model in terms of a series inductance and a shunt capacitance. Expressions for these elements are

$$L/h = 100 \left[4\sqrt{\left(\frac{w}{h}\right)} - 4.2 \right] \text{ nH/m} \tag{7.3}$$

The shunt capacitance in the equivalent circuit model is given as for $w/h < 1.0$:

$$C/w = \frac{(14\epsilon_R + 12.5)(w/h) - (1.83\epsilon_R - 2.25)}{\sqrt{w/h}} \text{ pF/m} \tag{7.4}$$

and for $w/h > 1.0$:

$$C/w = \left[(9.5\epsilon_R + 1.25)\left(\frac{w}{h}\right) + 5.2\epsilon_R + 7.0 \right] \text{ pF/m} \tag{7.5}$$

The Touchstone syntax for an uncompensated 90° bend is

 MBEND   1   2   W=   ANG=90°   M=0

7.2.3 Symmetric Step Changes in Microstrip Width

The equivalent circuit for an abrupt symmetric change in microstrip width, like the bend, is a series inductance and a shunt capacitance.

Following Garg, Ramesh, and Bahl [4] the expression for the series inductance of a symmetric microstrip step is

$$L/h = 40.5 \left(\frac{w_1}{w_2} - 1.0\right) - 75 \left(\frac{w_1}{w_2}\right) + .2 \left(\frac{w_1}{w_2} - 1.0\right)^2, \quad \text{nH/m} \tag{7.6}$$

and the shunt capacitance for $\epsilon_R \leq 10.0$; $1.5 < \frac{w_2}{w_1} < 3.5$ is

$$C/\sqrt{w_1 w_2} = (10.1 \log \epsilon_R + 2.33) \frac{w_2}{w_1} - 12.6 \log \epsilon_R - 3.17, \quad \text{pF/m} \tag{7.7}$$

The Touchstone syntax for a symmetric microstep is

    MSTEP 1 2 W1= W2=

7.2.4 Microstrip Tee Junction

The microstrip tee junction is much like the combination of a microstrip line and a 90° bend. It can be perceived as a three-port device with one input and two outputs, where one output is a straight through (line) and the second output is a 90° bend. The straight through port is coupled to the input by a series inductance and a shunt capacitance, similar to a bend or a step transition, but the 90° output is coupled to the input by a transformer of ratio N. Following Hammerstad and Bekkadal [5]:

$$N^2 = \left[\frac{\sin[\pi (w_{\text{eff1}}/\lambda_{g1})(Z_{01}/Z_{02})]}{\pi(w_{\text{eff1}}/\lambda_{g1})(Z_{01}/Z_{02})}\right]^2$$

$$\times [1 - [\pi (w_{\text{eff1}}/\lambda_{g1})(d2/W_{\text{eff1}})]^2] \tag{7.8}$$

where

$$w_{\text{eff1},2} = \frac{hN}{Z_{0(1,2)}\sqrt{\epsilon_e}}$$

Z_{01} and Z_{02} are the characteristic impedances of the main output, and the 90° output respectively.

The reference plane displacement, d_1, of the main output port is

$$\frac{d_1}{w_{\text{eff}_2}} = .05 \frac{Z_{01}}{Z_{02}} N^2 \tag{7.9}$$

The reference plane displacement, d_2 for the 90° output port is

$$\frac{d_2}{w_{\text{eff1}}} = .5 - \left\{ .076 + .2 \left(\frac{2w_{\text{eff1}}}{\lambda_{g1}} \right)^2 + .663 \exp\left(-1.71 \frac{Z_{01}}{Z_{02}} \right) \right.$$
$$\left. - .172 \ln \left(\frac{Z_{01}}{Z_{02}} \right) \right\} \frac{Z_{01}}{Z_{02}} \tag{7.10}$$

For $Z_{01}/Z_{02} \le .5$, the model's shunt capacitance, C, is determined by the following expression:

$$\frac{\omega C \lambda_{g1}}{Y_{01} w_{\text{eff1}}} = \left(\frac{2w_{\text{eff1}}}{\lambda_{g1}} - 1 \right) \frac{Z_{01}}{Z_{02}} \tag{7.11}$$

where

$$Y_{01} = 1/Z_{01}$$

For $Z_{01}/Z_{02} \ge .5$,

$$\frac{\omega C \lambda_{g1}}{Y_{01} w_{\text{eff1}}} = \left(\frac{2w_{\text{eff1}}}{\lambda_{g1}} - 1 \right) \left(2 - 3 \frac{Z_{01}}{Z_{02}} \right) \tag{7.12}$$

The Touchstone syntax for a microstrip tee junction is

MTEE 1 2 3 W1= W2= W3=

7.2.5 Microstrip Cross Junction

Unfortunately, good closed form expressions for the elements of a microstrip cross junction are not readily available. However, a good equivalent circuit model has been found useful by some designers. Akello, Easter, and Stephenson [6] have shown that there are a number of series inductors linking the four ports of the cross junction. A common shunt capacitor works in conjunction with these series inductors to form the model. Although no good description exists for the inductors, there is a rule of thumb for calculating the common shunt capacitor:

$$C_T = \frac{3}{4} C_M \tag{7.13}$$

for the range $1.32 \leq \frac{w_4}{h} < 3.0$, where

$$C_M = \frac{c}{Z_0\sqrt{\epsilon_e}}$$

Z_0 and ϵ_e refer to the common input line, and c is the speed of light in free space.
The Touchstone syntax for the microstrip cross junction is

    MCROS 1 2 3 4 W1= W2= W3= W4=

7.2.6 Bond Wires

Bond wires interconnect various elements in an MIC circuit. The wires themselves are parasitic inductors. The inductance of a bond wire in free space is given as

$$L = 2l\left[\ln\left(\frac{l}{d}\right) + .5 + .22\left(2\frac{d}{l}\right)\right] \text{ nH} \qquad (7.14)$$

where l is the wire length in cm, and d is the wire diameter in cm.

The mutual inductance between two parallel bond wires in free space, separated by a gap S (in cm), is

$$L_M = 2l\left[\ln\left(\frac{l}{S}\right) - 1 + \left(\frac{S}{l}\right) - .25\left(\frac{S}{l}\right)^2\right] \text{ nH} \qquad (7.15)$$

If the wire is close to a ground plane, it may be more accurate to model the wire as a free space microstrip line with a width equal to its diameter and a height equal to its separation from the ground plane.

The Touchstone syntax for a bond wire is

    WIRE 1 2 D= L= RHO=

where
    D is the wire diameter,
    L is the wire length,
    RHO is the metal resistivity relative to pure gold.

The Touchstone syntax for a ribbon (flat wire) is

    RIBBON 1 2 W= L= RHO=

where
> W is the ribbon's width,
> L is the ribbon's length,
> RHO is the ribbon's resistivity to pure gold.

7.2.7 Thin-Film Resistor

A thin-film resistor has a dc resistance equal to

$$R_{dc} = \left(\frac{l}{w}\right) R_s \tag{7.16}$$

where
> l is the length of the resistor trace,
> w is the width of the resistor trace,
> R_S is the sheet resistance (in Ω/square) of the resistive layer.

At microwave frequencies, the length of a resistor contributes series inductance and its width contributes shunt capacitance. The thin-film resistor really is a lossy microstrip transmission line. A good model for the thin film resistor is to place its dc resistance in series with a lossless microstrip line of width w and length l (and a substrate of height h, and relative dielectric constant, ϵ_R).

The Touchstone syntax for a thin film resistor is

> TFR 1 2 W= L= R_S= F=

where
> R_S is the sheet resistance in Ω/square,
> F is the reference frequency for skin effects.

7.2.8 Grounds

Ribbon and bond wire grounds are discussed in Section 7.2.6. Via hole grounds are modeled as inductors whose inductance is calculated using expression (7.14), with l as the height of the substrate, and d as the via hole's diameter.

A wrap-around is difficult to model because whether it is more nearly a microstrip line or a ribbon is not obvious. Often with wrap-around grounds, the trace metal becomes very thin at the substrate's edge as a natural result of the sputtering deposition process. Such metal thinning increases the inductance of the wrap-around ground and makes it unsuitable for microwave frequency operation. Therefore, this grounding technique is not recommended.

7.3 TRANSISTORS, DIODES, AND CAPACITOR CHIPS

Most of the semiconductor devices used in MIC circuits are in chip form. Capacitor chips also commonly are used. All chips are mounted into an MIC circuit using the conventional hybrid assembly techniques known as *die attachment*. Die attachment is done using either a gold-tin solder preform braze process performed in a forming gas atmosphere or with gold or silver epoxy. The chip's die attach must be mechanically strong, of low electrical loss, and of low thermal resistance. All three of these requirements can be achieved using the gold-tin preform technique. Silver epoxy die attachment should be used only when convenience and low temperature processing are of paramount importance, due to its relatively high electrical and thermal resistance.

Chips always are preferable to package devices, for high-frequency, wideband electrical performance. The detrimental effects of package parasitic elements cannot be tolerated in most high-frequency, broadband designs. However, to take full advantage of the performance potential of a chip, observing certain design strategies is important:

- Bond wire lengths always must be kept to an absolute minimum except when they act as intentional inductors. Excessive bond wire length is the single biggest "killer" of high frequency performance.
- If the backside of the chip is to be grounded, the chip should be mounted on a metal ground rib so that its topside is mechanically flush with the substrate's topside. This strategy will minimize the bond wire length between the substrate and the chip.
- Bypass capacitors that connect directly to a semiconductor, such as the source bypass capacitors of a FET, must be mounted right *next* to the semiconductor device so bond wire lengths are kept to a minimum.
- The dc blocking capacitors must be die attached flush with the edge of a gap in a transmission line. *Short* bond wires are used to bridge the gap from the line to the top surface of the capacitor.

Figure 7.6 shows various low-parasitic die attach configurations. New configurations can be developed by expanding the basic ideas shown in Figure 7.6.

Figure 7.7 shows the topside bonding pattern for some typical chip devices: Low-noise GaAs FETs, power GaAs FETs, bipolar transistors, diodes, capacitors. Also, each element's schematic symbol is shown next to its physical "footprint." Backside metalization and bonding pads are gold almost exclusively. For this reason, most microwave circuits use a gold metal system: with gold conductor patterns on the circuit substrates, gold-plated carriers and grounding ribs, and gold bond wire. Exclusively using gold increases long-term reliability by eliminating the growth mechanisms for certain destructive intermetallic compounds (such as "purple plague" and dendritic growth) associated with the other metals such as aluminum and silver.

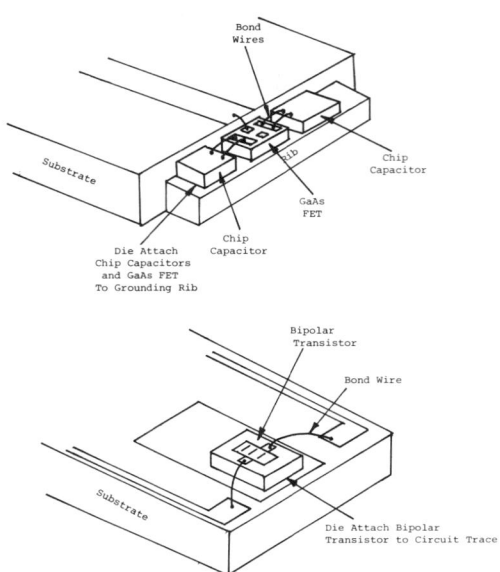

Fig. 7.6 Examples of low-parasitic die attachment configurations for GaAs FETs and bipolar transistors.

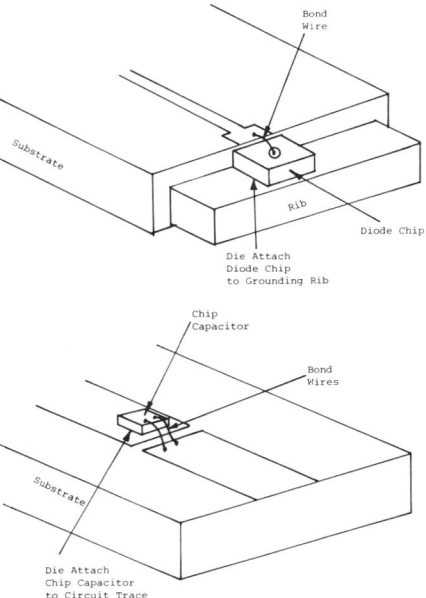

Fig. 7.6 cont'd. Examples of low parasitic die attachment configurations for diodes and chip capacitors.

7.4 THERMAL CONSIDERATIONS

GaAs FETs and bipolar transistors dissipate electrical energy. This electrical energy becomes heat and raises the temperature of the device and its immediate environment. High device temperatures can cause failures, or severely limit long-term lifetimes; thus, designing a semiconductor's thermal environment so that low device temperatures are maintained under all conditions is very important.

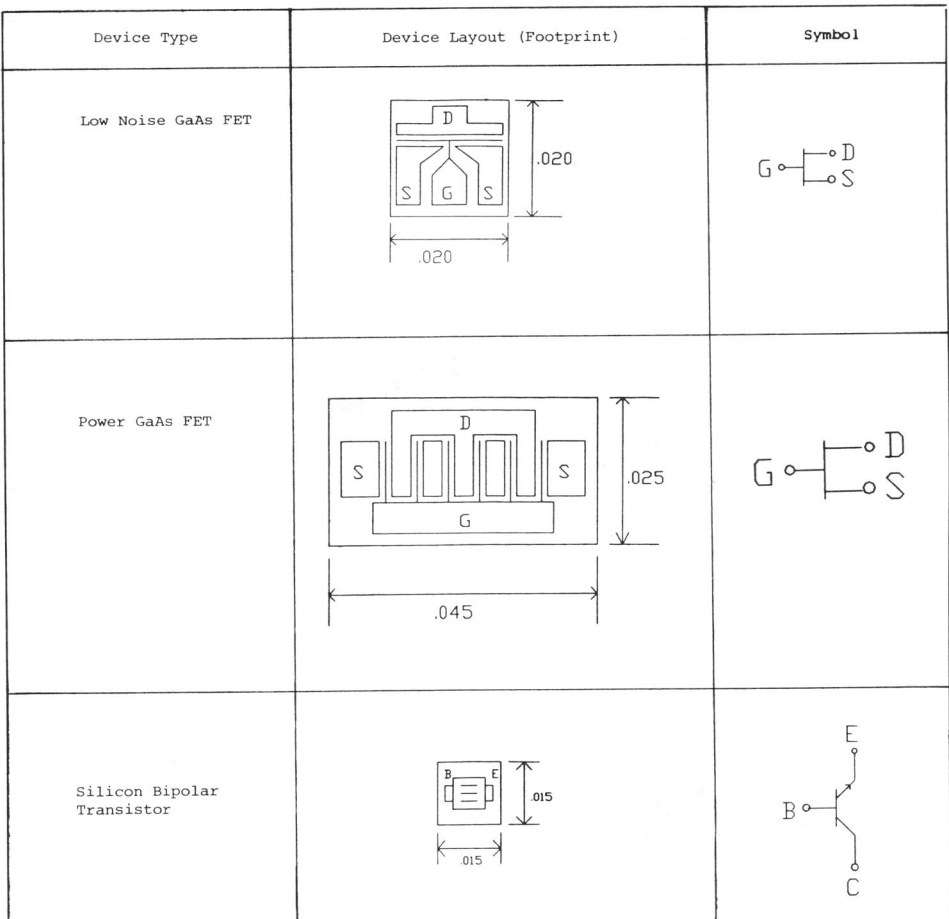

Fig. 7.7 Bonding pad pattern for commonly used MIC devices.

Device Type	Device Layout (Footprint)	Symbol
PIN Diode or Varactor Diode	⊡ 010 × 010	ANODE ▽ CATHODE
Chip Capacitor	☐ 020 × 020	─┬─ ─┴─

Fig. 7.7 cont'd.

The operating temperature of the active region of a semiconductor device, such as a GaAs FET, is calculated in the following way. In analogy to electrical resistance, a thermal resistance, θ_R, relates a temperature gradient, ΔT, to dissipated power, P_{diss}.

$$\Delta T = \theta_R P_{\text{diss}} \tag{7.17}$$

Let us calculate θ_R for the case of a GaAs FET, which is die attached to a copper grounding rib and brazed to a Kovar carrier mounted in an aluminum housing. This thermal configuration is diagrammed in Figure 7.8. We may calculate the individual thermal resistances as follows:

θ_{R1} = Thermal resistance from the active region to the bottom of the FET chip

$$= \frac{\text{Chip Thickness}}{(\text{Chip Area})K_{\text{GaAs}}} \tag{7.18}$$

θ_{R2} = Thermal resistance of die attach material

$$= \frac{\text{Die Attach Thickness}}{(\text{Chip Area})K_{\text{Die Attach}}} \tag{7.19}$$

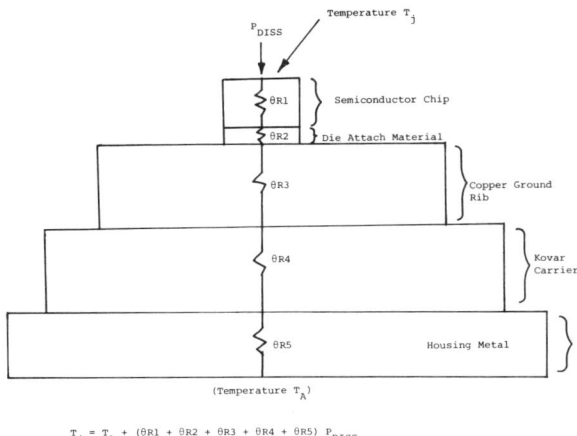

Fig. 7.8 The thermal resistance configuration of a semiconductor chip device mounted in an MIC circuit.

θ_{R3} = Spreading resistance of heat entering the copper rib (treated as a semiinfinite heat sink)

$$= \frac{2}{(\text{Periphery of Chip}) K_{\text{Cu}}} \qquad (7.20)$$

θ_{R4} = Spreading resistance of heat entering the Kovar carrier (treated as a semiinfinite heat sink)

$$= \frac{2}{(\text{Periphery of Rib}) K_{\text{Kovar}}} \qquad (7.21)$$

θ_{R5} = Spreading resistance of heat entering the aluminum housing (treated as a semiinfinite heat sink)

$$= \frac{2}{(\text{Periphery of Carrier}) K_{\text{Al}}} \qquad (7.22)$$

where
$K_{\text{Silicon}} = 1.5 \text{ W} - \text{cm}^{-1} - {}^\circ\text{C}^{-1}$
$K_{\text{GaAs}} = .5 \text{ W} - \text{cm}^{-1} - {}^\circ\text{C}^{-1}$
$K_{\text{Die Attach}} = 3.0 \text{ W} - \text{cm}^{-1} - {}^\circ\text{C}^{-1}$
$K_{\text{Alumina}} = .30 \text{ W} - \text{cm}^{-1} - {}^\circ\text{C}^{-1}$
$K_{\text{Cu}} = 3.9 \text{ W} - \text{cm}^{-1} - {}^\circ\text{C}^{-1}$
$K_{\text{Kovar}} = .17 \text{ W} - \text{cm}^{-1} - {}^\circ\text{C}^{-1}$
$K_{\text{Al}} = .97 \text{ W} - \text{cm}^{-1} - {}^\circ\text{C}^{-1}$

The total thermal resistance is

$$\theta_R = \theta_{R1} + \theta_{R2} + \theta_{R3} + \theta_{R4} + \theta_{R5} \tag{7.23}$$

These expressions for thermal resistance are easily solved by using the HP-41C program THERM R, which is discussed in Section 4.4.

For good semiconductor reliability, it is important that $T_j = T_A + \Delta T$ does not exceed 200°C in GaAs and 150°C in Si. This means that for a given ambient temperature, T_A, it is required that

$$T_j = 150°C \geq \theta_R P_{\text{diss}} + T_A \tag{7.24}$$

This can be a particularly difficult requirement if T_A is greater than 100°C. To design the thermal path for minimum θ_R becomes extremely important if P_{diss} and T_A are high.

7.5 ARTWORK

7.5.1 Rubyliths

The conventional approach to photomasking involves making a master artwork on a two color, two layer mylar sheet called a *rubylith*. The two colors usually are red and transparent. The circuit pattern is cut into the red layer, so that when red material is removed, an exactly scaled-up red replica of the circuit remains. Typical scale factors are 10, 20, 40, and 50. Photomasks are created by photographically reducing the rubylith to a 1 : 1 scale. Separate rubyliths are made for conductor and resistor patterns because each trace type requires a separate masking step.

If an etchback process is to be used, great care must be taken to ensure that proper etch factors are built into the rubylith. Rubyliths are cut on a precision machine called a *coordinatograph*. Some coordinatographs are computer controlled, whereas others are controlled manually.

Some CAD layout computer programs such as EEsof's MICAD will cut rubyliths using a conventional pen plotter in which the pen is replaced by a special cutting tool. Rubylith quality equivalent to that obtained with a coordinadograph is possible with this approach. An example of a rubylith is shown in Figure 7.9.

7.5.2 Digitizing

Highly complex machines, called *pattern generators,* create photomasks directly from a digital input characterizing the circuit's layout. The rubylith step, and even

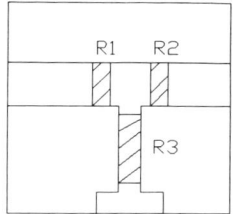

Fig. 7.9 An MIC "T" pad example showing how rubyliths are created for the conductor and resistor patterns of an MIC circuit.

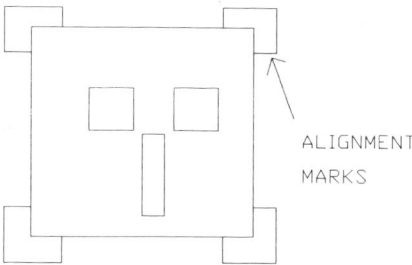

Fig. 7.9 cont'd. Resistor rubylith for "*T*" pad example.

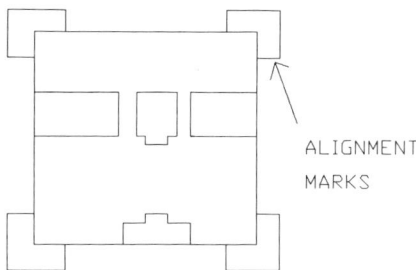

Fig. 7.9 cont'd. Conductor rubylith for "*T*" pad example.

the full-scale artwork step, is unnecessary with this technique. All that is required for pattern generation is a set of mathematical coordinates characterizing the layout.

Because of the high capital investment required to purchase a pattern generator, most microwave companies using this approach work with an independent mask-making shop. The masking shop will accept digital input from its customers, either as an encoded magnetic tape (if the customer has the capability to produce one) as fully dimensioned artwork, or even as artwork done to scale on a coordinate grid. Digitized mask-making has become a very attractive alternative approach for the following reasons:

- It saves time and money by avoiding the rubylith step;
- The circuit easily is modified by changing the digital input to the pattern generator and creating a new mask;
- Circuit masters, in the form of magnetic tapes or floppy disks, are easier to store than 50:1 scale rubyliths.

7.6 MASK-MAKING

7.6.1 Film or Chrome-Glass Masks

Photo masks can be made with two basic types of material: (1) flexible emulsions or films, or (2) rigid glass with a chrome pattern. There are advantages and disadvantages to both. In the past few years the industry has been moving toward more use of chrome-glass masks. Table 7.1 compares the two types of masks.

Table 7.1 Mask Types—A Comparison

Type of Mask	Accuracy	Lifetime	Cost	Mask-Making Lead Time	Storage
Film	Fair-Good	Short	Moderate	Short	Simple
Chrome-Glass	Excellent	Long	High	Longer	Complex

7.6.2 Step and Repeat

A photomask containing the pattern of only one circuit can expose only one circuit at a time. Step and repeat is a masking technique that allows the simultaneous exposure of a whole array of identical circuits. The step and repeat mask consists of a repeating array of circuit patterns in rows and columns across the mask. Each circuit pattern is separated from its neighbors by a narrow buffer zone exactly the width of the saw blade that ultimately will cut the exposed and etched substrate into individual circuits. For this reason, these buffer stripes are called *saw streets*. Figure 7.10 shows a typical step and repeat pattern. All the designer need do is

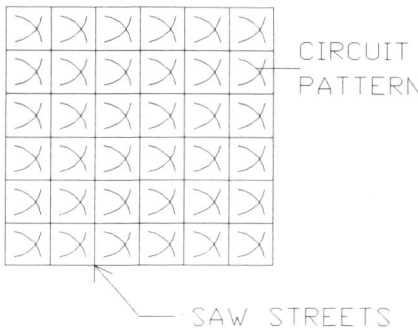

Fig. 7.10 A step and repeat photomask.

request a step and repeat mask from the mask shop, and the shop will take care of all the details. Of course, the ultimate substrate size must be specified. The same rubylith or digitized information is used for either a single shot mask or a step and repeat mask.

7.7 TYPES OF SUBSTRATES

7.7.1 Dielectric Choices

The most basic materials choice facing an MIC designer is choosing the type of substrate dielectric. Many types of material are available, and each has its own particular advantages and disadvantages. By far the most popular substrate material is alumina (Al_2O_3). Alumina is a hard, opaque, white, polycrystalline sapphire that can be polished to a high surface finish. Alumina's popularity is due to its ruggedness at high temperatures, its high dielectric constant (9.9), which leads to small circuits, and to its low electrical loss (loss tangent = .0001). Alumina is easily metalized with a number of popular metal systems. Table 7.2 compares a number of popular substrate materials. There are several excellent suppliers for each of these materials. Substrates may be purchased unmetalized, metalized, or if a circuit pattern has been provided as patterned circuit substrates.

A second class of circuit materials are the so-called soft boards, such as Duroid. Soft boards have a relatively low dielectric ($\epsilon_R = 2.2$) and very low loss, even at the highest microwave frequencies. Soft boards are a combination of Teflon and Fiberglas and generally are metalized with copper, as opposed to gold. Soft boards are important for applications using packaged devices (see Section 5.6).

Table 7.2 Properties of Hard Substrate Materials—A Comparison

Material	Dielectric Loss Tangent (10 GHz)	Dielectric Constant	Resistivity Ω-cm	Thermal Conductivity $(W - cm^{-1} - °C^{-1})$	Comments
Alumina (Al_2O_3)	.0001	9.9	10^{11}	.3	Low cost
Quartz (fused silica)	.000015	3.8	10^{17}	.01	Fragile
Beryllia (BeO)	.0003	6.6	10^{17}	2.5	High thermal conductivity, but its dust is toxic
Sapphire	$8.6 \cdot 10^{-5}$	11.5	10^{16}	.4	Expensive, anisotropic

7.7.2 Types of Metal Systems

Hard substrates are metalized by either a sputtering process (thin film) or by a silk-screen metal paste and cure process (thick film). A number of metal layers commonly are used in MIC circuit substrates. These layers are

- adhesion layer,
- barrier layer,
- conductor layer,
- resistor layer.

Soft boards use a simple single conductor layer of copper. Hard board substrates use some or all of the preceding metal layers. The purpose of these metal layers and the most commonly used metals is given in Table 7.3.

7.8 PHOTOLITHOGRAPHY

7.8.1 Photoresist

Photoresist is a waxlike photosensitive material applied as a liquid to metalized substrates during photolithography. Once cured by exposure to elevated temperature, the photoresist may be photographically exposed through a photomask to form a pattern of the circuit contained on the mask. After photographic exposure, the photoresist changes its chemical properties and becomes quite hard. During a developing step, the nonexposed regions of the substrate (i.e., those areas masked from the exposing light) remain unchanged in hardness and easily are washed

Table 7.3 Types of Metalization—A Comparison

Type of Layer	Purpose	Common Metals (soft board)	Common Metals (hard board)	Comments
Adhesion Layer	Provide adhesion between metal layers and substrate material		Chromium Titanium-Tungsten	Titanium-Tungsten more common on hard board
Barrier Layer	Prevent penetration to the adhesion layer of solders applied to the conductor layers		Nickel Palladium	Not often used
Conductor Layer	Provide primary electronic conductivity path	Copper, .5 to 1.0 mils thick	Copper, gold, aluminum	Gold most common on hard boards is .10 to .20 mils thick
Resistor Layer	Resistor material		Tantalum-nitride, nickel-chromium	Tantalum-nitride is the more popular

away, leaving a hard photoresist pattern defining the circuit elements. The hardened photoresist will serve to "mask" the circuit conductor and resistor regions against the etching step, which comes next. The resist-patterned substrate then is subjected to a series of chemical etches that remove first the unmasked conductor metal, then the unmasked barrier metal, and finally the unmasked adhesive metal.

Photoresist is available in both "positive" and "negative" forms. After development, positive photoresist turns hard once exposed to light. Negative photoresist works in the reverse.

Table 7.4 offers a detailed step-by-step list of the photolithographic process steps, together with a list of the equipment necessary to complete each step. Cleanliness is very important at all times because even minute dust particle contamination can cause defects in the circuit that could lead to electrical anomalies or failure.

7.8.2 Etching

As mentioned earlier separate etching may be necessary for each of the metal layers deposited on a substrate. Etching is a chemical process by which unmasked (by the patterned photoresist) metal is chemically removed to define a metallic circuit pattern on the substrate.

The two basic kinds of etch processes popular in the microwave industry are etchback, and plate up. The two processes are really quite similar. In the etchback process, work begins with a substrate containing a sputtered conductive layer of full thickness (.10–.20 mils). Etching must remove *all* of this material to define the circuit. The major drawback of the etchback process is that the unetched circuit element edges are sufficiently high to allow undercutting, sometimes by as much as the thickness of the metal layer. This is why an etch factor must be included in the artwork intended for etchback processing. The etch factor compensates for the loss of material at each conductor's edge by undercutting (see Figure 7.11 for details).

With the plate up process, a very thin conductive layer is initially sputtered onto the substrate. Because the element edges formed during etching are very low in this case, little or no undercutting occurs. After etching is completed and all photoresist is removed, the conductor traces are thickened by electroplating conductor metal on top of the original conductor metalization. In the plate up process, no etch factor is included in the artwork because significant undercutting does not occur.

Table 7.4 Steps in Photolithography

Step	Purpose	Equipment
Apply Photoresist	Spread a very uniform thin film of photoresist on the substrate	Clean room, spinner
Cure Photoresist	Harden photoresist from a liquid to a waxlike solid.	Temperature controlled oven, thermometer
Expose Photoresist	Optically project the photomask pattern onto the cured photoresist	Mask aligner, high-intensity UV light source
Develop Photoresist	Use developing chemicals to alter the chemical properties of the exposed photoresist to fix the circuit pattern; wash away unexposed photoresist	Chemicals, wet chemical processing bench with a fume hood
Etch	Apply etching chemicals to remove unmasked metal layer	Chemicals, wet chemical processing bench with fume hood
Etch Strip	Remove all remaining photoresist from etched circuit	Chemicals, wet chemical processing bench with fume hood

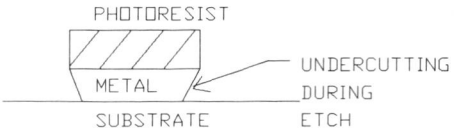

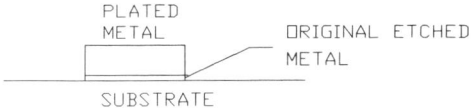

Fig. 7.11 Diagrams showing the *etchback* and *plate up* MIC etching processes.

7.9 RESISTOR STABILIZATION

Thin-film resistors require a heat treatment process to fix the value of sheet resistance on a particular substrate and stabilize the resistor against changes in sheet resistance over time. Heat treatment creates an oxide or nitride layer on the surface of the patterned resistor traces. Exposure to high temperatures thickens this oxide layer over time. Because the oxide layer is nonconducting, as it thickens, that portion of the resistive layer's cross-sectional area available for conduction decreases, which increases the net resistance. Heat treatment can only raise the value of thin film resistors, they cannot be lowered. To analyze a resistor's change during heat treatment, refer to Figure 7.12, which is a diagram of the resistor cross section.

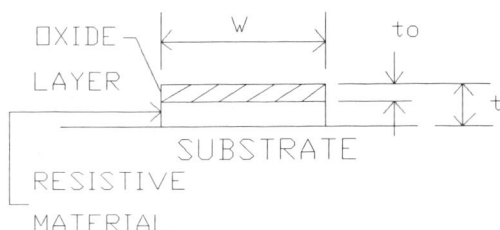

Fig. 7.12 How heat treatment affects thin-film resistors.

Let l be the length of the resistor, w its width, and t be its initial thickness. Let t_o be the thickness of the oxide layer, which increases with time during heat treatment. The resistance of this structure is

$$R = \left(\frac{w}{l}\right) R_S \left(\frac{t}{t - t_o}\right) \tag{7.25}$$

where R_S is the preheat treatment value of the resistive layer's sheet resistance.

Not uncommonly, heat treatment raises the value of all resistors on a substrate by 10 to 20%; therefore, initially design all resistors to be on the low side of their desired value by some percentage, and then heat treat them "up" to their proper value. Some heat treatment always is necessary because the oxide layer acts as a "cap" on top of the resistive layer, preventing any destabilizing changes over the long term once the circuit has been completed and is in the field. Heat treatment is something of an art, and each laboratory must develop a unique schedule that produces predictable results with their circuits.

7.10 SAWING

If a step-and-repeat photomask is used to process a substrate, the substrate must be subdivided into individual circuits after the photolithography-etch-resistor stabilization processes are completed. This usually is accomplished by a special "dicing" saw equipped with a circular blade with either a carborundum or diamond cutting edge. Many saws can be set on automatic control to cut along the saw streets running "east to west," and then the substrate is turned on its moveable stage by 90° and the saw automatically cuts along all saw streets running "north to south."

It is very important when creating the step-and-repeat pattern, as shown in Figure 7.10, that the width of the saw streets placed on the photomask exactly match the "cutting width" of the saw to be used for substrate dicing. If a single fabrication laboratory is taking care of both photomask-making and substrate fabrication, they will take care of the saw street width. However, if the mask maker is a separate organization from the fabrication laboratory, the designer must be sure to specify the proper saw street width, based on which saw ultimately will dice the substrate.

7.11 BRAZING PROCESS

Brazing is a hard soldering process by which circuit substrates and grounding ribs are attached to metal plates called *carriers*. The most common braze material used in the microwave industry is a gold-germanium eutectic. Gold-germanium flows at temperatures above 420°C and melts the gold on the back side of the substrate and the gold plating on the carrier and the ground rib, to form a eutectic bond between all three parts. Because new gold from the substrate and the carrier (and rib) liquifies during brazing, the percentage of gold in the gold-germanium eutectic is increased. The melting point of this eutectic increases rapidly with increased gold content, causing the material to solidify very quickly. This means the operator has a limited chance to change the placement of the parts before they are "frozen" in place. For this reason, brazing fixtures quite often are designed to ensure proper parts placement before and during heating and flow of the brazing material. The following is a list of important points to observe in setting up a successful brazing process.

- Use only substrate and carrier materials that have highly compatible thermal expansion coefficients. Otherwise, strains are built into the parts at brazing that can crack the substrate or warp the carrier. The most popular choice of materials in the microwave industry is alumina substrates and Kovar carriers. Because ground ribs are quite small, they can be made out of gold-plated copper (for thermal conductivity reasons) without causing expansion or cracking problems.

- Be sure that high-quality, high-process-temperature gold plating (mil-standard) is used on the carrier and the ribs. A sample of all plating lots should be subjected to at least 450°C and checked for blistering and peeling.

- The carrier thickness should be at least twice the substrate thickness to prevent warping. This requirement can be at odds with the electrical requirement for thin carriers to create very direct grounds at high frequencies. The designer will have to compromise these two considerations on a case-by-case basis to best satisfy the overall needs of a particular design project.

- Gold-germanium strip material should be cut into solder "preforms" that are about 75% of the surface area of the substrate to be brazed. Preforms larger than 75% can cause excessive flow of braze material. Those significantly less than 75 percent will not provide proper "wetting" between the substrate and the carrier, leaving voids between the substrate and the carrier. Such voids can cause both electrical performance and reliability problems.

- Brazing should be done in a reducing atmosphere to eliminate any organic contamination from the surface areas of the work. The most commonly used reducing atmosphere is forming gas (10% hydrogen, 90% nitrogen). Of course, pure hydrogen is an excellent reducing atmosphere, but it is too dangerous to use in its pure form. Many commercially available brazing stations make use of a system of ducts above the heated stage that surrounds the work with a "blanket" of forming gas.

- Although brazed parts can be positioned manually by an experienced operator, best results are obtained by specially designed brazing fixtures to fix the relative position of substrates, braze preforms, carriers, and ribs during the brazing process. Brazing fixtures ensure accurate dimensions on the completed circuit or carrier assembly.

- Make sure the carrier surface finish is smoother than $\sqrt{32}$.

7.12 DIE ATTACH PROCESS

Chip components are attached to circuit substrates, ground ribs, and carriers, by a process called die attach. There are two kinds of die attach: epoxy die attach, and eutectic die attach. Table 7.5 compares various types of die attach material, their cure or flow temperatures, and the relative advantages and disadvantages of each.

Table 7.5 Die Attach Processes—A Comparison

Die Attach Process	Material	Flow-Cure Temperature	Comments
Epoxy Die Attach	Gold epoxy	+150°C	Consistent with gold metal system, expensive
	Silver epoxy	+150°C	Possible reliability problems due to intermetallic compound growth
Eutectic Die Attach	Gold-tin	+290°C	Low thermal resistance, will not reflow earlier gold-Ge brazing
	Gold-germanium	+420°C	Low thermal resistance, may reflow gold-tin brazing
	Gold-silicon	+450°C	Good for attaching bare-back silicon chips. May reflow gold-Ge brazing

Eutectic die attachment is commonly done on the same kind of heated stage used in brazing stations; in fact, eutectic die attachment can be thought of, and treated as, brazing for small parts. The temperatures involved depend on the particular die attaching material (see Table 7.5). A forming gas atmosphere often is advisable to promote "wetting."

Eutectic die attaching materials may be purchased in the form of disk-shaped preforms. The operator places a preform on the work at the place where the die attach is to take place and waits a few seconds for it to melt. As soon as the preform melts, the operator, using a pair of tweezers (or using a vacuum chuck with an automatic die attach machine) lifts the chip, which is to be attached, onto the melted preform. Care must be taken to be sure the chip is properly oriented. The chip is carefully "scrubbed" back and forth by the operator to promote wetting on its backside. With the appearance of solder fillets around the base of the chip, the carrier or substrate is removed from the heated stage and allowed to cool.

Epoxy die attach is a very simple procedure. After preparing the appropriate epoxy, the operator places a small quantity of epoxy on the site of the die attach. The amount is critical, because too little epoxy will not secure the chip and too much will "bury" the chip in conductive epoxy and short it out. Once the right amount of epoxy is in place, the operator picks up the chip with tweezers and places it squarely in the middle of the epoxy (heading in the right direction, of course), and "pushes" the chip down into the epoxy. If the die attachment appears to be good, the work is placed in a curing oven (usually set to $+150°C$) for approximately one hour.

The thermal conductivity of gold-tin die attaching material is $3.0W\ cm^{-1} - °C^{-1}$. The thermal conductivity of silver epoxy is $.016W\ cm^{-1} - °C^{-1}$. This means that if everything else is equal, the thermal resistance of an epoxy die attachment is 188 times higher than a gold-tin eutectic die attachment. Eutectic die attachment is an insignificant contribution to the overall thermal resistance of a FET or bipolar transistor; however, an epoxy die attachment easily can become the dominant contributor to overall thermal resistance. Importantly, *always* use eutectic die attachment whenever significant power dissipation is anticipated in a chip under normal operation. Using epoxy for die attaching devices with significant power dissipation (such as power GaAs FETs) can reduce operating lifetime, or cause catastrophic failure. Epoxy die attachment is a fast, effective fabrication technique, but its effects must be factored carefully into the thermal calculation whenever significant power dissipation occurs.

7.13 WIRE BONDING

Semiconductor and capacitor chips, once die attached to a circuit substrate, are connected electrically to the circuit to create an electrically functioning component. The interconnection between the chip and the circuit substrate is done by a process technique called wire bonding. Two types of wire bonding commonly are performed in the microwave industry: wedge bonding, and ball bonding.

Bonding is a technique by which similar metals are "bonded" together under the influence of *temperature* and *pressure*, even though their melting points are never exceeded. For this reason, bonding sometimes is referred to as thermal compression bonding. Strictly speaking, both wedge and ball bonding are thermal compression techniques, because they both employ a combination of pressure and temperature to form the bond.

Some bonding machines have a capability called *ultrasonic scrubbing*, which is accomplished by attaching an ultrasonic transducer to the bonding tool, to vibrate the tool from side to side as the bond is made. This scrubbing motion enhances the bonding process. Scrubbing also can cause problems such as substrate cracking in fragile devices such as GaAs FETs. For this reason, ultrasonic scrubbing normally is avoided in the microwave industry.

7.13.1 Wedge Bonding

Wedge bonding is by far the most popular bonding technique in the microwave industry. This is because wedge bonds are very narrow (less than twice the wire diameter) and the wire is bonded in the plane of the bonding pad rather than at right angles to the pad, as in the case of a ball bond. The bonds themselves are very short and direct and have low parasitic inductance.

Wedge bonding uses a narrow, pointed tungsten carbide bonding tool (a wedge) to press a gold bond wire (usually .7 mil diameter) into a gold bonding pad on a chip or into a gold circuit trace. The "footprint" of the bonding tool is approximately 2 × 2 mils in area, not much larger than the wire's diameter but smaller than the chip's bonding pads. Both the bonding tool and the work are heated to facilitate bonding. Bonding machines have a system of counterweights that precisely control the amount of downward force the wedge tip exerts on the wire and the chip. Too little force causes the bond not to "stick," and too much force will crack the chip or flatten the wire. The stage and tip temperatures also must be set precisely to achieve consistent high-quality bonds. This combination of tool force, stage temperature, and tip temperature are called the *bonding schedule*. Operators setting up a machine for the first time must experiment for a considerable period to work out a schedule that works for the particular machine and the work to be run on it. If the work remains the same, a given schedule, once established, can be maintained over long periods of time. However, if the work should change (for example, if the diameter of the wire is increased or the thickness or quality of the pad metalization is changed), the schedule may have to be modified to accommodate the changes.

Bonding machines allow the precise position of the work stage under the bonding tip to be carefully controlled by the operator, through a high mechanical advantage manipulator. In some bonding machines, wire is fed automatically underneath the bonding tip, advancing by just the right length each time a bond is made. Figure 7.13 pictorially shows how a wedge bond is formed.

7.13.2 Ball Bonding

Ball bonding works on a slightly different principle from wedge bonding, although both types of bonding are thermal compression techniques. In the case of ball bonding, a small flame (from a hydrogen torch or an electric arc) melts the very end of the wire to form a very small diameter ball. This ball is forced onto the chip's bonding pad by a tungsten carbide bonding tip, which is called a *capillary* because of the small diameter hole that passes along its center line to accommodate the wire. When a bond is made, the capillary is instantaneously heated by an electric current. As the ball contacts the bonding pad, it is distorted by tool pressure into a mushroom shape.

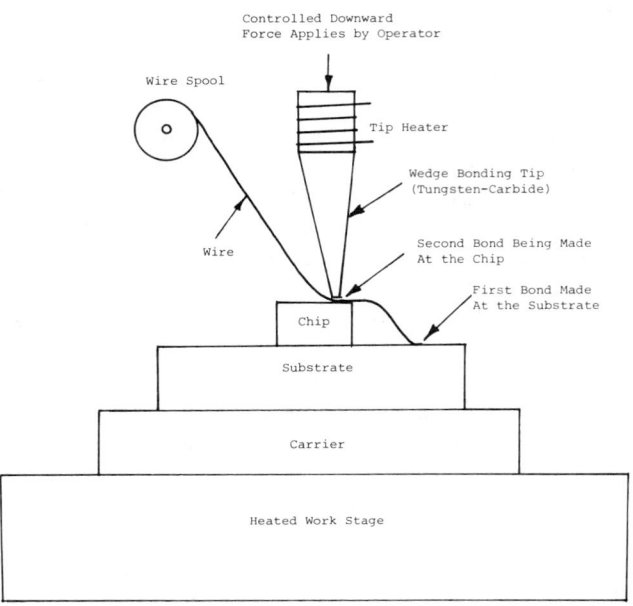

Fig. 7.13 The wedge wire bonding process.

The main advantage of ball bonding is that the technique does *not* require a heated work stage. Completed circuits with temperature sensitive components (such as tunnel diodes) may be ball bonded "cold," so as not to endanger the heat sensitive components. However, the price paid is a larger bond size, because the ball diameter is difficult to control to sizes less than twice the wire diameter. Also, the bond wire will be longer because the wire initially is pointing in a direction normal to the plane of the bonding pad and must be looped around into a "hairpin" shape to connect at the other end. See Figure 7.14 for some graphical details of the ball bonding process.

The designer must carefully plan for additional bond wire inductance if ball bonds are to be used. Ball bonds are too large for the small bonding pads that are available on GaAs FETs and bipolar transistors. Even small ball bonds often can short out these devices. Normally ball bonding is confined to interconnecting passive elements and conductor traces on the circuit substrate. The times when ball bonding is advantageous is for a few quick bonds to be made on a noncritical part of a circuit; and it is important *not* to heat the entire circuit.

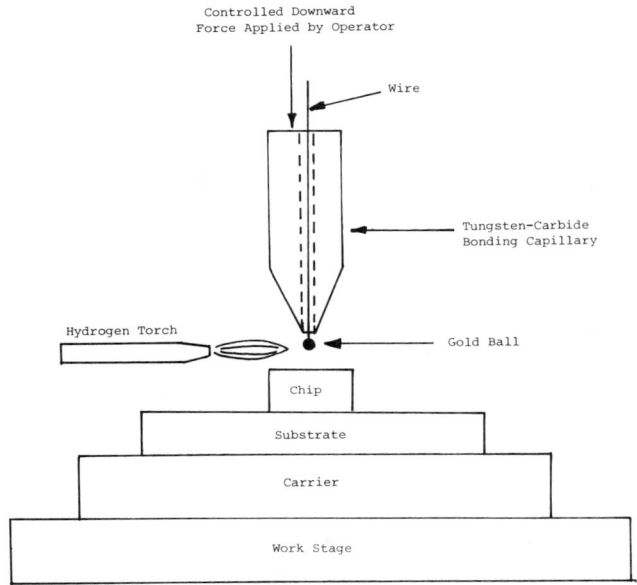

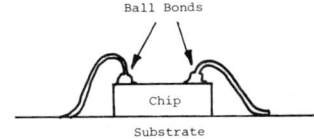

Fig. 7.14 The ball wire bonding process.

REFERENCES

1. T. C. Edwards, *Foundations for Microstrip Circuit Design,* John Wiley and Sons, New York, 1981.
2. P. Silvester and P. Benedek, "Equivalent Capacitance of Microstrip Open Circuits," *IEEE Trans. Microwave Theory Tech.*, Vol. MTT-20, August 1972, pp. 511–576.
3. K. C. Gupta, B. Garg, and I. J. Bahl, *Microstrip Lines and Slot Lines,* Artech House, Norwood, MA, 1979.
4. B. Garg, N. Ramesh, and I. Bahl, "Microstrip Discontinuities," *Int. J. Electronics,* Vol. 45, 1978, pp. 81–87.
5. E. Hammerstad and F. Bekkadal, "A Microstrip Handbook," ELAB Report, STF 44A74169, N7034, University of Trondheim—NTH, Norway, 1975.
6. R. Akello, B. Easter, and I. Stephenson, "Equivalent Circuit of the Asymmetric Crossover Junction," *Electronics Letters*, Vol. 13, February 17, 1977, pp. 117–118.

Chapter 8
Final Layout of MIC Amplifier and Oscillator Design Examples

The basic design of the MIC amplifier circuit examples is given in Chapter 5, and the basic design of the MIC oscillator circuit examples is given in Chapter 6. In this chapter, a final reduction to a buildable layout will be done for each example. The following final design steps were developed in Chapter 7 as the way of proceeding from the basic topology to the final layout:

- Perform an initial layout consistent with the process design rules.
- Modify the Touchstone-SuperCompact circuit model to include all layout parasitic elements.
- Optimize the circuit model to regain any "lost" simulated performance.
- Modify layout to include optimized element values.

This process will be followed with each of the MIC circuit examples. Actual dimensions are not shown on the layout drawings so that the same drawing may stand for both the initial and the final layouts. In some cases, the original Touchstone file had to be completely rewritten because the layout parasitic elements were sufficiently extensive to make simple file modifications very difficult. We address each example one by one.

Remember, especially at the layout stage, *there is no substitute for creative thinking.* The designer should "brainstorm" many possible layout options and arrive at his or her final choice through careful consideration of the pros and cons of each choice. Simplest is usually best concerning layouts. Be very careful not to overcomplicate a circuit layout because complications lead to parasitics and losses that degrade performance.

A word of caution is in order at this point. The designer must always remain aware of the inherent limitations of the available computer-aided design tools. Some element models, especially the parasitic element models (bends, tees, crosses), in both SuperCompact and Touchstone have certain built-in inaccuracies. Experience has taught many designers that these models are not completely accurate for all frequencies or all geometries. Generally, problems develop at high frequencies or high width-to-height ratios. A wise policy is to allow sufficient time in all design programs for the construction and empirical modification of at least one prototype unit. If significant empirical changes are warranted, the designer will want to incorporate these modifications into the artwork and the photomasks before manufacture begins.

CAD modeling has been of revolutionary help to the designer. In fact, today we could barely function without these tools, and the evolution continues. We anticipate, as time goes on, that the CAD models gradually will improve in accuracy, gradually removing the empirical element from the design process. This is particularly important in the case of MMIC design (see Chapters 9 and 10), where empirical modifications are impossible; if a circuit does not perform to expectation, the only recourse for the designer is to redesign it, a very costly process.

8.1 THE 6 TO 18 GHz BALANCED AMPLIFIER FINAL LAYOUT

The input and output circuits are patterned on separate alumina substrates (10 mils thick), shown in Figure 8.1. The two FETs along with various bypass capacitors and chip resistors are mounted on a metal grounding rib that runs between the input substrate and output substrate. Lange couplers on both substrates are terminated with deposited 50 Ω resistors grounded by welding a ribbon between a metal pad and carrier ground.

The principal parasitic elements in this circuit are the step discontinuities, microstrip bends, bond wires, and the deposited resistors. Figure 8.2 shows a reconfigured schematic diagram that includes all these elements. The Touchstone file modified to include all parasitic elements is shown in Table 8.1. This circuit requires very little reoptimization to regain its idealized performance. Table 8.1

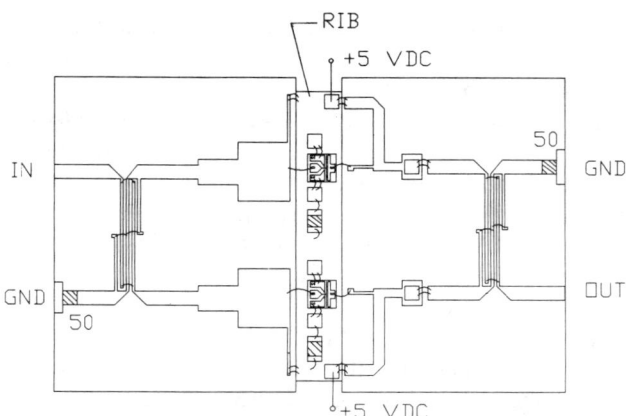

Fig. 8.1 Layout of the 6 to 18 GHz balanced FET amplifier.

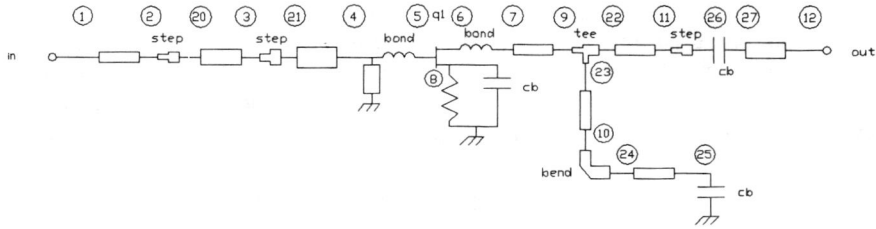

Fig. 8.2 The final schematic of the single ended 6 to 18 GHz amplifier, including all layout parasitic elements.

Table 8.1 Touchstone—Ver(1.30–May-17-85)—Ser(15738-1598-1000)—Con(1CCC22124P) 376.CKT 07-05-89 10:16:48

```
! THE BALANCED AMPLIFIER TOPOLOGY   FOR  6 TO 18 GHZ INCLUDING ALL LAYOUT
! PARASITIC ELEMENTS.
! THE GaAs FET IS A HARRIS HMF0310.
CKT                                              FREQ
    ! THE SINGLE ENDED AMPLIFIER                     SWEEP 2 26 2

    MSUB ER=9.9 H=10 T=.2 RHO=1 RGH=0            OUT
    MLIN 1 2 W=13.7 L=45.5
    MSTEP 2 20 W1=13.7 W2=31.9                       AMP DB[S21] GR1
    MLIN 20 3 W=31.9 L=46.2                          AMP DB[S11] GR1
    MSTEP 3 21 W1=31.9 W2=41.5                       AMP DB[S22] GR1
    MLIN 21 4 W=41.5 L=3.6                           BAL DB[S21] GR2
    MLSC 4 W=2.89 L=52.05                            BAL DB[S11] GR2
    IND 4 5 L=.10                                    BAL DB[S22] GR2
    S2PA 5 6 8 HMF0310                               BAL DB[S12] GR3
    IND 6 7 L=.10
    RES 8 0 R=20                                 GRID
    CAP 8 0 C=100
    MLIN 7 9 W=1.0 L=6.6                             RANGE 2 26 2
    MTEE 9 22 23 W1=1 W2=4.69 W3=6.27                GR1 -20 10 5
    MLIN 23 10 W=6.27 L=41.61                        GR2 -20 10 5
    MBEND 10 24 W=6.27 ANG=90 M=0                    GR3 -40 0 10
    MLIN 24 25 W=6.27 L=20.0
    CAP 25 0 C=50                                OPT
    MLIN 22 11  W=4.69 L=15.20
    MSTEP 11 26 W1=4.69 W2=12.8                      RANGE 17 19
    CAP 26 27 C=50                                   BAL DB[S21]=7.0 40
    MLIN 27 12 W=12.8 L=6.36                         RANGE 6 17
    DEF2P 1 12 AMP                                   BAL DB[S21]=7.0 20
                                                     AMP DB[S22]<-15   5
! CONNECT TWO AMPS IN A BALANCED CONFIGURATION

    MLANG 1 2 3 4 W=1.4 S=.8 L=80
    MLANG 7 8 5 6 W=1.4 S=.8 L=80
    AMP 4 5
    AMP 2 7
    TFR 1 0 W=10 L=10 RS=50 F=0
    TFR 6 0 W=10 L=10 RS=50 F=0
    DEF2P 3 8 BAL
```

also gives the final values for the circuit elements. The final simulated performance is shown in Figure 8.3. Were this amplifier to be constructed, an artwork master would be created for each circuit substrate, which would be fabricated separately (paying especially careful attention to maintaining the coupler gap dimensions).

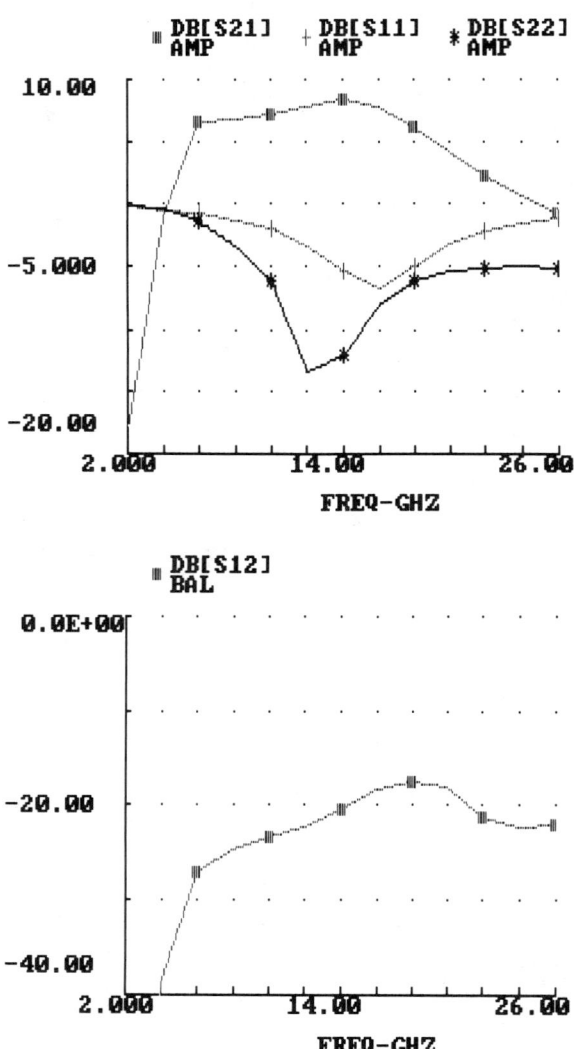

Fig. 8.3 Final simulated performance of the 6 to 18 GHz balanced amplifier.

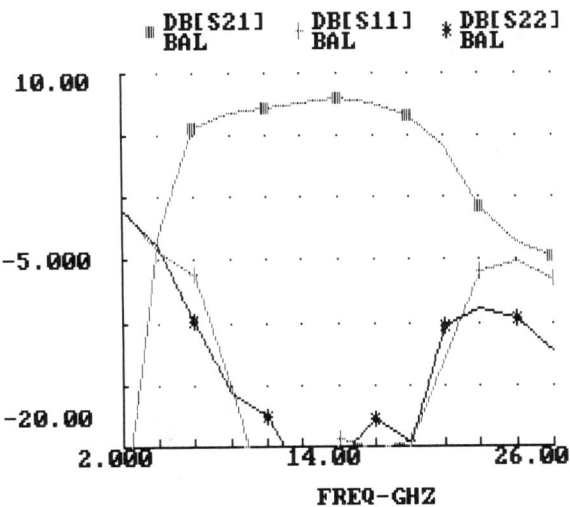

Fig. 8.3 continued.

During assembly, the input and the output substrates would be brazed to the carrier with the metal rib positioned between them. The brazing, die attaching, and bonding processes would follow the plan already discussed in Chapter 7.

8.2 THE 2 TO 6 GHz TWO-STAGE LOSSY MATCH AMPLIFIER FINAL LAYOUT

The 2 to 6 GHz lossy match amplifier is fabricated on three alumina substrates (each 10 mils thick) separated by two metal grounding ribs. The first substrate is the input matching circuit for the first stage. The second substrate is an interstage matching circuit; and the third substrate is the output matching circuit for the second stage.

Many kinds of parasitic elements arise in the initial layout, which is shown in Figure 8.4. These elements include

- microstrip tees,
- microstrip bends,
- bond wires,
- capacitor pads acting like parasitic open stubs.

A circuit diagram including all of the layout parasitic elements is given in Figure 8.5. A new Touchstone file including all of the new elements is shown in Table 8.2.

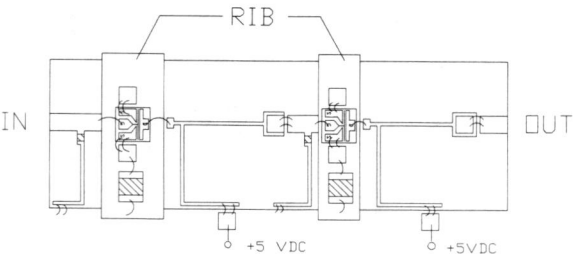

Fig. 8.4 The layout of the lossy match 2 to 6 GHz amplifier.

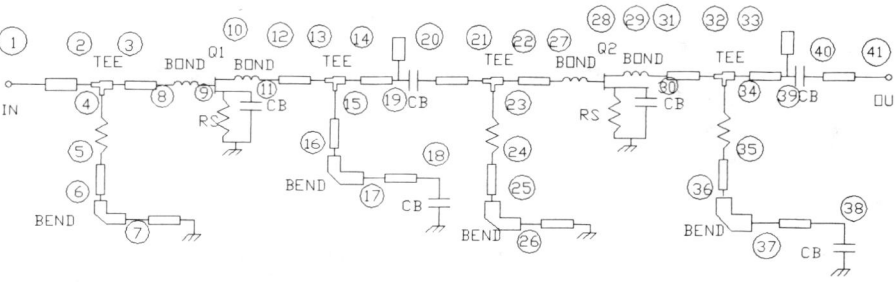

Fig. 8.5 The final schematic diagram of the 2 to 6 GHz lossy match amplifier including all layout parasitic elements.

Table 8.2 Touchstone—Ver(1.30–May-17-85)—Ser(15738-1598-1000)—Con(1CCC22124P) 380.CKT 07-26-89 11:08:51

```
! THE TOPOLOGY OF A TWO STAGE 2 TO 6 GHZ LOSSEY MATCH AMPLIFIER, INCLUDING
! ALL LAYOUT PARASITIC ELEMENTS.
! THE GaAs FETS ARE LITTON/DEXCEL DXL3501'S
CKT
      MSUB ER=9.9 H=10 T=.2 RHO=5 RGH=0
      MLIN 1 2 W=10 L\11.17452
      MTEE 2 3 4 W1=10 W2=10 W3=2.0
      TFR 4 5 W=2.0 L\0.29028 RS=50 F=0
      MLIN 5 6 W=1.6 L\14.94141
      MBEND 6 7 W=1.6 ANG=90 M=0
      MLSC 4 W=1.63410    L\77.03024
      MLIN 3 8 W=10 L\31.99817
      IND 8 9 L=.15
      S2PA 9 10 11   DX3501
      IND 10 12 L=.15
      MLIN 12 13 W=1 L\4.94162
      RES 11 0 R=20
      CAP 11 0 C=100
```

Table 8.2 cont'd.

```
MTEE 13 14 15 W1=1 W2=1.5 W3=1
MLIN 15 16 W=1 L\212.68260
MBEND 16 17 W=1 ANG=90 M=0
MLIN 17 18 W=1 L\120.47170
CAP 18 0 C=50
MLIN 14 19 W=1.5 L\24.88184
MLEF 19 W=20 L=20
CAP 19 20 C=50
MLIN 20 21 W=10 L\21.09626
MTEE 21 22 23 W1=10 W2=10 W3=2
TFR 23 24 W=2 L\2.27900 RS=50 F=0
MLIN 24 25 W=1 L\284.31050
MBEND 25 26 W=1 ANG=90 M=0
MLSC 26 W=1 L\48.15461
MLIN 22 27 W=10 L\8.62156
IND 27 28 L=.15
S2PA 28 29 30   DX3501
RES 30 0 R=20
CAP 30 0 C=100
IND 29 31 L=.15
MLIN 31 32 W=1 L\34.44980
MTEE 32 33 34 W1=1 W2=1 W3=2
TFR 34 35 W=2 L=2         RS=50 F=0
MLIN 35 36 W=1 L\167.10250
MBEND 36 37 W=1 ANG=90 M=0
MLIN 37 38 W=1 L\0.34895
CAP 38 0 C=50
MLIN 33 39 W=1 L\164.77850
MLEF 39 W=20 L=20
CAP 39 40 C=50
MLIN 40 41 W=10 L\7.93163
DEF2P 1 41   AMP

FREQ

    SWEEP 1 8 .5

OUT

    AMP DB[S21] GR1
    AMP DB[S11] GR1
    AMP DB[S22] GR1

GRID

    RANGE 1 8  .5
    GR1 -20 20 5

OPT

    RANGE 2 6.5
    AMP DB[S21] =10      40
    AMP DB[S11] <-10     10
    AMP DB[S22] <-10     10
```

This circuit required a fair amount of reoptimization to regain the same level of simulated performance as the ideal circuit. The final optimized simulation is shown in Figure 8.6. The optimized circuit element values are given in Table 8.2.

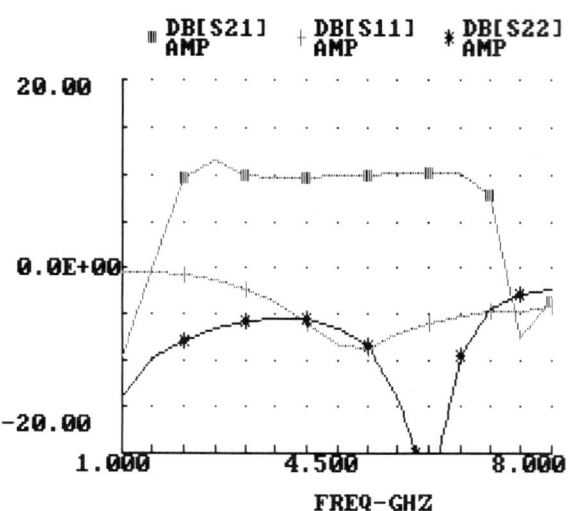

Fig. 8.6 Final simulated performance of the 2 to 6 GHz lossy match amplifier.

8.3 TWO-STAGE 4.5 TO 5.0 GHz LOW-NOISE AMPLIFIER FINAL LAYOUT

The layout of the 4.5 to 5.0 GHz low-noise amplifier is unique among the design examples in that it is intended for fabrication on a low-dielectric Teflon-Fiberglas board. The commercial name for this type of board material is Duroid™. Duroid™ comes in various thicknesses and usually is metalized with copper. This low-noise amplifier uses 25 mil thick Duroid™ material.

Both stages are fabricated onto a single circuit board. The GaAs FETs are used in packaged form to allow presorting for noise figure. The source contacts of the FETs are grounded through via holes in the board. Gate and drain bias voltage is applied to each stage through four identical high-impedance, low-impedance choke networks. The amplifier's layout is shown in Figure 8.7.

The major layout parasitic elements are bends, tees, and crosses as shown in the final schematic diagram in Figure 8.8. Because the GaAs FETs and chip capacitors are soldered in place (rather than wire bonded, as is the case on alumina circuits), no bond wire parasitic elements appear in this circuit.

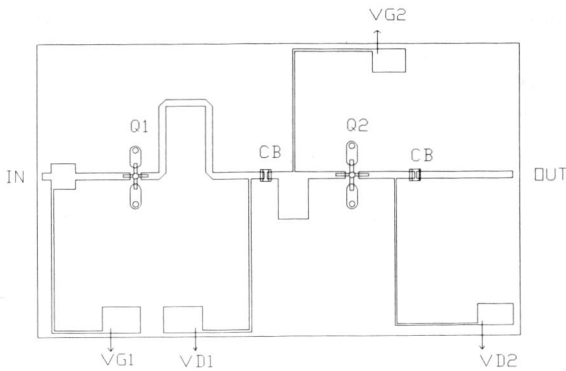

Fig. 8.7 The layout of the two-stage low-noise amplifier for 4.5 to 5.0 GHz.

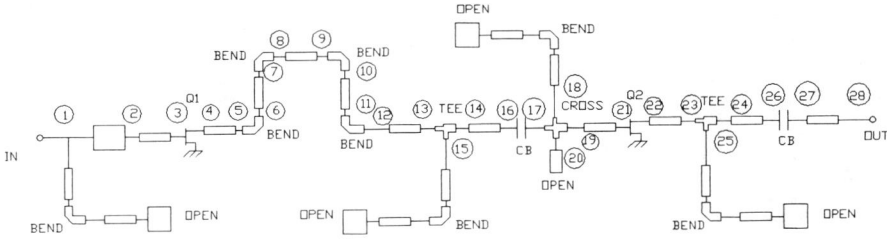

Fig. 8.8 The final schematic diagram of the two-stage 4.5 to 5.0 GHz low-noise amplifier including all layout parasitic elements.

A final Touchstone file, shown in Table 8.3, is created from the schematic diagram. The circuit element dimensions are adjusted by further optimization to account for the layout parasitic elements. Nearly all of the circuit's basic performance is recovered by this process. The final Touchstone simulations are shown in Figure 8.9.

8.4 THE 4 TO 10 GHz ONE WATT AMPLIFIER FINAL LAYOUT

The 1 W, 4 to 10 GHz amplifier's layout, shown in Figure 8.10, is very similar to the layout of the 6 to 18 GHz balanced amplifier described in Section 8.1. The circuit is divided into input and output substrates (25 mil thick alumina), and the power FETs are mounted on a metal grounding rib that separates the substrates. Bypass

Table 8.3 Touchstone—Ver(1.30–May-17-85)—Ser(15738-1598-1000)—
Con(1CCC22124P) 392.CKT 08-11-89 01:05:48

```
! THE FINAL TOPOLOGY OF A THE TWO STAGE LOW NOISE AMPLIFIER FOR 4.5 TO 5.0 GHZ
! INCLUDING ALL LAYOUT PARASITIC ELEMENTS.
! THE GaAs FETS ARE NE04583.

CKT                                              ! CASCADE THREE STAGES

    MSUB ER=2.2 H=25 T=1 RHO=1 RGH=0                 NAIN 1 2
    MLIN 1 2 W=5.0        L\589.44340                NA2P 2 3
    MBEND 2 3 W=5 ANG=90 M=0                         NAOUT 3 4
    MLIN 3 4 W=5 L\189.91890                         NBIN 4 5
    MLEF 4    W\31.21907  L\93.68091                 NB2P 5 6
    DEF1P 1   BIAS                                   NBOUT 6 7
                                                     DEF2P 1 7    AMP
! FIRST STAGE
                                                 FREQ
    BIAS 1
    MLIN 1 2 W\270.14390 L\166.88640                 SWEEP 4.0   5.5   .1
    MLIN 2 3 W=5.0       L\133.77400
    DEF2P 1 3 NAIN          ! INPUT NETWORK      OUT

    S2PA 3 4 0    NE04583                            AMP DB[S21]   GR1
    DEF2P 3 4   NA2P       ! FET #1                  AMP DB[NF]    GR2
                                                     AMP DB[S11]   GR3
    MLIN 4 5 W=35 L\132.42040                        AMP DB[S22]   GR3
    MBEND 5 6 W=35 ANG=90 M=0
    MLIN 6 7 W=35 L\128.89780                    GRID
    MBEND 7 8 W=35 ANG=90 M=0
    MLIN 8 9 W=35 L\1.52298                          RANGE 4.3   5.2   .1
    MBEND 9 10 W=35 ANG=90 M=0                       GR1 0 25 5
    MLIN 10 11 W=35 L\91.12880                       GR2 0 5 .5
    MBEND 11 12 W=35 ANG=90 M=0                      GR3 -30 0 5
    MLIN 12 13 W=35 L\52.18962
    MTEE 13 14 15 W1=35 W2=35 W3=5               OPT
    BIAS 15
    MLIN 14 16 W=35 L\63.97654                       RANGE 4.3   5.2
    CAP 16 17 C=50                                   AMP DB[NF]<1.5      50
    DEF2P 4 17   NAOUT     ! OUTPUT NETWORK          AMP DB[S21]=20.0    40
                                                     AMP DB[S11]<-10     10
! SECOND STAGE                                       AMP DB[S22]<-15     10

    MCROS 17 18 19 20 W1=35 W2=5 W3=11.39 W4=54.03
    BIAS 18
    MLEF 20 W=54.03136    L\21.24822
    MLIN 19 21 W=11.39021 L\129.46620
    DEF2P 17 21   NBIN     ! INPUT NETWORK

    S2PA 21 22   0    NE04583
    DEF2P 21 22   NB2P      ! FET #2


    MLIN 22 23 W=5.0      L\15.29126
    MTEE 23 24 25 W1=5.0     W2=5.0    W3=5
    BIAS 25
    MLIN 24 26 W=5.0             L\77.65469
    CAP 26 27 C=50
    MLIN 27 28 W=5.0      L\27.72465
    DEF2P 22 28   NBOUT     ! OUTPUT NETWORK
```

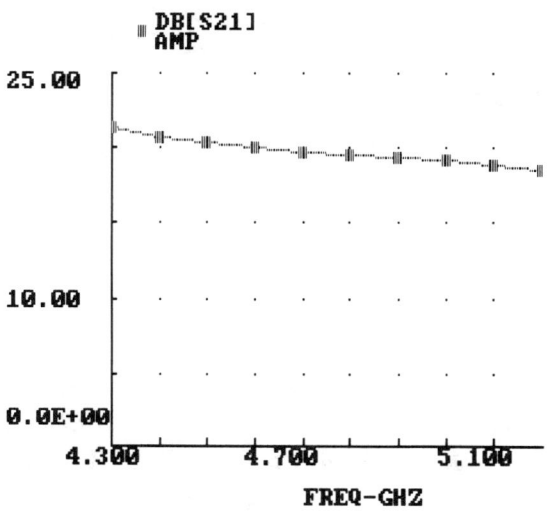

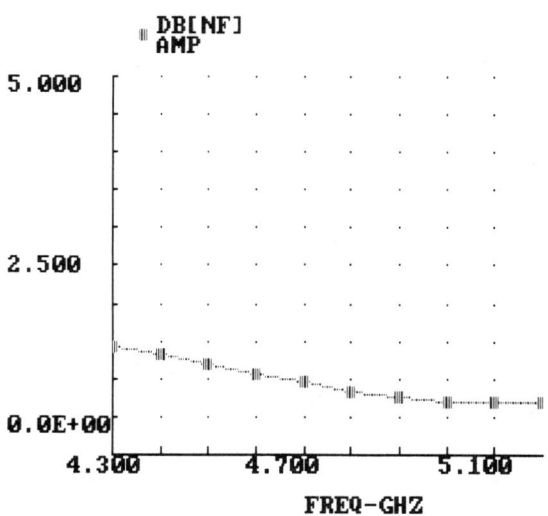

Fig. 8.9 Final simulated performance of the 4.5 to 5.0 GHz low-noise amplifier.

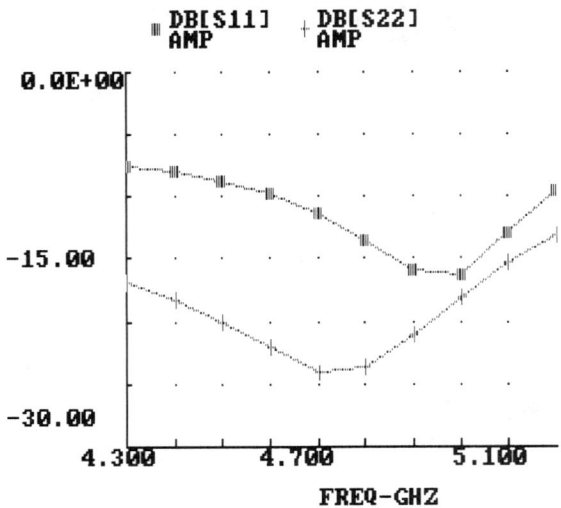

Fig. 8.9 continued.

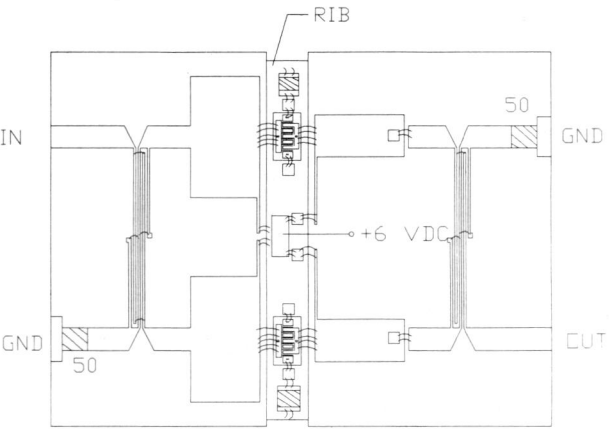

Fig. 8.10 Layout of a 4 to 10 GHz 1 W amplifier.

capacitors (100 pF) and source resistors are mounted onto the rib in close proximity to the FETs. Because the power FETs have such large contact pads, a large number (up to five) bond wires connect it into and out of the circuit. The advantage of multiple bond wires is reduced wire inductance. Both substrates contain Lange couplers and deposited termination resistors. The coupler's .8 mil gaps present the greatest challenge to the photolithic fabrication but they are well within the capabilities of present technology.

For thermal reasons, it is very important that the power FET chips be mounted on a gold-plated copper grounding rib. Because each chip dissipates about 2 W of heat, each increase of 1°C/W in thermal resistance will increase the chips active area temperature by 2°C. See Section 7.4 for details on thermal design.

The layout is very straightforward and contains only a few parasitic elements. These are steps, bond wires, and thin-film resistors. The amplifier's schematic diagram including these layout parasitic elements, is shown in Figure 8.11. Table 8.4 shows the final Touchstone file with the optimized element values accounting for layout parasitics. The final performance simulations are given in Figure 8.12. Nearly all of the ideal performance is preserved with the final layout as a direct result of its simplicity. Whenever possible, simplest *always* is best.

8.5 THE 6 TO 18 GHz BALANCED PIN ATTENUATOR FINAL LAYOUT

Like the balance amplifiers already discussed, the balanced PIN attenuator's layout uses two alumina substrates (each 10 mils thick) separated by a metal grounding rib. The PIN diode chips are die attached to this grounding rib. This very simple, straightforward layout contains few parasitic elements. Therefore, it differs from the basic topology only by the inclusion of bond wires and 50 Ω lines connecting the Lange coupler to the PIN diodes and the deposited terminations.

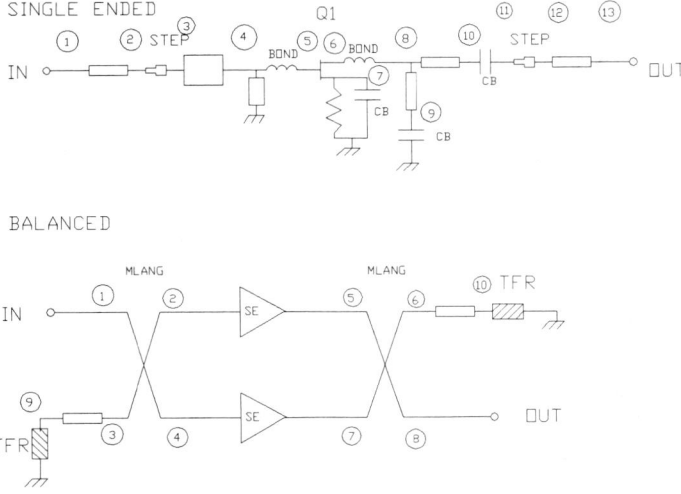

Fig. 8.11 The final schematic diagram of the 4 to 10 GHz 1 W amplifier, including layout parasitics.

Table 8.4 Touchstone—Ver(1.30–May-17-85)—Ser(15738-1598-1000)—
Con(1CCC22124P) 381.CKT 07-26-89 11:44:23

```
! THE FINAL TOPOLOGY OF THE 4 TO 10 GHZ BALANCED POWER AMPLIFIER,INCLUDING ALL
! LAYOUT PARASITIC ELEMENTS.
! THE GaAs FETS ARE MITSUBISHI MGF2116'S.
! LARGE SIGNAL S PARMETERS ARE CALCULATED USING THE LOAD LINE METHOD.
CKT                                          OPT
    ! SINGLE ENDED
    MSUB ER=9.9 H=25 T=.2 RHO=5 RGH=0        RANGE 4 10
    MLIN 1 2 W=25          L\4.85575         BAL DB[S21]=6   50
    MSTEP 2 3 W1=25 W2=80                    BAL DB[S11]<-15  10
    MLIN 3 4 W=80 L\4.64629                  BAL DB[S22]<-15  10
    MLSC 4 W=2.0      L\157.63930            AMP DB[S22]<-7   30
    IND 4 5 L=.10
    S2PA 5 6 7  MG2116L
    IND 6 8 L=.10
    CAP 7 0 C=200
    RES 7 0 R=10
    MLIN 8 9 W=2.04447 L\30.37491
    CAP 9 0 C=50
    MLIN 8 10 W=50        L\109.99360
    CAP 10 11 C=50
    MSTEP 11 12 W1=50 W2=25
    MLIN 12 13 W=25 L\31.33295
    DEF2P 1 13 AMP

    ! BALANCED
    MLANG 1 2 3 4   W=1.2  S=.8  L=170
    MLANG 5 6 7 8   W=1.2  S=.8  L=170
    AMP 2 5
    AMP 4 7
    MLIN 3 9 W=25 L=30
    TFR 9 0 W=25 L=25 RS=50 F=0
    MLIN 6 10 W=25 L=30
    TFR 10 0 W=25 L=25 RS=50 F=0
    DEF2P 1 8   BAL

FREQ

    SWEEP 3 11 1

OUT

    AMP DB[S21] GR1
    AMP DB[S11] GR1
    AMP DB[S22] GR1
    BAL DB[S21] GR2
    BAL DB[S11] GR2
    BAL DB[S22] GR2

GRID

    RANGE 3 11 1
    GR1 -15 15 5
    GR2 -30 15 5
```

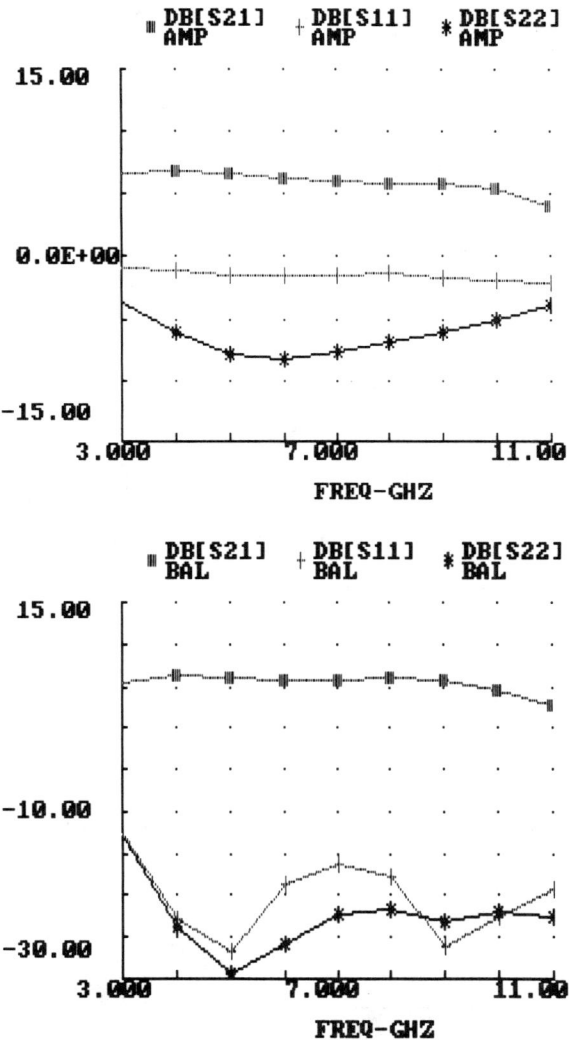

Fig. 8.12 Final simulated performance of the 4 to 10 GHz 1 W power amplifier.

The attenuator's layout is shown in Figure 8.13. The final schematic diagram is given in Figure 8.14. The final Touchstone file is given in Table 8.5, and the final simulated performance, including layout parasitic elements is shown in Figure 8.15.

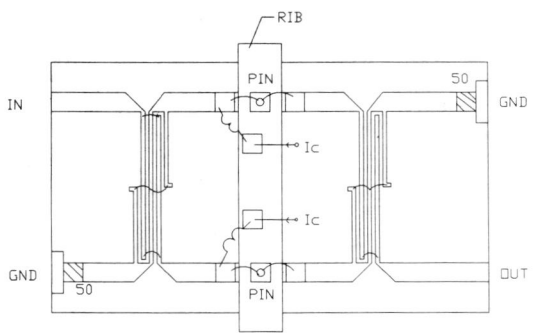

Fig. 8.13 The layout of a 6 to 18 GHz balanced PIN diode attenuator.

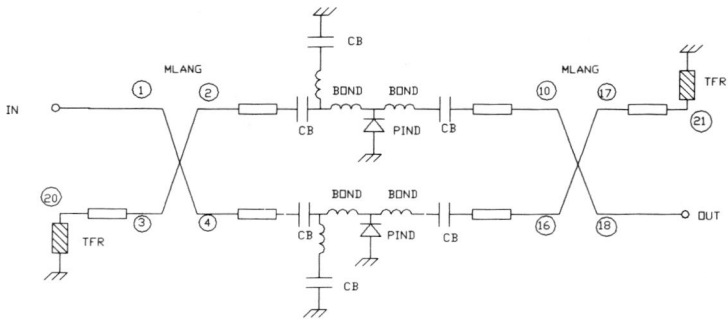

Fig. 8.14 The final schematic diagram of the 6 to 18 GHz balanced PIN diode attenuator, including layout parasitic elements.

Table 8.5 Touchstone—Ver(1.30–May-17-85)—Ser(15738-1598-1000)—Con(1CCC22124P) 382.CKT 07-26-89 22:29:23

```
! THE FINAL TOPOLOGY OF A 6 TO 18 GHZ BALANCED PIN ATTENUATOR, INCLUDING LAYOUT
! PARASITIC ELEMENTS.

VAR
    RS=.50
    RD=2
    CD=.08
    L1=50
    WC=1.4
    SC=.8
    LC=80
    CB=50
```

Table 8.5 cont'd.

```
CKT

    MSUB ER=9.9 H=10 T=.15 RHO=5 RGH=0

    ! PIN DIODE MODEL
    RES 1 2 R^RS
    PRC 2 3 R^RD C^CD
    DEF2P 1 3 PIND

    ! SINGLE ENDED ATTENUATOR

    CAP 5 6 C^CB
    MLIN 2 5 W=10 L^L1
    IND 6 7 L=.15
    PIND 7 0
    IND 7 19 L=16
    CAP 19 0 C^CB
    IND 7 8 L=.15
    MLIN 9 10 W=10 L^L1
    CAP 8 9 C^CB
    DEF2P 2 10 SE

    ! BALANCED ATTENUATOR

    MLANG 1 2 3 4 W^WC S^SC L^LC
    MLIN 3 20 W=10 L=25
    TFR 20 0 W=10 L=10 RS=50 F=0
    SE 2 10
    SE 4 16
    MLANG 10 17 16 18 W^WC S^SC L^LC
    MLIN 17 21 W=10 L=25
    TFR 21 0 W=10 L=10 RS=50 F=0
    DEF2P 1 18 BAL

FREQ

    SWEEP 5 20 1

OUT

    BAL DB[S21] GR1
    BAL DB[S11] GR2
    BAL DB[S22] GR2

GRID

    RANGE 5 20 1
    GR1 -30 0 5
    GR2 -30 0 5
```

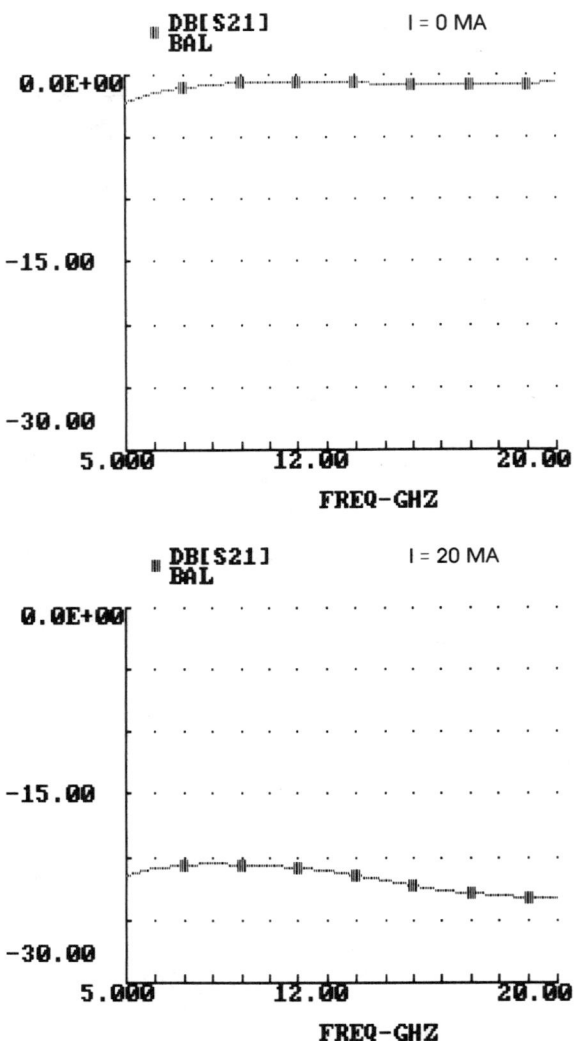

Fig. 8.15 Final simulated performance of the 6 to 18 GHz balanced pin attenuator.

8.6 THE 12 TO 18 GHz FET VARACTOR VCO FINAL LAYOUT

The 12 to 18 GHz FET varactor VCO is fabricated on a single 10 mil thick alumina board. The varactor diode chip is die attached to a large value (at least 100 pF) blocking capacitor, which is mounted off, but next to, the alumina board. Both the alumina board and the varactor-blocking capacitor are brazed to a common gold-plated Kovar carrier.

The GaAs FET chip is die attached to a metallized pad on the alumina board that acts as the source feedback capacitor. Therefore, the GaAs FET's sources are bonded to this pad to complete the circuit. The gate bias choke and the drain bias choke are simple long-bond wires of the proper length. These wire chokes may be modeled on Touchstone as wire elements if they are suspended above the circuit, or they may be treated as microstrip transmission line elements if they are in close contact with the substrate. We will model them as *wire* elements in this example.

The layout of the 12 to 18 GHz VCO is shown in Figure 8.16. The VCO's final schematic diagram is given in Figure 8.17. The principal layout parasitic elements in this circuit are bond wires, bends, and steps. A final Touchstone file, created from the schematic diagram, is shown in Table 8.6. The circuit elements are adjusted by the Touchstone optimizer to account for the effects of the layout parasitics. Figure 8.18 gives the final simulated performance, which is very similar to the performance of the basic circuit.

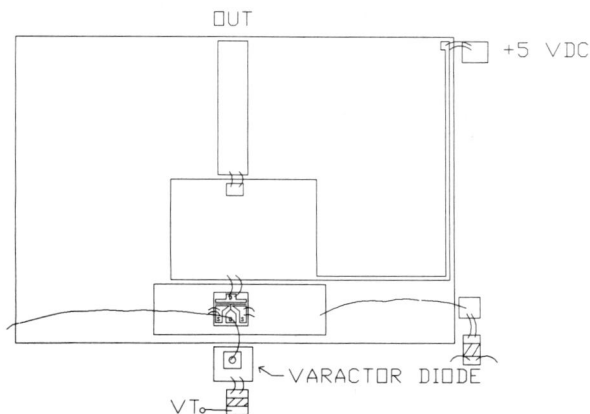

Fig. 8.16 The layout of the 12 to 18 GHz FET-varactor VCO.

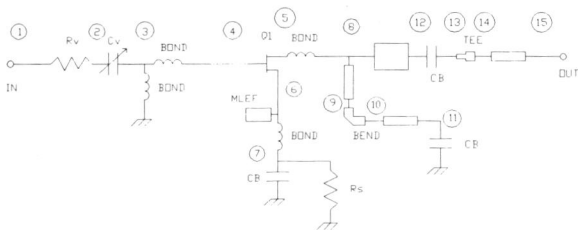

Fig. 8.17 The final schematic diagram of the 12 to 18 GHz FET-varactor VCO.

Table 8.6 Touchstone—Ver(1.30–May-17-85)—Ser(15738-1598-1000)—Con(1CCC22124P) 393.CKT 08-11-89 13:05:49

```
! FINAL TOPOLOGY FOR 12 TO 18 GHZ FET \ VARACTOR  VCO CIRCUIT INCLUDING ALL
! LAYOUT PARASITIC ELEMENTS.

VAR

   R1=2
   CV=1

CKT

   MSUB ER=9.9 H=10 T=.2 RHO=5 RGH=0
   RES 1 2 R^R1
   CAP 2 3 C^CV
   WIRE 3 0 D=.7 L=280.51290 RHO=1.0
   WIRE 3 4 D=.7 L=23.99218 RHO=1.0
   S2PA 4 5 6   NEC71000
   MLEF 6 W=40        L=25
   CAP 7 0 C=50
   RES 7 0 R=20
   WIRE 5 8 D=.7 L=17.98854 RHO=1.0
   MLIN 8 9 W\6.72440 L\124.10560
   MBEND 9 10 W=5 ANG=90 M=0
   MLIN 10 11 W\8.36883 L\10.12712
   CAP 11 0 C=50
   MLIN 8 12 W=113.94130 L=158.16200
   CAP 12 13 C=50
   MSTEP 13 14 W1=113.9 W2=10
   MLIN 14 15 W=10 L=10.25656
   DEF2P 1 15   VCO

FREQ

   SWEEP 10 20 1

OUT

   VCO DB[S21] GR1
   VCO RE[Z1] GR2
   VCO IM[Z1] GR2
   VCO   S11

GRID

   RANGE 10 20 1
   GR1 -10 10 5
   GR2 -100 100 10

OPT

   RANGE 12 18
   VCO RE[Z1] < -20    50
   RANGE 12 15
   VCO IM[Z1] <0.0     30
   RANGE 15 18
   VCO IM[Z1] >0.0     30
```

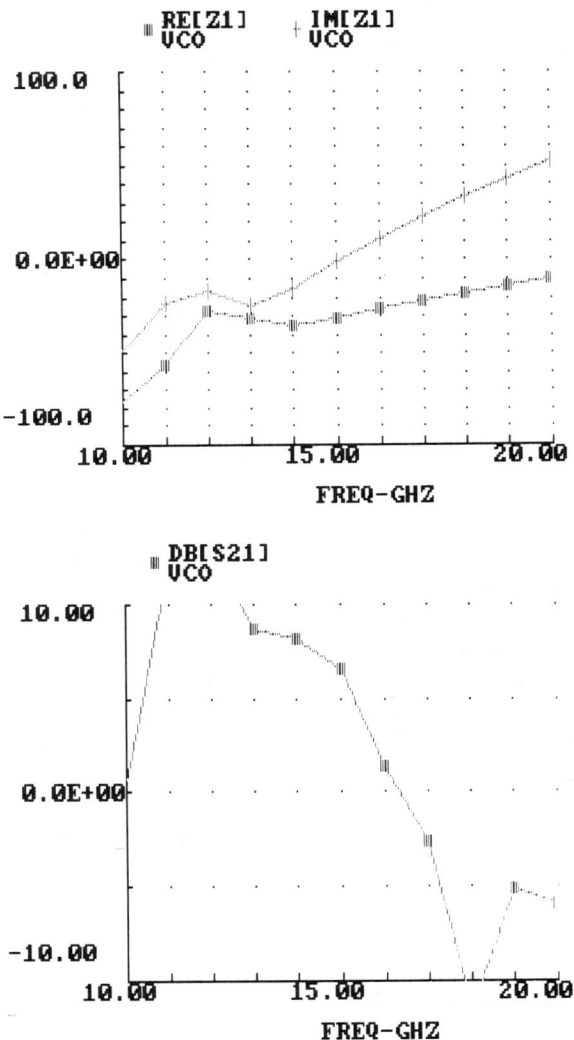

Fig. 8.18 Final simulated performance of the 12 to 18 GHz FET-varactor VCO.

8.7 THE 4 TO 8 GHz BIPOLAR VARACTOR VCO FINAL LAYOUT

The layout of the 4 to 8 GHz bipolar varactor VCO is shown in Figure 8.19. Its layout and construction techniques are very similar to previous examples. The circuit board is 10 mil alumina material. The varactor diode chip is attached to a

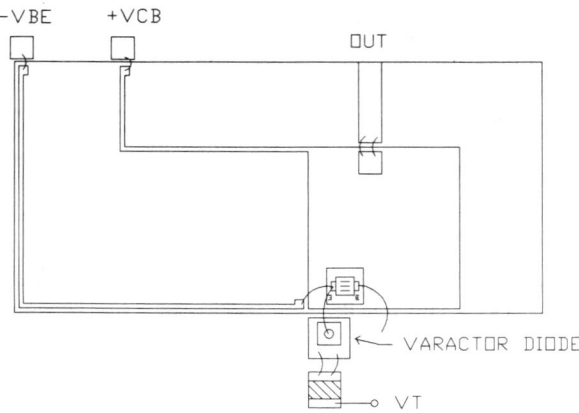

Fig. 8.19 The layout of a 4 to 8 GHz bipolar VCO.

blocking capacitor, which is mounted next to the circuit board. Both the board and blocking capacitor are brazed to a common gold-plated Kovar carrier. The bipolar transistor chip is die attached to a gold pad on the circuit board. Both the board and the blocking capacitor are brazed to a common gold-plated Kovar carrier. The bipolar transitor chip is die attached to a gold pad on the circuit board. The varactor is connected to the transistor by a short bond wire. The feedback inductor is formed by another bond wire that runs from the transistor's base pad to ground.

The circuit's principal layout parasitic elements are bond wires and bends. The VCO's final schematic diagram, including these parasitic elements is shown in Figure 8.20. A Touchstone file created from this diagram is shown in Table 8.7. The circuit elements are adjusted by the Touchstone optimizer to account for the effect of the parasitic elements. Figure 8.21 gives the final simulated performance for the 4 to 8 GHz VCO. The final performance is very similar to the performance of the basic circuit.

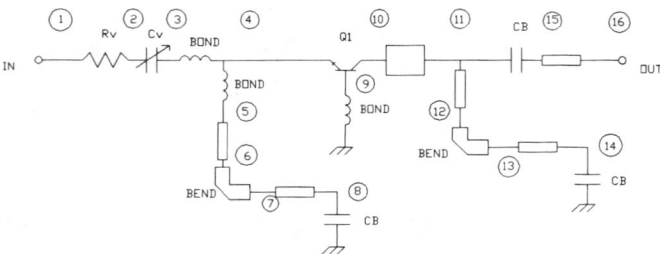

Fig. 8.20 The final schematic diagram of the 4 to 8 GHz bipolar VCO including all parasitic elements.

Table 8.7 Touchstone—Ver(1.30–May-17-85)—Ser(15738-1598-1000)—
Con(1CCC22124P) 386.CKT 07-28-89 16:27.59

```
! THE FINAL TOPOLOGY OF THE 4 TO 8 GHZ BIPOLAR VCO INCLUDING BOND WIRES AND
! LAYOUT PARASITICS.

VAR

    RV=2.0
    CV=.50

CKT

    MSUB ER=9.9 H=10 T=.2 RHO=5  RGH=0

    ! OSCILLATOR CIRCUIT

    RES 1 2 R^RV
    CAP 2 3 C^CV
    WIRE 3 4 D=.7 L=15 RHO=1
    WIRE 4 5 D=.7 L=15 RHO=1
    MLIN 5 6 W=2 L\126.46700
    MBEND 6 7 W=2 ANG=90 M=0
    MLIN 7 8 W=2 L\21.08020
    CAP 8 0 C=50
    S2PA 9 10 4  NE64400
    WIRE 9 0 D=.7 L\55.51556 RHO=1
    MLIN 10 11 W=70         L\27.40002
    MLIN 11 12 W=2 L\11.42306
    MBEND 12 13 W=2 ANG=90 M=0
    MLIN 13 14 W=2 L\70.48246
    CAP 14 0 C=50
    CAP 11 15 C=50
    MLIN 15 16  W=10.0     L\24.18758
    DEF2P 1 16 OSC

FREQ

    SWEEP 3 9  1

OUT

    OSC RE[Z1] GR1
    OSC IM[Z1] GR1
    OSC DB[S21] GR2
GRID

    RANGE 3 9  1
    GR1 -100 100 10
    GR2 -20 20 5


OPT

    RANGE 4 9
    OSC RE[Z1]<-30  50
    OSC IM[Z1]>10   30
    OSC DB[S21]>0   20
```

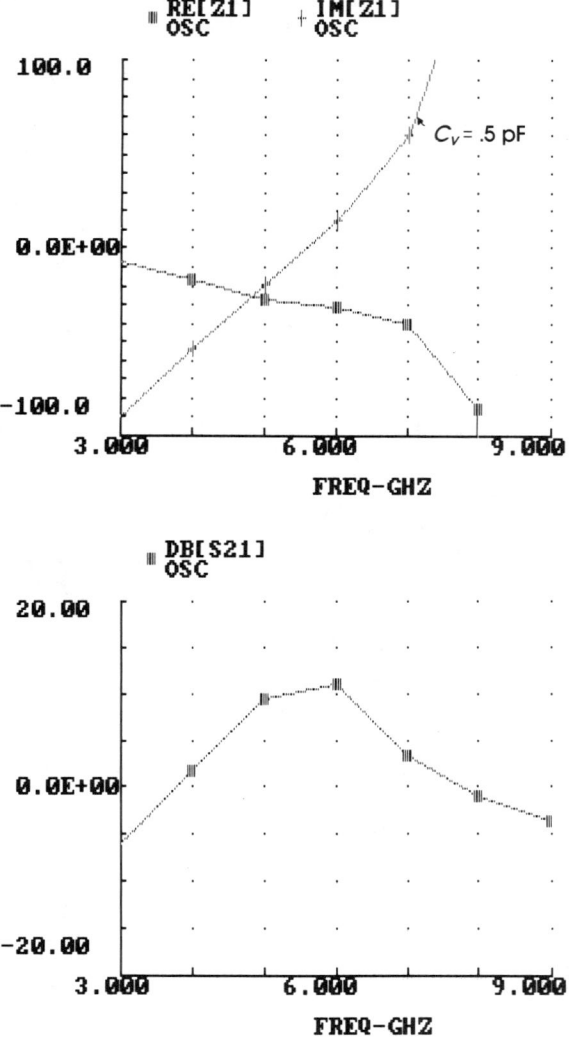

Fig. 8.21 Final simulated performance of the 4 to 8 GHz bipolar/varactor VCO.

8.8 THE 2 TO 8 GHz BIPOLAR YIG-TUNED OSCILLATOR FINAL LAYOUT

The final layout for the 2 to 8 GHz YIG-tuned bipolar oscillator example is shown in Figure 8.22. The oscillator circuit uses a single 10 mil thick alumina board. The YIG

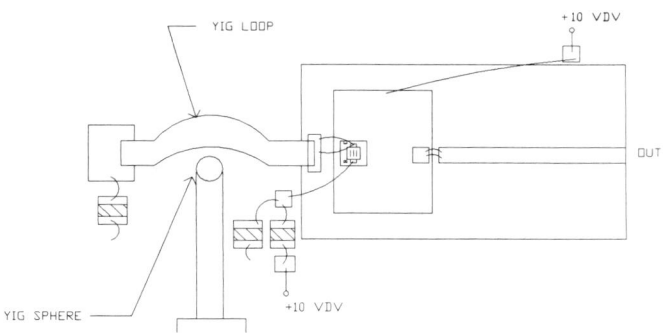

Fig. 8.22 The layout of a 2 to 8 GHz bipolar YIG-tuned oscillator.

loop (usually 1/2 turn 10 mil gold ribbon) is welded to a pad on the far left-hand side of the circuit board and extends up and around the YIG sphere. The other end of the loop ribbon is welded to a bypass capacitor. The YIG sphere is fixed to the end of a ceramic rod, which is held in a heater or temperature controller. The YIG sphere and its rod may be rotated and moved back and forth within the coupling loop to allow setting the sphere on its temperature stable axis and optimize the coupling between the sphere and the coupling loop.

The bipolar transistor is die attached to a gold pad on the circuit board, which we will model as an open stub. The backside of the transistor is the collector, so the contact pad is an important part of the oscillator output matching network. The feedback inductor is simply a bond wire between the transistor's base pad and ground. A second bond wire between the collector pad and a bypass capacitor forms the collector bias choke. Bond wires between the transistor's emitter pad and the pad to which the coupling loop is attached, connects the transistor to the YIG resonator.

The circuit's layout parasitic elements are just the bond wires. The final schematic diagram of the 2 to 8 GHz YTO is given in Figure 8.23. A Touchstone file created from this diagram is shown in Table 8.8. The circuit elements are adjusted by the Touchstone optimizer to account for the effects of the parasitic elements.

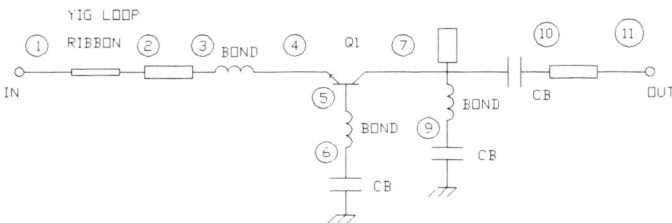

Fig. 8.23 The final schematic diagram of the 2 to 8 GHz bipolar YTO, including all layout parasitic elements.

Table 8.8 Touchstone—Ver(1.30–May-17-85)—Ser(15738-1598-1000)—Con(1CCC22124P) 387.CKT 07-28-89 19:24:43

```
! THE FINAL TOPOLOGY OF THE 2 TO 8 GHZ YIG TUNED BIPOLAR OSCILLATOR USING
! COMMON BASE\INDUCTIVE FEEDBACK.
! MODEL INCLUDES ALL DISCONTINUITIES AND WIRES.


CKT

    MSUB ER=9.9 H=10 T=.2 RHO=1 RGH=0
    RIBBON 1 2 W=5 L=100 RHO=1
    MLIN 2 3 W=20 L=10
    WIRE 3 4 D=1.4 L=15 RHO=1
    S2PA 5 7 4 NEC56710
    WIRE 5 6 D=.7 L\39.14231 RHO=1
    CAP 6 0 C=50
    MLEF 7    W\19.82010 L\108.09470
    WIRE 7 9 D=.7 L\149.00550       RHO=1
    CAP 9 0 C=50
    CAP 7 10 C=50
    MLIN 10 11 W=10 L=100
    DEF2P 1 11 OSC


FREQ
SWEEP 1 10 1

OUT

    OSC RE[Z1]   GR1
    OSC IM[Z1]   GR1
    OSC DB[S21]  GR2

opt
range 2 8
OSC RE[Z1] <-20     50
OSC IM[Z1] > 20     10
OSC DB[S21] >0       5

GRID
RANGE 1 10 1
GR2 -10 10 5
GR1 -100 100 10
```

Figure 8.24 gives the final simulated performance of the 2 to 8 GHz YTO. The final simulation is very similar to the simulation of the basic circuit.

8.9 THE FINAL LAYOUT OF THE 6 TO 18 GHz FET YIG-TUNED OSCILLATOR

This YIG-tuned oscillator is fabricated on two separate 10 mil thick alumina substrates, as shown in Figure 8.25. The first substrate contains the oscillator GaAs

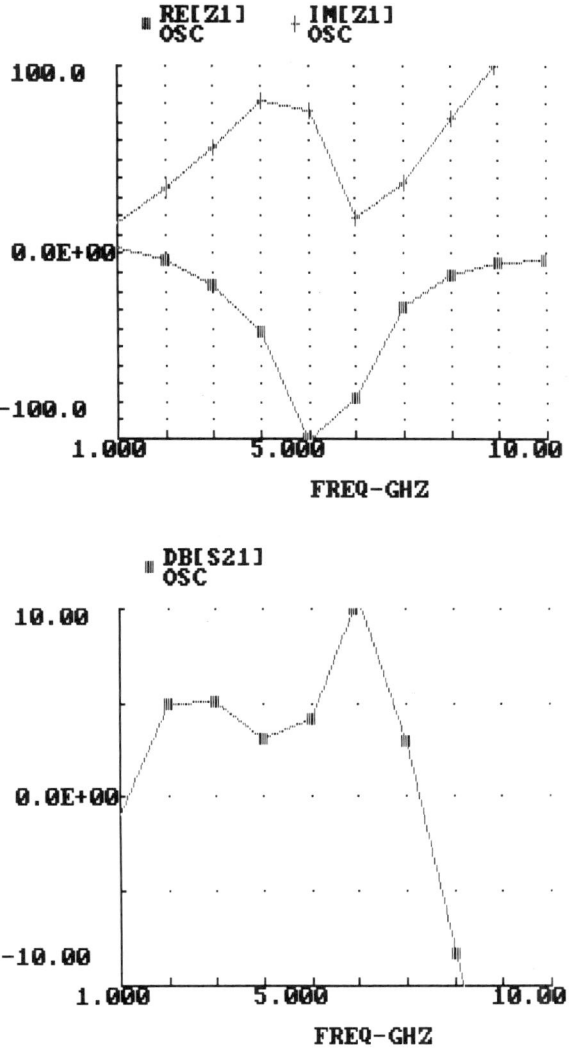

Fig. 8.24 Final simulated performance of the 2 to 8 GHz bipolar/YIG-tuned oscillator.

FET circuit and is composed mainly of a gold pad that serves as the source feedback capacitor, to which the oscillator FET is die attached. The second substrate contains the buffer amplifier's output circuit and is not much more than a 50 Ω line with a blocking capacitor. A metal grounding rib separates the two substrates. The buffer amplifier GaAs FET is die attached to the grounding rib.

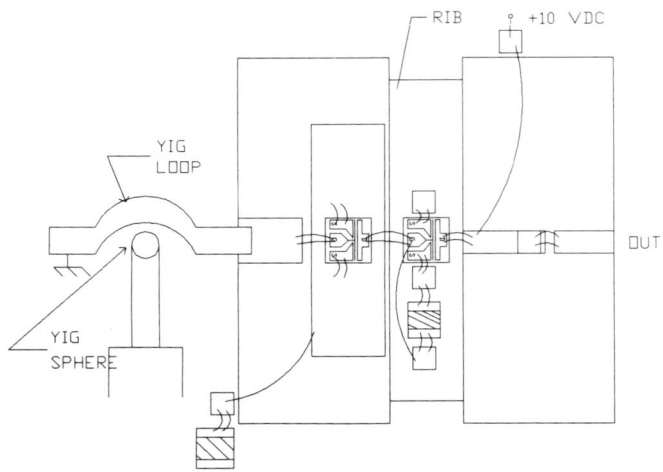

Fig. 8.25 The layout of a 6 to 18 GHz FET YIG-tuned oscillator.

As in the previous example, bond wires serve as bias chokes. The principal parasitic elements are the gold ribbon that serves as the YIG loop and the bond wires. It may be necessary to use a two- or three-turn YIG loop to increase the loop's inductance so that the circuit's self-resonance is moved below the low end of circuit's negative resistance band, to prevent spurious oscillations. As in the last example, the YIG sphere is mounted on a ceramic rod, which may be adjusted in position within the loop to control coupling and rotated to set the sphere on a temperature stable axis.

A final schematic diagram of the circuit is given in Figure 8.26. A Touchstone file created from this diagram is shown in Table 8.9. The circuit elements are adjusted by the Touchstone optimizer to account for the effects of the parasitic elements. The final simulation is given in Figure 8.27. The final simulated performance is quite similar to the performance of the basic circuit.

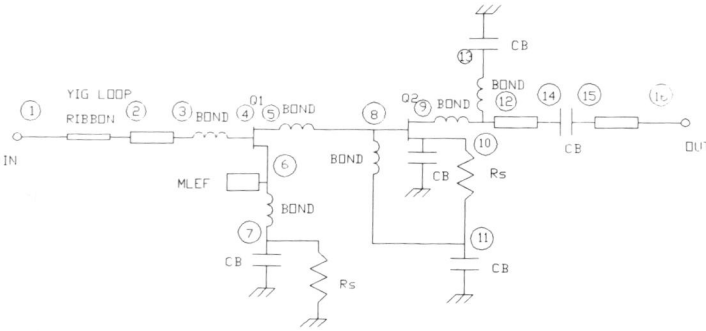

Fig. 8.26 The final schematic diagram of the 6 to 18 GHz FET YTO, including all layout parasitic elements.

Table 8.9 Touchstone—Ver(1.30–May-17-85)—Ser(15738-1598-1000)—
Con(1CCC22124P) 388.CKT 07-29-89 19:25:47

```
! THE FINAL TOPOLOGY OF THE 6 TO 18 GHZ YIG TUNED FET OSCILLATOR.
! ALL WIRES AND LAYOUT PARASITIC ELEMENTS ARE INCLUDED.
! ALL MATCHING ELEMENTS ARE DISTRIBUTED.


CKT
    MSUB ER=9.9 H=10 T=.2 RHO=5 RGH=0
    RIBBON 1 2 W=5 L=400 RHO=1
    MLIN 2 3 W=35      L\18.72797
    WIRE 3 4 D=1.4 L\44.31247 RHO=1.0
    S2PA 4 5 6   NEC71000
    MLEF 6 W\34.37332 L\31.31019
    WIRE 6 7 D=.7 L\256.37790 RHO=1.0
    CAP 7 0 C=50
    RES 7 0 R=20
    WIRE 5 8 D=1.4 L\32.68973 RHO=1.0
    S2PA 8 9 10 NEC71000
    WIRE 8 11 D=.7 L\212.68000 RHO=1.0
    RES 10 11 R=20
    CAP 11   0 C=50
    WIRE 9 12 D=1.4 L\5.27952 RHO=1.0
    WIRE 12 13 D=.7 L\189.97830 RHO=1.0
    CAP 13 0 C=50
    MLIN 12 14 W=10 L\27.20001
    CAP 14 15 C=50
    MLIN 15 16 W=10 L\23.18612
    DEF2P 1 16   OSC

FREQ
    SWEEP 2 19 1

OUT
    OSC RE[Z1]    GR1
    OSC IM[Z1]    GR1
    OSC DB[S21]   GR2
    OSC S11
    OSC S22

GRID
    RANGE 2 19 1
    GR1 -50 50 10
    GR2 -20 20 5

OPT

    RANGE 6 19
    OSC RE[Z1] <-20
    OSC DB[S21] >-5
    RANGE 5.0 5.1
    OSC IM[Z1]=0
```

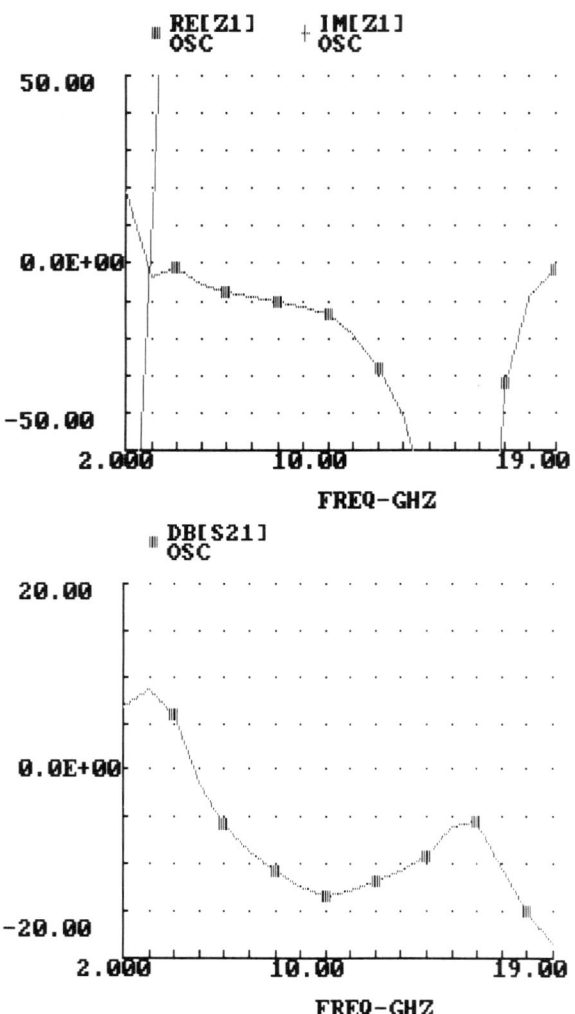

Fig. 8.27 Final simulated performance of the 6 to 18 GHz FET YIG-tuned oscillator.

8.10 DIELECTRICALLY TUNED 8.4 GHz OSCILLATOR FINAL LAYOUT

The layout of the dielectrically tuned 8.4 GHz FET oscillator is shown in Figure 8.28. The oscillator consists of two 10 mil thick alumina substrates: the first contains the resonator puck that is coupled to a deposited 50 Ω line; the second

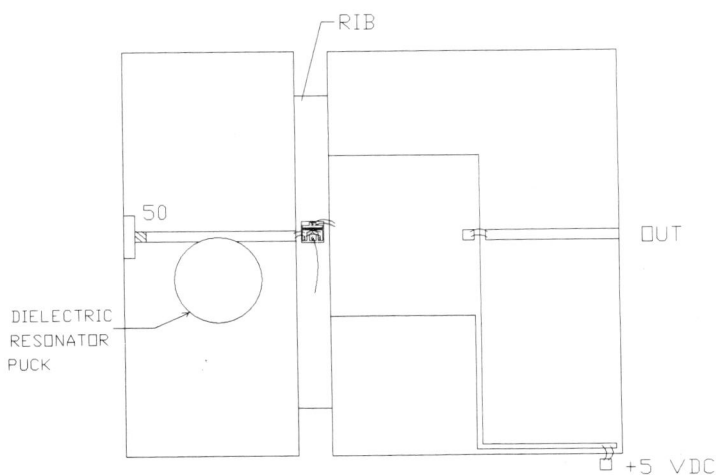

Fig. 8.28 The layout of a dielectrically tuned 8.4 GHz FET oscillator.

substrate contains the output matching elements. Between the two substrates is a metal grounding rib. The GaAs FET device is die attached to this ground rib.

A final schematic diagram for the dielectrically tuned oscillator is shown in Figure 8.29. The principal layout parasitic elements are bond wires and bends. The common gate feedback inductance is supplied by one of the bond wires. A Touchstone file based on the final schematic diagram is given in Table 8.10. The elements of the final circuit are adjusted by Touchstone's optimizer to achieve acceptable performance in the presence of the layout parasitic elements. The final simulated performance is given in Figure 8.30.

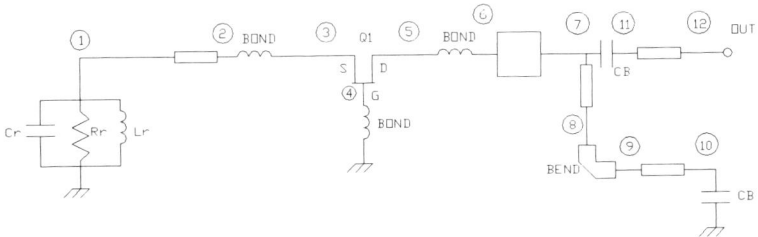

Fig. 8.29 The final schematic diagram of the 8.4 GHz DRO, including all layout parasitic elements.

Table 8.10 Touchstone—Ver(1.30–May-17-85)—Ser(15738-1598-1000)—
Con(1CCC22124P) 389.CKT 07-30-89 19:28:15

```
! THE FINAL TOPOLOGY OF AN 8.4 GHZ DIELECTRICLY TUNED FET OSCILLATOR USING
! COMMON GATE / INDUCTIVE FEEDBACK.
! MODEL INCLUDES ALL WIRES AND LAYOUT PARASITIC ELEMENTS.

CKT
    MSUB ER=9.9 H=10 T=.20 RHO=5 RGH=0
    PRLC 1 0 R=500 L=8.8E-4 C=412
    MLIN 1 2 W=10 L=10
    WIRE 2 3 D=1.4 L=10 RHO=1.0
    S2PA 4 5 3 NEC71000
    WIRE 4 0 D=.7 L\34.42538 RHO=1.0
    WIRE 5 6 D=1.4 L=10 RHO=1.0
    MLIN 6 7 W\11.66422 L\58.17238
    MLIN 7 8 W=2 L\19.56925
    MBEND 8 9 W=2 ANG=90 M=0
    MLIN 9 10 W=2 L\146.44980
    CAP 10 0 C=50
    CAP 7 11 C=50
    MLIN 11 12 W=10 L=50
    DEF1P 12 DRO

FREQ

    SWEEP 8.300 8.500 .01

OUT

    DRO DB[S11] GR1
    DRO RE[Z1] GR2
    DRO IM[Z1] GR2

GRID

    RANGE 8.300 8.500 .01
    GR1 -10 10 1
    GR2 -50 50 5

OPT

    RANGE 8.360 8.370
    DRO RE[Z1]<-10   50
    DRO IM[Z1]=0     30
    DRO DB[S11]>5.0  20
```

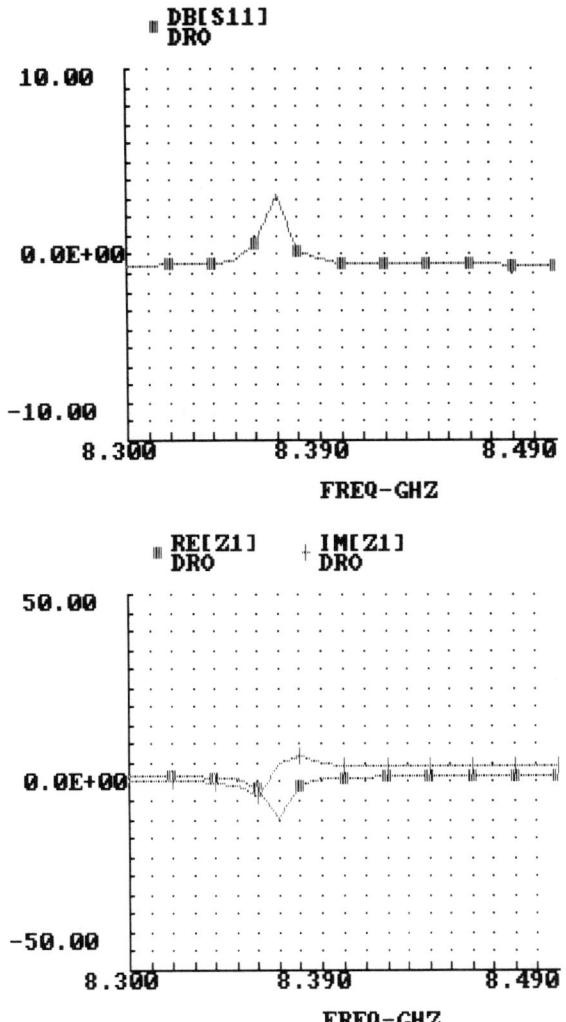

Fig. 8.30 Final simulated performance of the dielectrically tuned 8.4 GHz oscillator.

Chapter 9
MMIC Layout and Fabrication

9.1 MMIC ECONOMICS

Monolithic microwave integrated circuitry (MMIC) is a technique by which both GaAs FET devices and their associated matching circuitry are fabricated onto a single chip of GaAs. This technique, which currently is gaining considerable momentum in the industry, can produce single-chip amplifiers, oscillators, switches, phase shifters, mixers, and in an ultimate application, whole subsystems such as receiver front ends.

The development cost of a MMIC circuit is considerably higher than the cost of developing a comparable MIC circuit. To understand the motivation for MMIC, we first must examine the economics of the MMIC. This section will examine the various MMIC cost factors, one at a time, to present a clear picture of the origins of cost. Only with a clear picture of the cost in mind can the designer judge when an MMIC makes sense in a given application.

9.1.1 Cost of Design Rules

Most MMIC chips are manufactured at "foundry" companies, which operate as fabrication services for both external and internal customers. The MMIC fabrication process is very complex (containing up to fourteen different masking levels), and each foundry has its own way of carrying out these process steps. This means each foundry uses a process that is consistent but completely unique to itself. For this reason, foundries draw up a complex, yet very complete set of design rules, which enable the designer to produce a successful design within the physical limits of that foundry's process. All the physical and electrical characteristics of the materials used in the process are presented in the design rules.

Foundries sell these design rules to their clients. The cost (in 1988 US dollars) can vary from $5,000 to $30,000, depending on the foundry and the process details.

For example, the rules for a .5 μm FET gate length process may cost more than the design rules for a 1.0 μm FET gate length process. Because the designer must have a set of design rules, the first step after selecting a foundry is to purchase the set. As the cost of the design rules is quite high, most foundries will waive the design rule cost if the client processes some minimum number of wafers at that foundry (about ten).

9.1.2 Mask Set Costs

MIC circuits require two photomasks, one for the resistor layer and one for the conductor layer. MMIC circuits require a mask set containing up to fourteen masks. This means that a set of MMIC masks can be very expensive. The average mask set cost is $20,000.00, which is a lot of money to commit to an untested design. Mask turn-around time depends on a number of factors, and it can be anywhere from eight weeks to fourteen weeks. Although this is a one-time charge, the mask set turns out to be one of the major cost factors associated with a MMIC circuit. Mask sets also are a major area of risk because some or all of the masking levels may have to be replaced if the circuit does not perform as expected.

In addition to the cost of the masks themselves, a foundry client will have to pay the costs of designing the circuit and generating the CALMA GDSII tape that contains all the digitized pattern information from which the mask set is made.

9.1.3 Fabrication Costs

MMIC circuits are fabricated by a step-and-repeat process that completely covers the surface area of a GaAs wafer with a mosaic of identical circuits (separated by a grid of saw streets). The maximum number of circuits that can come from a wafer of diameter, d, is

$$N_{max} = \frac{\pi d^2}{4 a_c} \tag{9.1}$$

where a_c is the area of one MMIC circuit.

If we call w the width and l the length of each circuit,

$$a_c = w \cdot l$$

and

$$N_{max} = \frac{\pi d^2}{4w \cdot l} \tag{9.2}$$

Figure 9.1 shows a drawing of such a wafer or chip pattern. Unfortunately, the chips lying near the wafer's edge may be lost due to incompleteness during fabrication. Some material also is lost during the "dicing" process. The number of chips realistically available from a wafer is the number of chips that lie within the largest square that can be located within the wafer. This square will have four sides of length $d/\sqrt{2}$. So,

$$N = \frac{(d/\sqrt{2})^2}{w \cdot l} \qquad (9.3)$$
$$= \frac{d^2}{2w \cdot l}$$

Example

A 3-inch diameter wafer is patterned with MMIC chips of length .100 inch and width .070 inch. How many chips will this wafer reasonably produce?

$$N = \frac{3^2}{2(.100)(.070)} = 643$$

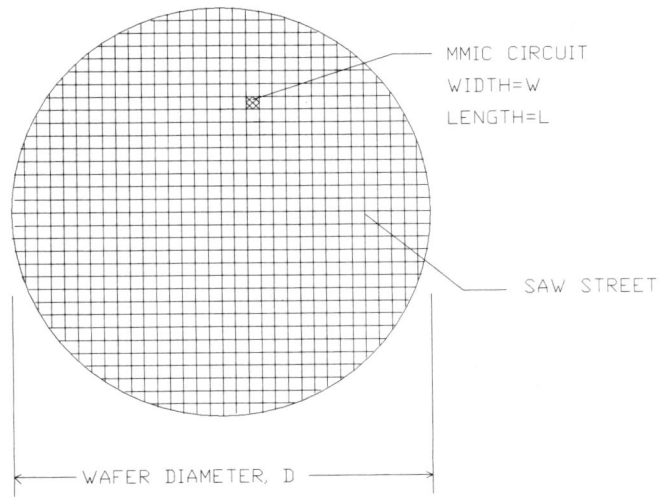

Fig. 9.1 GaAs wafer patterned with MMIC circuits.

After dicing, the wafer is tested on a wafer probe station to determine which chips are electrically good, relative to the specification. Very early in the processing sequence, the material parameters of the wafer are carefully monitored to identify whole wafers that are not likely to meet specifications. These wafers are scrapped early to minimize the labor investment. Because of cost and lead time considerations, only "good" wafers should complete fabrication. Final wafer probing may be a full RF probe test, or it may be a simpler dc probe test. In any event, the wafer will "yield" a certain percentage of chips that will test electrically good. The number of good chips available from a wafer is

$$N_{\text{good chips}} = \frac{d^2}{2w \cdot l} \times (\text{Yield}) \tag{9.4}$$

Yield is a highly variable quantity depending upon the difficulty of the specification, the nature of the circuit design, and the amount of process variability during fabrication. Generally, yields above 50% are considered good, and yields below 50% are considered less desirable; although what constitutes an economically viable yield is largely a question of market supply and demand. Sometimes very unique chips may be run successfully at 30% yield. The lower yield simply means a higher cost of manufacturing, implying a higher selling price, assuming the customer is willing to pay it. Of course, a high yield always is desirable and should be the ultimate goal of any circuit design efforts.

At 50% yield, using (9.4), the number of good chips in this example is found to be

$$N_{\text{good chips}} = 643(.50)$$
$$= 321$$

Fabrication costs usually are charged by the foundry on a per-wafer basis. Typically, the fabrication charges (in 1988 US dollars) per wafer are $10,000 to $20,000 for one or two wafers (some foundries ask for a two-wafer minimum), dropping to $5,000 to $10,000 per wafer for a run of ten or more wafers. These charges, plus the other fixed charges such as design rules, mask set, wafer probing and dicing, must be amortized over the number of good chips to determine the cost per chip.

9.1.4 Probing, Thinning, Dicing, and Packaging Costs

After fabrication, several more process steps occur before the chips are ready for the end user. These steps all add to the total cost of fabricating a chip. The sequence of back-end processing steps is

- Wafer probing to identify electrically good chips;
- Wafer thinning to reduce the overall chip thickness, thus thermal resistance, and establish the proper microstrip dielectric thickness. This step applies only a nonsubstrate "via" process;
- Backside metalization;
- Dicing;
- Sawing the wafer into individual chips;
- Visual inspection, rejecting chips with visual imperfections and any chips electrically marked "bad" during wafer probing;
- Packaging, placing good chip into "waffle packs" for shipment to user.

Total cost for all back-end process steps is $1,000 to $2,000 per wafer and is relatively independent of the number of wafers.

9.1.5 Cost Per Chip as a Function of Volume

Let us create a chart listing the total cost of MMIC fabrication for a given number of wafers. As before, we assume typical US dollar values for 1988. Then, let us calculate how many good chips will be obtained from these wafers. Finally, we divide the total cost by the total number of good chips to arrive at the cost per chip as a function of quantity. Table 9.1 presents these costs organized according to number of wafers.

Table 9.1 Total Cost of an MMIC Chip as a Function of Volume

No. of Wafers	Design Rules Cost	Mask Set Cost	Fabrication Cost	Back-End Process Cost	Total Cost	Number of Good Chips*	Cost Per Chip
1	$20,000	$20,000	$15,000	$2,000	$57,000	321	$177.5
2	$20,000	$20,000	$20,000	$3,000	$63,000	642	$98.13
5	$20,000	$20,000	$35,000	$5,000	$80,000	1,605	$49.84
10	—	$20,000	$50,000	$10,000	$80,000	3,210	$24.92

* Assume a 3-inch wafer with 50 percent die yield.

The cost per chip clearly is a strong function of volume because a high percentage of total MMIC cost is fixed, constant for any volume (i.e., design rules, mask sets, *et cetera*). If it is known beforehand what a chip *must* cost, then we can determine what volume is necessary to make this chip work economically. For instance, if the cost of the chip analyzed in Table 9.1 is not to exceed $50, the quantity must be greater than 1,600 chips (or, five wafers) to meet that goal.

Remember, we have not included any design costs associated with the initial development of the circuit. We are treating development costs as nonrecurring engineering costs, which are accounted separately. If this is not the case, the engineering costs also must be included as a fixed cost to be amortized over the total number of chips.

9.1.6 A Comparison Between MIC and MMIC Costs

Often the question arises: Does it make more sense to make a circuit in an MIC or a MMIC format? Although, in practice, this is a very application-specific question, depending on a number of factors including turn-around time, electrical performance, volume, and weight, we will investigate how to answer this question on purely economic grounds.

First, we must analyze the cost of a typical MMIC circuit as shown in Table 9.2. Although admittedly the numbers are "guesses" for a hypothetically "typical" circuit, they provide a feeling for how the costs divide up and behave with volume. The amounts are typical of 1988 US dollars.

Table 9.2 Per Circuit Cost Analysis of a Typical MIC Circuit

Volume	Semi-conductor	Substrate	Carrier	Miscellaneous	Total Material	Labor	O.H. (Assume 100%)	Total Cost per Circuit
1	$100	$50	$25	$10	$185	$70	$70	$325
10	$80	$35	$20	$9	$144	$60	$60	$264
100	$60	$25	$15	$8	$108	$55	$55	$218
1000	$40	$10	$10	$7	$67	$50	$50	$167

Typical MIC and MMIC costs are compared as a function of volume in Figure 9.2. For the minimum MMIC volume (one wafer, or 321 chips), the MMIC cost is less than the MIC costs. If we project the MMIC cost to lower volume (i.e., fabricate one wafer for a requirement of less than 321 chips and amortize the cost of one wafer over the required number of chips), the break-even point between the two technologies is 200 circuits. Clearly, from Figure 9.2, MMIC costs descend with volume much faster than the MIC costs. The reasons are

- MIC costs are labor intensive.
- MIC labor costs change little with volume because it takes a certain fixed time to manually process, tune, and test a circuit. We have assumed that this work is done manually. Automated assembly might significantly reduce assembly

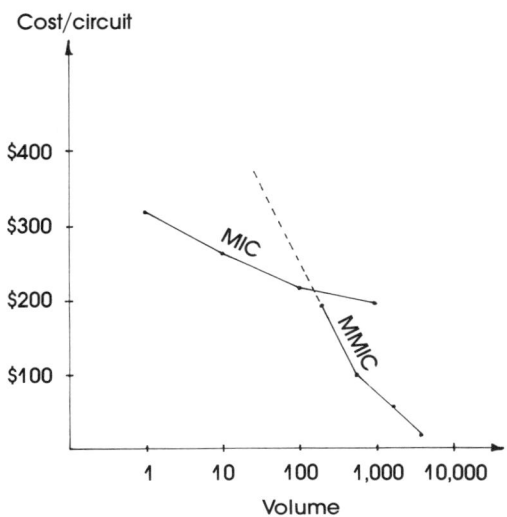

Fig. 9.2 A comparison of the cost per circuit for MIC *versus* MMIC fabrication techniques.

labor, but the test and tune labor, which is about 60% of the total, would remain the same.
- MMIC circuits require no tuning. Either they work or they do not work. Meticulously detailed designing is required in the beginning, but, once a successful design reaches the manufacturing phase, its labor content is very low and its cost drops rapidly with volume.

In general economic terms, if 1 to 100 circuits are to be manufactured, MIC is the most practical approach. If 100 to 500 circuits are to be manufactured, the decision falls into grey area and many factors must be considered. If more than 500 circuits are to be manufactured, MMIC has a clear economic advantage. Remember that many other factors influence the MIC *versus* MMIC decision, cost is rarely the only, or even the most significant, factor. If a MMIC design can achieve a very desirable volume or weight reduction or if it can achieve a special level of performance, a higher cost may be justified.

Many (more than 100 in December 1988) commercially available MMIC circuits are now on the market, with more becoming available all the time. Check which chips already are available to meet a given requirement before embarking on a custom design. There is no reason to "reinvent the wheel," if a chip is already available that meets your requirements. A commercial chip may cost more per chip than the foundry charges for a custom chip run, but there are no nonrecurring engineering charges (someone else has already done the design work) to be amortized over the number of the foundry chips needed for the requirement. It is very important to make a careful estimate of the design costs when facing a decision of this kind. Also, turn-around time can be a very important factor in a make-or-buy decision.

9.2 MMIC PROCESS DESCRIPTION

9.2.1 MMIC Structure

MMIC chips contain all the circuit elements necessary to make an entire functioning component. Both amplifiers and oscillators may be fabricated by MMIC techniques. The circuit elements available to the MMIC designer are

- GaAs FETs,
- GaAs diodes,
- microstrip transmission lines,
- thin-film and GaAs resistors,
- MIM capacitors,
- spiral inductors,
- metal interconnects and crossovers,
- grounds.

Each circuit element is formed "monolithically" as an integral part of the total GaAs chip, and then fabricated *en masse,* using step-and-repeat wafer processing techniques.

A wealth of literature exists on GaAs MMIC techniques. The reader is referred to the excellent books by Williams [1], Ghandi [2], Mun [3], Ferry [4], and Purcel [5] for additional background information and detail in this rapidly growing field.

MMICs are processed in up to fourteen steps, compared to MIC circuits that are processed in two steps. Each process step requires its own mask, which is why MMIC mask sets are so costly. Table 9.3 lists the most common process steps, the circuit elements they help to form, and their MIC equivalent. These process steps are typical, although each foundry has developed its own unique process with its own combination of steps. A basic graphic representation of a MMIC structure created by these process steps is shown in Figure 9.3.

Note that no bipolar transistor presently is available as part of a GaAs MMIC process. However, Si bipolar MMIC processes do exist and are very useful for amplifiers operating below 4 GHz.

The next few sections provide information about each process step. For more details about a particular process, check the foundry's design rules.

Table 9.3 MMIC Process Steps Compared with MIC Process

MMIC Process Step (or level)	Circuit Element	MIC Equivalent
N^- Implant	GaAs resistor	Thin-film resistor,
	FET channel	Discrete FET
N^+ Implant	GaAs resistor	Thin-film resistor,
	FET source-drain contact regions	Discrete FET
Annealing	FET channel (repair crystal damage caused by implant)	Discrete FET
Ohmic Metal	FET source-drain ohmic contacts	Discrete FET
Gate Metal	FET gate	Discrete FET
Thin-Film Resistor Metal	Thin-film resistor	Thin-film resistor
First-Level Metal	Microstrip lines interconnects	Conductor metallization
Airbridge Metal	Interconnects, crossover, spiral inductor	Bond wires, wound inductors
Dielectric Layer	Capacitor (airbridge metal forms top plate, first-level metal forms bottom plate)	Chip capacitor
Airbridge Via Holes	Direct vertical interconnect between airbridge metal, and first-level metal	Bond wire
Dielectric Via Holes	Direct vertical interconnect between airbridge metal and first-level metal that passes through the dielectric layer	No MIC equivalent
Substrate Via Holes	Grounds (direct connection through the GaAs substrate from the topside to the bottomside)	Via grounds, wrap-around grounds, grounding ribbons, grounding ribs
Passivation	Protective outer coating of hard dielectric material	No MIC equivalent

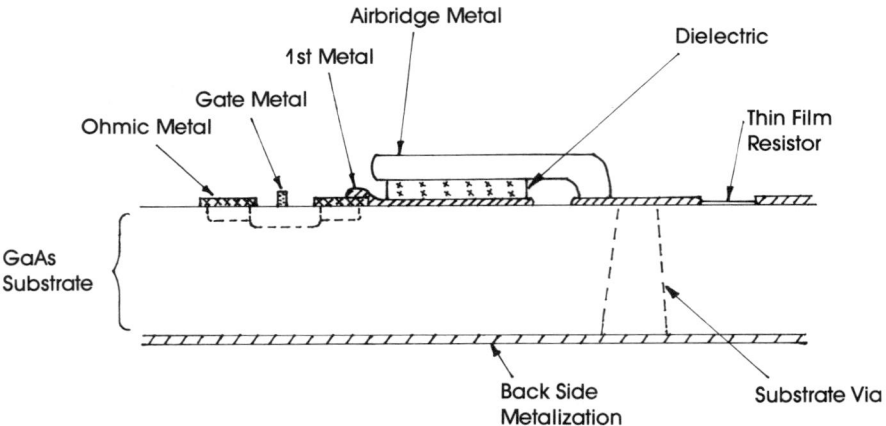

Fig. 9.3 The structure of a GaAs MMIC chip.

9.2.2 GaAs Material

Wafers are round, thin sheets of GaAs cut from a grown crystal ingot. Wafers of GaAs must be low in impurities and defects and semiinsulating. The semiinsulating requirement is a direct result of the need for a low-loss dielectric substrate to form low-loss metal microstrip lines. Sometimes impurities such as chrome intentionally are added to the crystal "melt" to compensate other unintentional impurities such as silicon and make the material more nearly semiinsulating.

The two principal methods for GaAs crystal growth are the horizontal Bridgeman [6] technique, and the liquid encapsulated Cyochralski [7] technique. Each approach has advantages and disadvantages, although the liquid encapsulated Cyochralski material customarily is used for material to be doped by ion implantation, the most common doping technique for MMIC devices. A great deal of study has been done on these crystal growth techniques over the years. Materials researchers are continually working toward a purer, more uniform material because pure material translates into high-performance MMIC chips.

9.2.3 Doping by Ion Implantation

Semiinsulating GaAs must be intentionally doped to create the active channel of FETs, diodes, and GaAs resistors. Doping can be accomplished by epitaxial techniques such as liquid phase epitaxy, vapor phase epitaxy, and molecular beam epitaxy. However, doping by ion implantation has become the preferred technique for MMIC devices because of its natural adaptation to high-volume manufacturing.

Ions are implanted using a linear accelerator to accelerate doping ions (silicon or selenium for N^- type material) to energize in the range of 30 to 400 keV. These doping ions are directed onto the surface of the GaAs wafer, penetrating to some depth before coming to rest within the wafer's crystal structure. The ion implantation process does considerable crystal damage to the GaAs material, which must be "healed" by a high-temperature (about 850°C) annealing process before the doping profile introduced by ion implantation is "activated." Figure 9.4 shows a typical ion implanted doping profile. To encapsulate the GaAs wafer in a material such as silicon nitride often is necessary to prevent arsenic atoms from escaping during the annealing process. Performing the implant right through this encapsulation is possible.

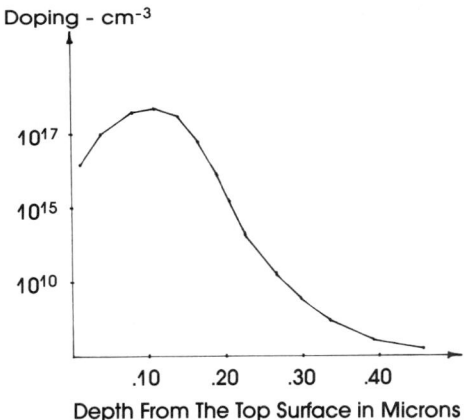

Fig. 9.4 A typical ion implanted doping profile.

Individual device channels are defined by either mesa etching or implant isolation techniques. In implant isolation, a second ion implant (usually hydrogen) is directed specifically to the region between devices. This implant damages the crystal lattice in these areas, spoiling conductivity. No annealing is done after the isolation implant, so the damaged regions between the devices are highly insulating, effectively isolating the devices.

9.2.4 Ohmic Contact Metal

Ohmic contacts are used to form the source and drain contacts of FETs and diodes. They also are used to form the contacts for GaAs and thin-film resistors. Ohmic contacts must be nonrectifying, low-resistance contacts, if these devices are to function at full performance. FET performance (gain and noise figure) in particular is degraded by poor ohmic contacts.

Ohmic contacts consist of highly doped regions within the GaAs, below a deposited metal contact (usually gold-germanium). The metal is patterned and then alloyed into the GaAs by a high-temperature process, producing a very high doping level right at the surface of the GaAs, which is the prerequisite for a good ohmic contact. Often an N^+ implant just below the ohmic metal enhances the low resistance characteristic of the contact. Figure 9.5 shows an example of an ohmic contact.

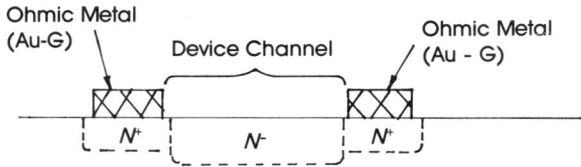

Fig. 9.5 Example of ohmic metal contacts applied to the surface of an ion-implanted GaAs device (FET channel).

9.2.5 Resistive Layers

MMIC resistors are formed from a number of different materials. Some of the common resistor materials are

- N^- implant doped GaAs: 500–1000 Ω/square;
- $N+$ implant doped GaAs: 100–500 Ω/square;
- NiCr thin film: 50–100 Ω/square;
- Alloyed ohmic metal: 2 Ω/square.

Resistors may be patterned like any other layer. GaAs resistors must be designed with ohmic contacts at either end. The value of the contact resistance must be factored into the overall resistance value.

Like the MIC resistor, the MMIC resistor has a dc resistance value of

$$R = \left(\frac{l}{w}\right) S \tag{9.5}$$

where S is the material's sheet resistance in Ω/square.

9.2.6 Gate Metal

An FET gate is a long, thin strip of metal that forms a Schottky barrier contact along the middle of an FET's GaAs channel. Choices of gate metal are aluminum, chromium, titanium, or molybdenum. These metals form the best gates on GaAs because of a combination of good adhesion properties and their relatively slow diffusion into GaAs. Except for aluminum, which has sufficient low resistivity for making low resistance gates on its own, gold often is used as an overlay metal to reduce the electrical resistance of the gate strip. However, a barrier metal such as platinum must be deposited between the adhesion metal and the gold to prevent diffusion of the gold into the adhesion metal and through it into the GaAs. A typical gate metal system might have the following thickness:

- titanium Schottky barrier—1500 Å,
- platinum diffusion barrier—1000 Å,
- gold—5000 Å.

The gate usually is formed by a sequence of processing steps starting with the creation of a slot in the middle of the channel surface. This slot is called the *gate recess*. The gate recess is etched to a depth of about 1500 Å. Because gates normally are .5 μm in length (the narrow direction), nonoptical, short wavelength, lithographic techniques, such as electron-beam lithography often are used to pattern the gate recess in the photoresist. Some undercutting occurs during etching, making the recess wider than the opening in the photoresist. The next step is to evaporate gate metal into the recess (and on top of the resist). Because of the recess undercutting, the gate metal does not fill the recess slot but is self-aligned in the middle of the recess. The self-alignment of the gate in the center of the recess is very important for the proper operation of the FET and therefore is a critical process step. Figure 9.6 shows graphically the details of gate formation.

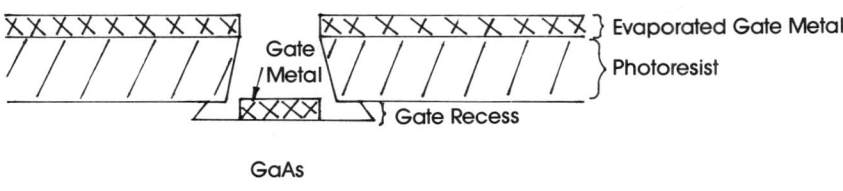

Fig. 9.6 The formation of a self-aligned recessed gate.

9.2.7 First Metal

Gold is the most popular metallization on GaAs because of its high electrical conductivity and its chemical inertness. However, gold does not adhere well to GaAs, so it must be used in conjunction with some other metal or metals. Like the gate metal system described earlier, other metal layers on GaAs also must have an adhesion layer and a barrier layer to prevent the diffusion of gold into the GaAs. A titanium-platinum-gold metal system is used extensively for the first-level metal, as is chromium-platinum-gold.

First-level metal is the first layer of interconnect metal in MMIC circuits, and it is in contact with the GaAs substrate. First-level metal is useful for making connections between devices (much like bond wires in MIC circuits), as the bottom plate for capacitors, and as microstrip transmission lines. First-level metal usually is quite thin, so transmission lines formed from it may be high in loss. First-level metal often is used to "overcoat" ohmic and gate metals to increase their conductivity. The thickness of the first level metal is determined by skin depth and current carrying considerations. At 5 GHz, the skin depth in gold is about 1.0 μm, so the first-level metal should be at least this thick.

First-level metal normally is patterned by a lift-off process because etching usually is not practical due to the multiple metals involved. Referring to Figure 9.7, in the lift-off process, photoresist is applied to the wafer and patterned to remove resist in regions where the metal is to be applied. Next metal is either evaporated or

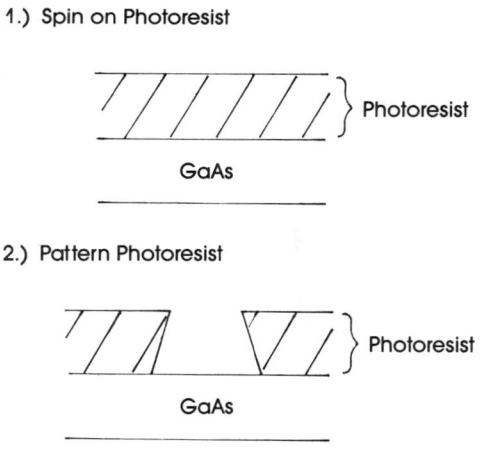

Fig. 9.7 Steps of the lift-off process for patterning metal layers (first metal in this example).

3.) Deposit Metal

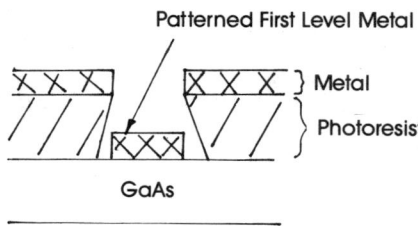

4.) "Lift-off" Excess Metal, Strip Photoresist

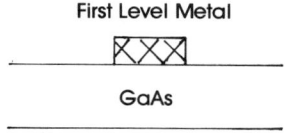

Fig. 9.7 cont'd.

sputtered onto the photoresist and the GaAs. Evaporation is the preferred technique because it is much less likely to coat the edge of the photoresist. Next, the photoresist is dissolved by a solvent, and the excess metal on the top of the photoresist is washed away with the photoresist, leaving behind the patterned first-level metal.

9.2.8 Dielectric

Dielectrics are used to form MIM capacitors, but they also are useful as crossover insulation and as passivation layers over completed circuits to prevent mechanical damage and contamination. The most common dielectric material for GaAs MMIC devices is silicon nitride (Si_3N_4) a wideband gap insulator with a dieletric constant of about 6.0. It is a barrier to alkali ion migration; it is not hygroscopic; it is dense; it can be applied easily to a wafer by sputtering or by *plasma enhanced chemical vapor deposition* (PECVD).

The thickness of the Si_3N_4 layer is a compromise between potential pinhole damage if the film is too thin and parasitic effects (such as increased C_{dg} in FETs coated with Si_3N_4) if the film is too thick. The compromise usually is about 2000 Å thickness. Sometimes the deposition of Si_3N_4 can cause crystal damage on the surface of the GaAs, which can effect the operation of devices such as FETs. Process engineers must take care to ensure that dielectric deposition does not affect the device's performance significantly. Other materials such as SiO_2, SiN, and various polyimides also are used as dielectrics. Each has its own advantages and disadvantages. Overall however, Si_3N_4 has become an industrial standard.

MIM capacitors are formed by creating a sandwich of the first-level metal on the bottom, the dielectric in the middle, and the second-level metal on the top. The capacitance per unit area of this type of capacitor is

$$C/a = \frac{\epsilon_R \cdot \epsilon_0}{d} \tag{9.6}$$

where d is the dielectric's thickness, and

$$\epsilon_0 = 8 \cdot 854 \cdot 10^{-15} \text{ F/mm},$$

For Si_3N_4, $\varepsilon_R = 6.0$, and let us assume $d = 2000$ Å, or

$2 \cdot 10^{-4}$ mm

$$C/a = 6 \cdot 0 \left(\frac{8 \cdot 8540^{-15}}{2 \cdot 10^{-4}} \right) = 265.6 \text{ pF/mm}^2$$

This is a typical capacitance per unit area for GaAs MMIC MIM capacitors.

9.2.9 Second-Level Metal

Second-level metal is used to provide circuit crossovers, low-loss conductors with high current-carrying capability, and the top plates of MIM capacitors. Second-level metal usually is quite thick, often 2 μm or more, and made of gold with an adhesion layer of Ti or Cr. Second-level metal either can be deposited on top of a permanent dielectric layer or the dielectric layer can be washed out with solvents, leaving the metal unsupported except at specific locations called *vias* or *posts*. This unsupported second metal is called *airbridge metal*. Second metal may be thickened by electroplating to increase current carrying capability and decrease electrical resistance. When plated, the thickness of the second metal can be increased to as much as 8 μm.

Second metal usually is formed by a lift-off process. The final process step is to wash out the temporary dielectric, if the second metal is to be an airbridge. Airbridges have very low capacitance to the first metal at crossover as a result of the low ($\varepsilon = 1.0$) dielectric constant of air, and the relatively wide spacing between the first metal and the airbridge metal. Airbridges are much like freeway overpasses, providing crossovers above existing conductors. Because of their noncontact with the GaAs, air-bridges can be used to synthesize very high-impedance microstrip lines.

9.2.10 Dielectric and Airbridge Vias

Vias are vertical connections between various metal layers. The two most commonly connected layers are first-level metal and second-level metal. Vias are formed by first patterning holes in the photoresist that coats the dielectric where a via is desired. Next, a hole is etched wherever there is a hole in the photoresist. The etchant used will etch the dielectric layer at an obtuse angle. This angle is important for maintaining metal thickness through the via. Once a hole is etched in the dielectric, the next process step defines the location of the second-level metal. Second metal then is deposited over the dielectric and, where holes exist, the second-level metal will contact the first-level metal through the via hole. Figure 9.8

1.) Etch Hole in Dielectric Layer

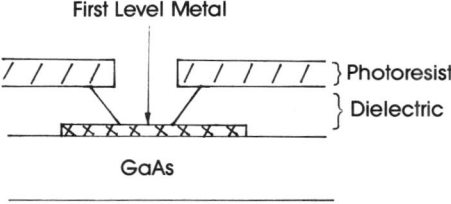

2.) Pattern Photoresist for Second Level Metal

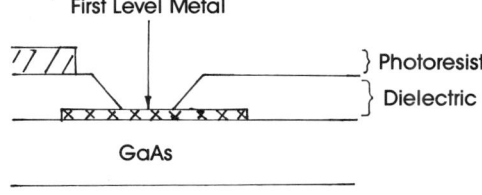

3.) Deposit Second Level Metal, and Lift-Off

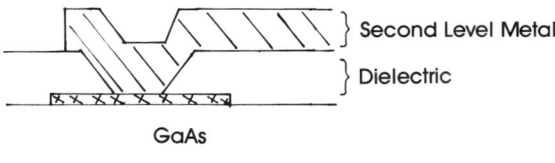

Fig. 9.8 Steps in the process for forming dielectric (and air-bridge; after removing the dielectric) vias.

shows the process steps used in forming a via of this type. The final lift-off step leaves the second metal and the vias in place. If the dielectric is permanent, the vias are called *dielectric vias,* if the dielectric is washed out as a final process step, the vias are called *airbridge vias*.

9.2.11 Substrate Vias

Substrate vias provide direct, low-inductance grounds from the backside of the substrate to any point on the frontside of the substrate. Although substrate vias pose a certain technological challenge to the foundry, they often can make a tremendous difference in terms of a circuit's electrical performance.

The substrate is first thinned by lapping the backside; typically a GaAs substrate is thinned to a thickness of 5 mils. Next, the backside is coated with photoresist which is patterned with the desired via holes. Via hole diameter typically is 2 to 5 mils. The resist is removed at these holes, and the substrate is etched through these holes in the resist. An etchant is used that attacks GaAs, but not the frontside metal. When the via holes have been etched through to the frontside, the resist is stripped and the backside of the wafer is metallized. An initial metallization is performed by sputtering, ensuring good coverage throughout the etched via holes. Next, the backside metal is plated up to a thickness of 5 μm or more. See Figure 9.9 for a diagram of the substrate via process.

1.) Pattern Backside Photoresist with Via Holes

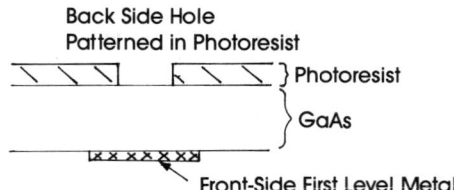

2.) Etch Via Hole in GaAs

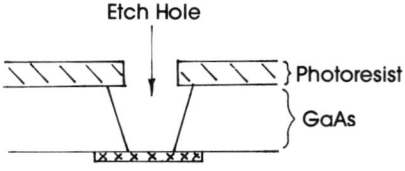

Fig. 9.9 Steps in the process for forming substrate via holes.

3.) Strip Photoresist, Deposit and Plate Up Backside Metal

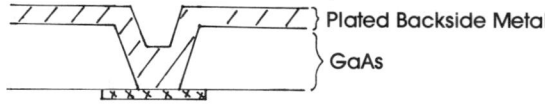

Fig. 9.9 cont'd.

9.2.12 Final Wafer Process Steps

After completion of backside processing, a wafer is dc-wafer probed, RF-wafer probed if possible, and then diced into individual chips, which are packaged in "waffle packs" for shipment.

The dicing process is accomplished in one of three possible ways:

1. The wafer is cut by a diamond circular saw along saw streets to separate the chips. This process is identical to the sawing step in processing MIC substrates, except that it requires a narrower saw blade. Some problems with sawing include chipping the GaAs material, which is very brittle, and metal wearing out the saw blades (especially plated metal).
2. Scribing the wafer along the saw streets with a diamond stylus and breaking it into die along the "scratches." For scribing to work, the wafer must be oriented so that the scribe lines are along wafer cleavage directions, which are 90° apart. With proper wafer orientation, this technique works fine for scribe scratches parallel to and perpendicular to each other.
3. The wafer may be patterned with photoresist, and etched along the saw streets to separate the chips. The etching must not undercut the chips too much, or the chips will be difficult to grip with tweezers during die attach.

9.3 MMIC DESIGN AND LAYOUT STRATEGY

This section will focus on the steps a MMIC designer must perform between achieving a basic circuit design (Chapters 5 and 6) and creating the digitized layout information (usually CALMA CDSII tapes) necessary to make a MMIC mask set. The design effort flows as a process, and we treat it as such by including the feedback loops necessary to certain refinements.

MMIC design always is conducted in meticulous detail. The reason is very simple. Unlike MIC circuits that, once fabricated, may be modified easily by adding small pieces of gold ribbon where needed, MMIC circuits *cannot* be modified. If an MMIC circuit does not perform to expectations, the designer must change the design, modify or remake the mask set, and fabricate new wafers. Such modifications may cost $20,000 or more and may take weeks or even months to carry out. Needless to say, mistakes in MMIC design are extremely painful. Therefore, the MMIC designer must make every effort to get it right the first time. This usually translates into very detailed modeling of all the circuit elements— nothing is left to chance. However, the fabrication process itself has a certain randomness to it. Materials parameters vary from wafer to wafer and sometimes even within a single wafer. The foundries deal with these variations by informing the designers of the statistical mean value of each material parameter and its standard deviation. Armed with these statistical parameters, the designer must question the behavior of the design in the face of random fluctuations in design parameters. This question is answered by either making computer calculations of performance sensitivity to individual parameters, or by using the Monte-Carlo analysis capability of Touchstone or SuperCompact to predict overall yield of a group of chips with statistical variations in the specified parameters. Figure 9.10 is a MMIC design sequence flow diagram. The following subsections address individual design events leading to a digitized final layout.

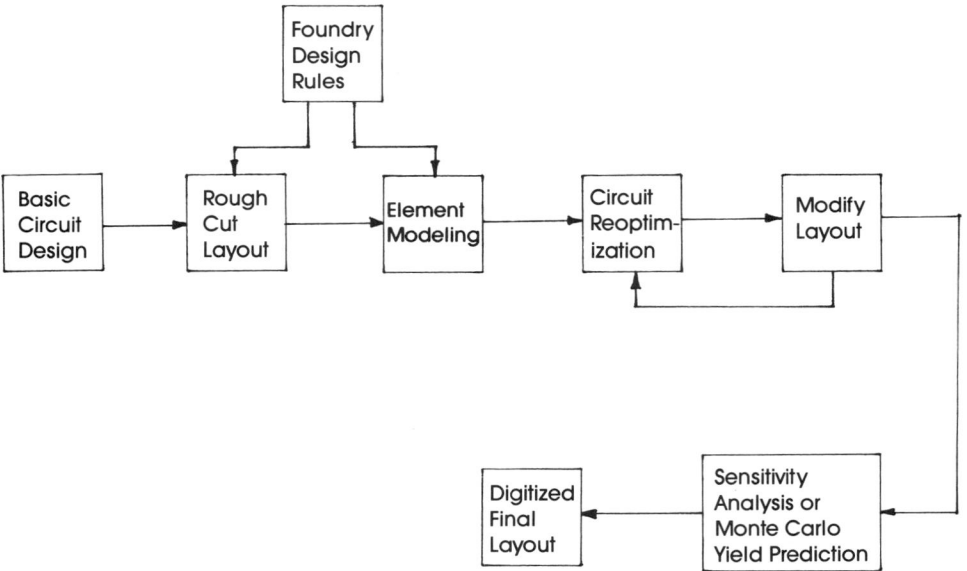

Fig. 9.10 Flow chart of the MMIC design process.

9.3.1 Foundry Design Rules

Obtaining complete information about a foundry's design rules is the first step in any MMIC design-layout effort. Each foundry has its own specialized GaAs MMIC process, which has evolved throughout the development of the foundry. This process is highly specialized and represents the combined thinking and experience of all the workers at the foundry. For a circuit design to be successfully fabricated, the designer must have complete knowledge of the foundry's process and be able to incorporate its features into the design. This knowledge is communicated to the designer by way of a set of *design rules*.

Design rules usually come in three-ring binder notebooks, and they are quite lengthy, have considerable detail, and are often three hundred pages or more in length. The designer must become intimately familiar with the design rules because he or she will constantly need to refer to them throughout the design process. Quite often, foundries offer classes for designers to teach them how to use their design rules. The following general areas are covered by a typical set of design rules:

- Process overview and specifications.
- Fabrication sequence and details.
- Device element models.
- Layout rules, mask requirements.

9.3.1.1 Process Overview and Specifications

Because so many special features compose a MMIC foundry's process specifications, simply writing down such a specification is difficult. For this reason, you should use a MMIC process survey sheet to organize all the design rule information into an easily accessible summary. Such a survey sheet provides a quick reference during the design process and allows a systematic approach to comparing foundry processes. One such survey sheet is shown in Table 9.4. Notice that space is provided to compare up to three foundries on a single sheet. All electrical parameters are normalized per unit length, or per unit area to facilitate easy comparisons and scaling.

9.3.1.2 Fabrication Details and Sequence

The information contained in this subsection of the design rules relates to the fabrication process details of the foundry. This material is very similar to the material presented earlier in Section 9.2.

Table 9.4 Foundry Process Survey Sheet

dc FET Parameters	Foundry A	Foundry B	Foundry C
I_{DSS} (mA/mm)			
g_m (mS/mm)			
V_p (V)			
V_S (V)			
V_{BGD} (V)			
V_{BGS} (V)			

RF FET Parameters			
g_m (mS/mm)			
C_{GS} (pF/mm)			
RI (Ω-mm)			
C_{DG} (pF/mm)			
R_{DS} (Ω-mm)			
C_{DS} (pF/mm)			
R_G (Ω-mm)			
R_R (Ω-mm)			
R_D (Ω-mm)			

Sheet Resistance	Foundry A	Foundry B	Foundry C
GaAs N^+ (Ω/sq)			
GaAs N^- (Ω/sq)			
Thin-Film Resistor (Ω/sq)			
Ohmic Metal (Ω/sq)			
First Metal (Ω/sq)			
Second (airbridge) Metal (Ω/sq)			
Gate Metal (Ω/sq)			

Layer Thickness			
Ohmic Metal (μm)			
Gate Metal (μm)			
First Metal (μm)			
Thin-Film Resistor (μm)			
Second (airbridge) Metal (μm)			
MIM Dielectric (Å)			
GaAs Substrate (mils)			

Table 9.4 cont'd.

Maximum Safe Current per Width	Foundry A	Foundry B	Foundry C
Gate Metal (mA/μm)			
Thin-Film Resistor (mA/μm)			
First Metal (mA/μm)			
Second (airbridge) Metal (mA/μm)			
Minimum Line Width			
Ohmic Metal (μm)			
Gate Metal (μm)			
First Metal (μm)			
Thin-Film Resistor (μm)			
Second (arbridge) Metal (μm)			
Minimum Line Spacing			
Ohmic Metal (μm)			
First Metal (μm)			
Thin-Film Resistor (μm)			
Second (airbridge) Metal (μm)			
Capacitance/Inductance			
MIM Capacitance (pF/mm^2)			
Substrate Via Inductance (nH)			

FET RF Performance	Foundry A	Foundry B	Foundry C
Power Output (W/mm)			
Efficiency (%)			
MAG at 18 GHz $-$ 300 μm (dB)			
S_{21} at 18 GHz $-$ 300 μm (dB)			
$NF_{min} - 300\ \mu$m (dB)			

Statistical Info Standard Deviation	Foundry A	Foundry B	Foundry C
g_m (mS/mm)			
C_{GS} (pF/mm)			

Table 9.4 cont'd.

Statistical Info Standard Deviation	Foundry A	Foundry B	Foundry C
RI (Ω-mm)			
C_{DG} (pF/mm)			
R_{DS} (Ω-mm)			
C_{DS} (Ω-mm)			
C_G (Ω-mm)			
R_G (Ω-mm)			
R_S (Ω-mm)			
R_D (Ω-mm)			
Sheet Resistance Thin-Film Resistor (Ω-Sq)			
Sheet Resistance First Metal (Ω/Sq)			
Line Width First Metal (mm)			
Line Width Thin-Film Resistor (mm)			
Line Width Second Metal (mm)			
MIM Capacitance (pF/mm^2)			
Substrate Via Inductance (nH)			

9.3.1.3 Device Element Models

As previously pointed out, it is very important to create the most accurate model possible for each device in a MMIC circuit. Foundry design rules often spell out these models in great detail. Let us consider the models in some detail.

FET

The customary small-signal model for a FET is shown in Figure 9.11. Many other models are possible, but this one has become very popular within the MMIC community. Model element values are scaled with gate width in the following way.

The following elements scale as w (directly with gate width): g_m, C_{ds}, C_{GS}, C_{DG}, and R_G. The following elements scale as $1/w$ (inverse with gate width): RI,

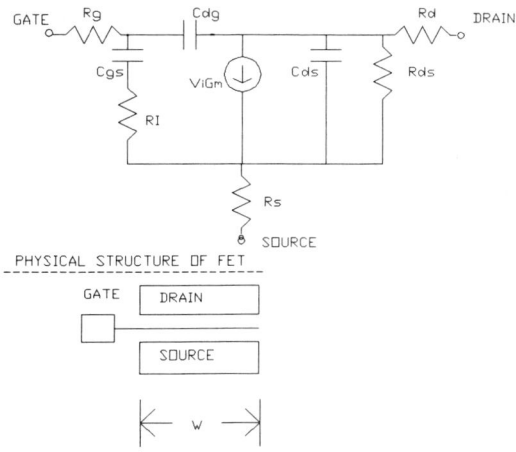

Fig. 9.11 The small-signal equivalent circuit model for a GaAs FET.

R_{DS}, R_s, and R_d. Figure 9.11 shows a small signal model and the physical structure of a FET.

MIM Capacitors

An MIM capacitor has parasitic capacitance between its bottom plate (first-level metal) and ground at the backside of the chip. For this reason, an MIM capacitor is modeled as a lumped capacitor with one terminal connected to an open circuited microstrip transmission line, where the open circuited stub models the parasitic capacitance to ground (see Figure 9.12).

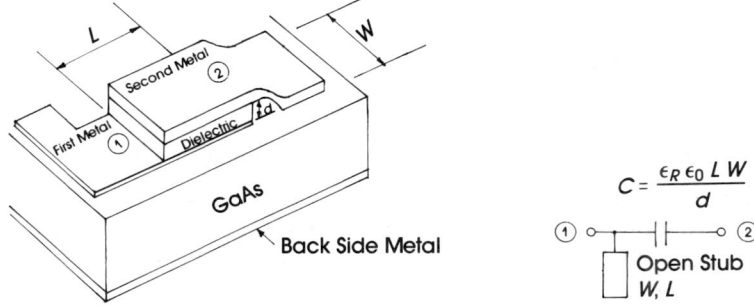

Fig. 9.12 The circuit model of an MIM capacitor.

The value of the lumped MIM capacitor is given by equation (9.7)

$$C = C' \cdot w \cdot l \tag{9.7}$$

where C' is the capacitance per unit area for the particular foundry process (9.6). The terms l and w are the length and width of the capacitor. The open circuited stub has dimensions $l \times w$, and a substrate thickness equal to the chip's thickness.

Thin-Film or GaAs Resistors

Thin-film or GaAs resistors have a dc resistance given by (9.5):

$$R = \left(\frac{l}{w}\right) S$$

where

l is the length of the resistor,
w is the width of the resistor,
S is the sheet resistance of the resistive material in Ω/square.

At microwave frequencies, a thin-film resistor has a complex impedance because its length acts as a transmission line. In its simplest form, this complex impedance can be modeled as a microstrip transmission line of length l and width w in series with the resistor's dc resistance. When modeling circuits, it is best to treat a physical resistor using special distributed resistor models available in Touchstone or SuperCompact.

In Touchstone, this model has the syntax

TFR 1 2 W=   L=   RS=   F=

where

W is the resistor's width,
L is the resistor's length,
RS is the sheet resistance of the material,
F is the frequency above which skin effects begin to occur.

The metal connections to the resistor must also be modeled carefully, because even the contact metallization has physical length and produces transmission line effects of its own. See Figure 9.13 for a diagram of the physical resistor.

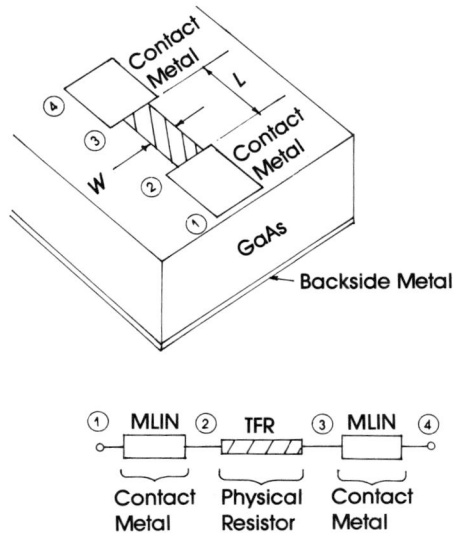

Fig. 9.13 The circuit model of an MMIC resistor (either thin-film or implanted).

Spiral Inductor

Quite often in MMIC designs, ideal inductances must be realized with physical elements such as microstrip lines and narrow lengths of airbridge metal. A very convenient and compact way to create inductance is using a structure called a *spiral inductor*. The most popular spiral inductor is the multiturn rectangular structure shown in Figure 9.14. Spiral inductors customarily are fabricated from airbridge metal. Their inductance is the sum of the self-inductance of the airbridge metal and the mutual inductances between parallel line sections.

To completely characterize a spiral inductance, a model must take into account certain parasitic elements. First, the frequency at which the total length of the coiled airbridge metal becomes a quarter wavelength determines the parallel resonance for the entire structure. This resonance usually is modeled as a capacitance, C_p, in parallel with the main inductance L_1, which are in resonance at the quater wave length resonance frequency. Second, the entire spiral structure has shunt capacitance through the substrate to ground. Barna [8] has shown that this capacitance is equal to

$$C_{SH} = \left(\frac{\epsilon_R \epsilon_0}{h}\right)(A + 1.6h)(B + 1.6h) \tag{9.8}$$

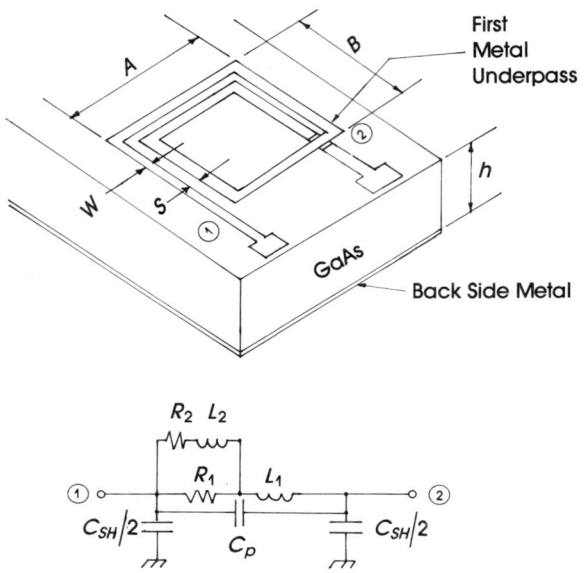

Fig. 9.14 The circuit model of a spiral inductor.

where
    A and B are the outer dimensions of the spiral inductor,
    C_{SH} is equally divided between the two terminals of the inductor,
    h is the substrate thickness.

The third parasitic is the skin effect that causes the airbridge conductor's effective thickness to decrease at high frequencies, with the effect of increasing both resistance and inductance at high frequencies. This effect usually is modeled as a series R-L circuit (R_2 and L_2) in parallel with a resistance R_1. As frequency increases, the inductance L_2 becomes a significant contributor to overall inductance and the overall resistance approaches R_1 in the high-frequency limit. The resistance approaches ($R_1 \cdot R_2/R_1 + R_2$) as the low-frequency limit.

Following Parisot [9], the equivalent circuit of a spiral inductor is shown in Figure 9.14. This model applies only to the spiral structure proper. Outside of the spiral structure, any additional metal lines must be modeled as transmission lines.

Normally, spiral inductors are not used near their resonance frequency because the inductive reactance of a spiral inductor is no longer linear with frequency for frequencies higher than 80% of its resonance frequency. To determine the resonance frequency of a spiral inductor is important before planning its use in the matching structure of a MMIC circuit.

9.3.1.4 Standard Cell Library

Some foundries make extensive use of a collection of designs for certain specific device elements, which are called *standard cells*. Usually, these standard cells are stepped in physical size and often can be combined in a way that allows the creation of any desired device element.

Standard cells are provided as both layout and circuit models. The standard cell layouts are extremely useful because they contain all the masking levels necessary to create a device within the foundry's MMIC process rules. For instance, an FET standard cell contains the following masking levels: N^+ implant, N^- implant, ohmic metal, and gate metal. Standard cells may be introduced into preliminary layouts by using so-called paper doll appliques, and into final layouts by using GDSII standard cell data supplied on tape by the foundry.

Standard cell models representing many foundries are available for both Touchstone and SuperCompact. These models may be read into the program from a disk or a tape and then called up as part of the total MMIC model, like any other circuit element, by using special syntax. The use of standard cells within the CAD model and the layout of a MMIC circuit can greatly enhance the accuracy and efficiency of GaAs MMIC circuit design. The designer will have to contact potential foundries to learn about the availability of standard cells. Some examples of standard cells are shown in Figure 9.15. The exact elements needed by the designer can be created by stretching, shrinking, and combining standard cells in creative ways.

9.3.1.5 Layout Rules, Mask Requirements

Because of the highly specialized nature of their process, each foundry has a specific set of rules pertaining to the preparation of MMIC circuit layouts. It is beyond the scope of this book to go into great detail about these rules, because each set is linked to a particular foundry (or to a particular process if the foundry offers more than one process). However, the following list summarizes the areas addressed by the layout rules:

- Minimum width, various levels;
- Maximum width, various levels;
- Minimum spacing, given levels;
- Minimum spacing, different levels;
- Overhang and inclusion, different levels;
- Bonding pad spacing;
- Maximum circuit size;
- Aspect ratio, total circuit;
- Direct-write E-beam registration markers;

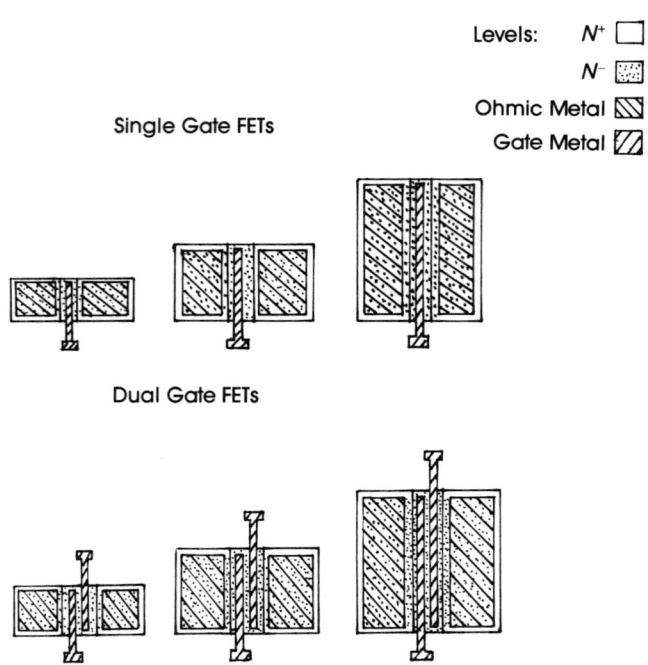

Fig. 9.15 Examples of standard FET cells.

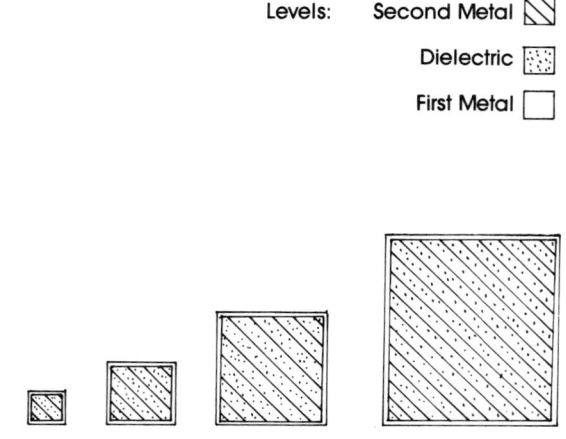

Fig. 9.15 (continued) Examples of standard MIM capacitor cells.

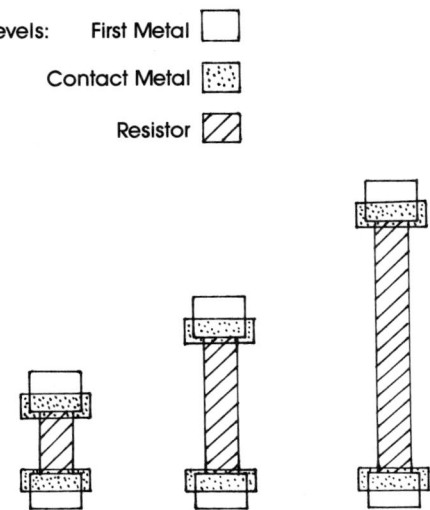

Fig. 9.15 (continued) Examples of standard thin film resistor cells.

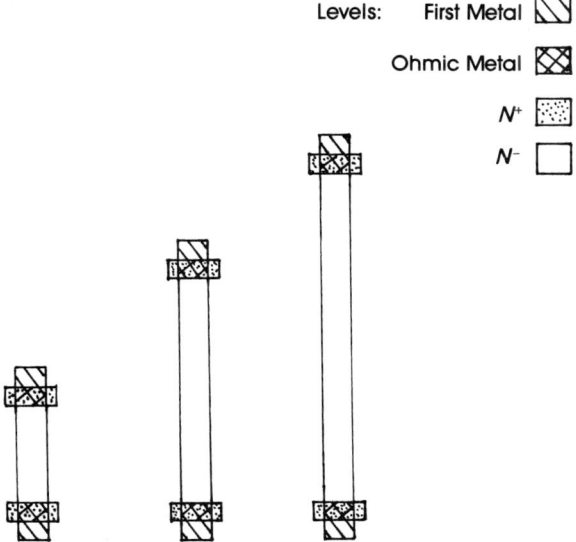

Fig. 9.15 (continued) Examples of standard GaAs N^- resistor cells.

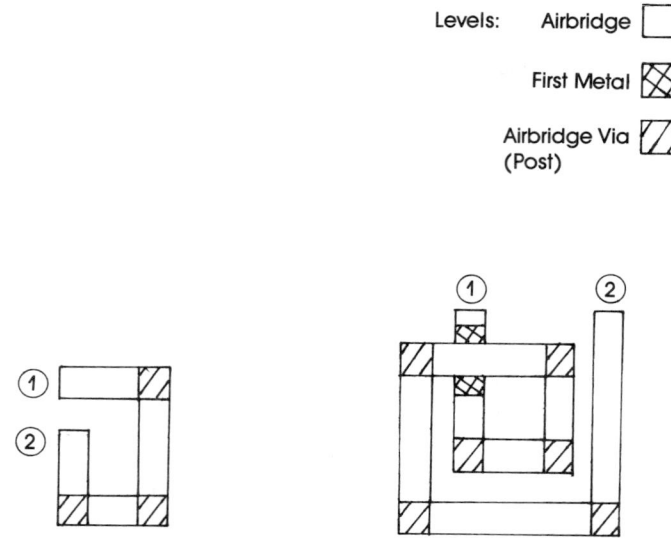

Fig. 9.15 (continued) Examples of standard spiral inductor cells.

- Maximum unsupported airbridge length;
- Maximum current per unit line width, various levels.

9.3.2 Rough Cut Layout

The rough cut layout transforms a basic electrical circuit model into a first-pass MMIC layout. This layout is used to establish an advantageous layout configuration in the available space with input, output, and bias pads appropriately placed at predetermined locations. The designer must be very careful to maintain a judicious balance between the desires for a very compact layout (to reduce cost by yielding more chips per wafer) and for an expanded (large) layout to reduce various performance inhibiting parasitic elements such as bends and cross coupling. Of course, the smallest possible chip area is very desirable from a cost viewpoint; however, small size is a false economy if a very compact layout introduces parasitic elements, compromising performance. It does no good to make a lot of chips with low yield. There is no substitute for highly creative thinking in making the rough cut layout. However, some established practices will facilitate this layout.

9.3.2.1 FET Layout

The most common type of MMIC FET structure is the so-called multifinger or interdigitated FET. This FET structure is created from several of the standard FET cells shown in Figure 9.15. The standard cells are placed side by side so that the ohmic contacts of adjacent FETs are touching. Ohmic areas ultimately will be source and drain contacts. The total gate width of such an FET is the gate width of one cell times the number of cells. The equivalent circuit scaling rules discussed in Subsection 9.3.1.3 are used to calculate the model of an interdigitated FET. However, the gate resistance, R_G, will be the parallel combination of the four gate strips providing a significant improvement over one long gate strip. An even number of standard cells always is used so that the outside ohmic regions are source contacts. Interior source contacts are connected to the outside source contacts with airbridge metal. Alternating drain contacts are connected together at their ends with first-level metal. The gate "fingers" are connected at a common gate contact with first-level metal. If an FET is intended to have dc-grounded source contacts (use with a two polarity bias supply), the source contacts are connected directly to a substrate via. If the FET is to be self-biased using a source resistor, its source will be connected to the top plate of a MIM bypass capacitor. The bottom plate of the MIM capacitor is grounded by a substrate via. Usually two MIM capacitors are used, each connected to an outside source contact. In most applications, a value of about 10 pF per MIM capacitor is sufficient, although this value directly affects the FET's lowest operating frequency. The layout of a typical four-finger FET using source capacitors is shown in Figure 9.16.

If the foundry process to be used does not have substrate vias, the designer must be sure to locate all FETs as close as possible to the edge of the chip to allow for low inductance grounding of the FET's sources. Indirect source contacts will add parasitic source inductance to the FET circuit. In amplifiers, such inductance can act as a feedback element, causing "peaking" or "suck outs" of gain at certain frequencies.

A key to good FET design is to be sure all ground contacts are as direct as possible. During layout, keep in mind that most foundries require that all the gate strips on a MMIC chip be parallel to one another.

9.3.2.2 Microstrip Bends

MMIC circuits always contain at least a few microstrip bends. Each bend represents a parasitic element that may affect performance and must be included in the overall circuit model. In general, tight bends such as square bends or mitered bends introduce more parasitics than circular bends with large radii. Of course, large radius bends will increase overall chip size, so the designer must weigh these

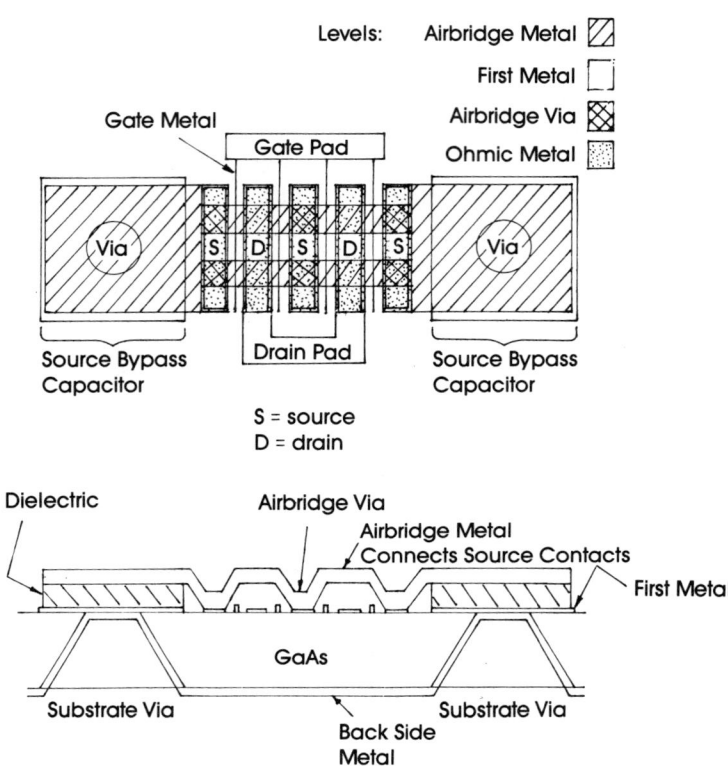

Fig. 9.16 The layout of a four-finger interdigitated FET structure, including source bypass MIM capacitors and substrate via grounds.

trade-offs. If the radius of a bend can be made five to ten times a line width, the bend becomes essentially parasitic free. This is a very desirable situation (especially at high frequencies) if the real estate is available. Figure 9.17 compares the three kinds of bends.

9.3.2.3 MIM Capacitors Versus Open Circuited Microstrip Stubs

When designing grounded shunt capacitors, the designer must choose between an MIM capacitor connected to ground through a substrate via and an open circuited microstrip stub. MIM capacitors become unrealistically small for values less than .1 pF; therefore, shunt capacitors below .1 pF are best realized by an open circuited microstrip stub. Above .1 pF, shunt capacitors are realized as MIM

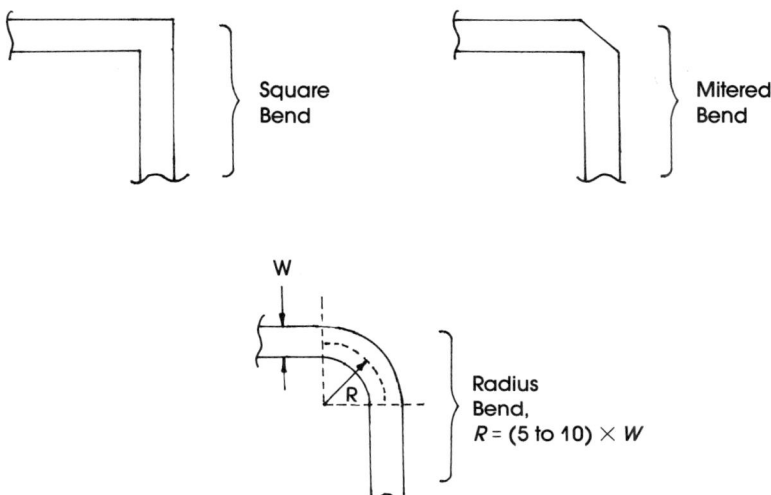

Fig. 9.17 Three types of microstrip bends.

capacitors. All series capacitors must be MIM type, which means that series capacitors cannot be less than .1 pF.

9.3.2.4 Spiral Inductors Versus Microstrip Lines

Most MMIC circuits require inductors at various points in the circuit topology. The two choices for realizing these inductors are spiral inductors and straight lengths of microstrip line. Spiral inductors are best suited to foundry processes that support very thin airbridge lines (10 μm or less). Usually, spiral inductors are not very useful for inductance below .2 nH. If all the inductors in a circuit are less than .5 nH or the foundry process does not support thin airbridge lines, the designer should consider using only straight microstrip lines as inductors.

If the decision is made to use a spiral inductor, it is important to check a particular inductor's resonance frequency to be sure resonance is well above the band of interest. Choke inductors are an exception to this rule because the impedance of a choke will remain high at and above resonance.

9.3.2.5 Coupled Line Sections

Two microstrip lines running parallel to each other will couple, causing crosstalk that can have an impact on performance of the MMIC. Coupling can be kept at a minimum by maintaining a space at least a three times line width's between adjacent lines. Of course, tradeoffs must be made constantly between this requirement and the desire for a compact circuit to minimize overall size.

9.3.2.6 On-Chip Versus Off-Chip Components

In some applications, it makes good sense to locate certain passive components such as inductors, capacitors, and resistors off chip. Some examples are

- A very high value choke inductance is needed for low frequency operation. If the necessary inductance requires an unusually large spiral inductor, to use a wound choke mounted off-chip may make sense.
- If a large value dc blocking capacitor is needed at the input or output of a MMIC, it may best be mounted off-chip.
- For nonsubstrate via hole processes, source bypass capacitors may be best mounted off-chip.
- Source resistors are often mounted off-chip to control the FET's drain current.

9.3.2.7 Active Loads

A very compact technique for biasing an FET circuit is to use a so-called active load. An active load is just an FET with its gate contact connected to its source contact. This FET always operates at I_{DSS} for any V_{DS} above V_S. The active load's impedance is very high for all voltages above V_S (see Figure 9.18) and at all

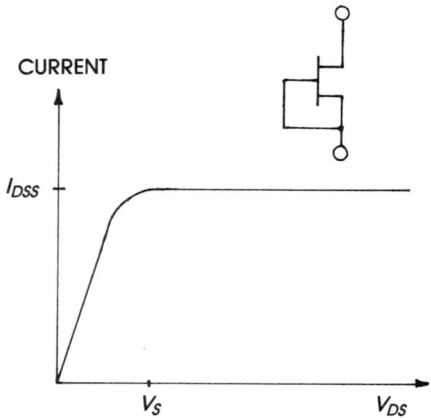

Fig. 9.18 The I-V characteristic of an active load.

frequencies. However, four compromises may have to be made to use an active load:

- Bias current is fixed at I_{DSS} of the active load.
- Considerable dc voltage (3 or more V), must be doped across the active load, dissipating dc power and decreasing overall efficiency.
- At high frequencies, the active load's impedance may decrease as a result of C_{DS}.
- In some circuits, the small but finite loss of the active load may be objectionable.

9.3.2.8 Paper Doll Techniques

A convenient way to initiate a rough cut circuit layout is to use "paper dolls," which are scaled appliques of standard cells arranged as on a scaled grid layout paper. The scale factor of this layout usually is between 200 and 1000. The paper dolls can be fastened with double stick tape and easily are moved around on the grid until an optimum configuration is achieved. Figure 9.19 shows an example of a paper doll rough cut layout for a two stage FET amplifier.

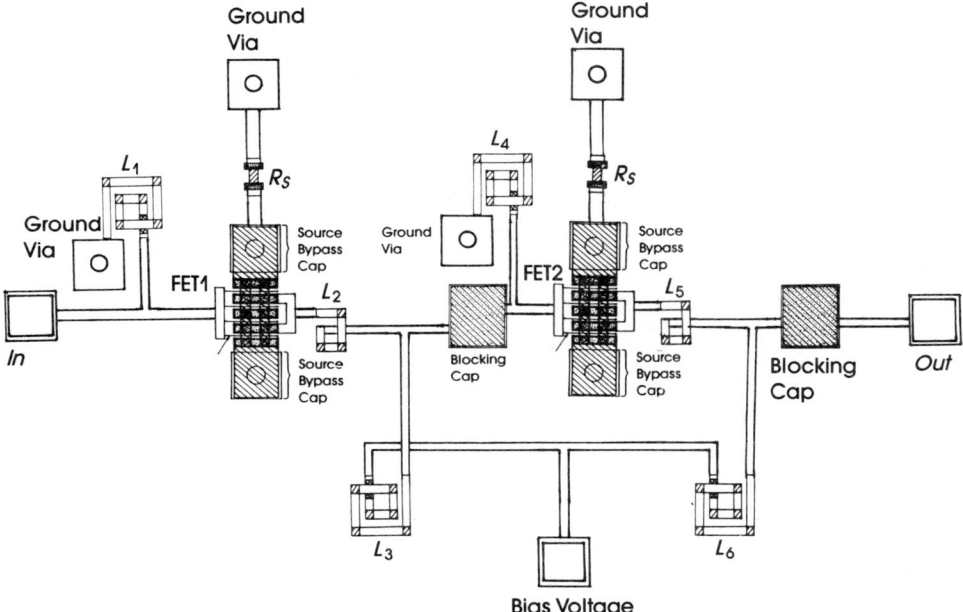

Fig. 9.19 A paper doll layout of a two-stage MMIC FET amplifier.

9.3.3 Circuit Reoptimization Including Element Models, Layout Parasitics, and Layout Modifications

Once the rough cut layout has been completed, the next step in the design process is to incorporate all the device element models and parasitic models into the Touchstone or SuperCompact circuit file. The following elements must be included in the overall circuit file at this time:

- FET models based on total gate width;
- MIM capacitor models;
- Spiral inductor models;
- Thin-film and GaAs resistor models;
- Microstrip lines (all metal lines and pads should be treated as microstrip lines);
- All microstrip parasitic elements as discussed in Section 7.2;
- Any off-chip device elements such as chokes, bias resistors, or bypass capacitors;
- Bond wires connected between the chip and the external circuit, both RF lines and bias lines.

The designer may need to "re-node" completely the basic circuit model to include all these additional elements. For the overall circuit file to become very long at this time is not unusual. Software packages and computers that analyze and optimize very large files without slowing significantly begin to have a real advantage with this type of circuit file. Large MMIC circuits occupy very large computer files.

The next step is to analyze the new circuit file and note any changes in overall performance resulting from the introduction of device and parasitic element models into the circuit file. In the rare event that no performance change has occurred, the design process can now advance to the sensitivity and yield analysis step. However, if performance needs to be improved, the circuit may have to be reoptimized to recover the lost performance.

Reoptimization can be performed either using the Touchstone or SuperCompact optimizer to vary designated circuit parameters (i.e., microstrip lines, capacitors, inductors, resistors) automatically, or it can be performed by manually varying elements a little bit at a time to regain lost performance. This "by hand" optimization procedure is very time consuming and should not be rushed.

Once reoptimization has recovered any lost performance (or at least brought performance up to acceptable levels), the rough cut layout must be modified to reflect these changes. Reoptimization and modification of the rough cut layout now becomes part of a feedback loop that interactively modifies the circuit, reintroduces these changes into the layout, and then reanalyzes the circuit's performance to see whether performance has been affected in any other ways.

This process may require several iterations to achieve the desired level of performance and requires considerable patience and creativity on the part of the designer. There are no "pat" or "sure-fire" approaches to these steps, all circuits are different and designers must treat each new circuit as an independent challenge.

It is very important at some point in the overall process to check all metals and resistor films for current carrying capability. *Imperative for reliability reasons is that no current per unit width design rules be violated at any point in a circuit.*

9.3.4 Sensitivity Analysis and Yield Predictions

Foundries have found that all process parameters experience statistical variations over the course of many wafer fabrication runs. These statistical process variations ultimately manifest as statistical variations of the element values in the circuit models. The designer must be aware of these statistical variations and plan a circuit's design so that an acceptable percentage of a quantity of circuits will meet the electrical specifications. A design can be tested analytically for performance changes with statistical process variations in one of two ways. The first method is called a *sensitivity analysis* and the second is called a *Monte-Carlo yield prediction*. Both methods perform calculations using either Touchstone or SuperCompact.

9.3.4.1 Process Variation and Sensitivity Analysis

Many of the parameters of the foundry process have statistical variations associated with them. Some examples are
- all FET model parameters,
- all sheet resistance parameters,
- all line widths,
- MIM capacitance per unit area.

Statistical variations in these parameters are communicated by the foundry to the designer in terms of a standard deviation about the nominal value of each parameter, which is equivalent to the statistical mean value. From the mathematics of statistics, the mean of a parameter, X, is defined as [10]

$$\overline{X} = \frac{X_1 + X_2 + \ldots X_N}{N} = \frac{\sum_{i=1}^{N} X_i}{N} \tag{9.9}$$

where N is the number of different values X assumed.

The standard derivation of X is defined as

$$\sigma_X = \sqrt{\frac{\sum_{i=1}^{N}[(x_i-\bar{x})^2]}{N}} \qquad (9.10)$$

Figure 9.20 shows graphically how $\bar{X}$ and σ_X are interpreted from a normal distribution of a parameter X; 68 percent of all values of X are contained within $\pm\sigma_X$ of $\bar{X}$; 95 percent of all values of X are contained within $\pm 2\sigma_X$, and over 99 percent of all values of X are contained within $\pm 3\sigma_X$ of $\bar{X}$.

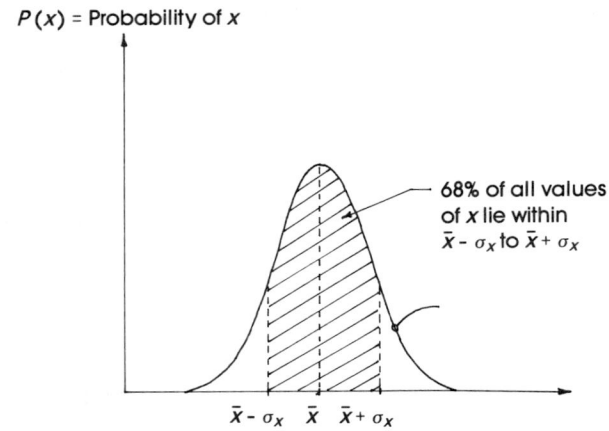

Fig. 9.20 The normal distribution of quantity X, showing the locations of $\bar{X}$ and σ_x.

The purpose of a sensitivity analysis is to investigate how sensitive a given performance parameter is to variations in a given device or process parameter. Sensitivity analysis is somewhat limited because it must be performed one parameter at a time, and very little sense of overall performance variations can be calculated from it. However, a sensitivity analysis can indicate any "sore spots," where the performance of a circuit is extremely sensitive to variations in a particular process parameter. To analyze sensitivity, let the performance parameter be called X (examples are gain, return loss, noise figure, phase shift, *et cetera*); and let the process variable be called Y (examples are the FET parameters, line widths, *et cetera*).

The sensitivity of X to Y will be defined as

$$S_X = \frac{\Delta X(\Delta Y)}{\bar{X}} \qquad (9.11)$$

where
$$\Delta X = X - \overline{X}$$
$$\Delta Y = Y - \overline{Y}$$
$\overline{X}$ = mean value of the performance parameter

A sensitivity analysis is carried out in the following way:
- Establish a value for ΔY. This typically is $\pm 10\%$ of $\overline{Y}$.
- Using Touchstone or SuperCompact, calculate the frequency response of X, first for the device or process parameter Y set to $\overline{Y} + \Delta Y$, and then for Y set to $\overline{Y} - \Delta Y$.
- For many frequencies across the band of interest, calculate $2\Delta X = X$ at $(\overline{Y} + \Delta Y) - X$ at $(\overline{Y} - \Delta Y)$.
- Divide each value of ΔX by $\overline{X}$.
- Plot $\Delta X/\overline{X}$ across the range of frequencies.

Figure 9.21 shows the steps in constructing a sensitivity plot. Typically, many such plots will be generated for a given circuit to investigate the circuit's performance sensitivity to a number of device and process parameters. If the circuit is found inordinately sensitive to one or two device parameters, modification may be necessary to reduce the sensitivity, because an inordinate sensitivity to any parameter always leads to poor yield. There is no formula for reducing sensitivity, so the designer must rely on creativity and sleuthing.

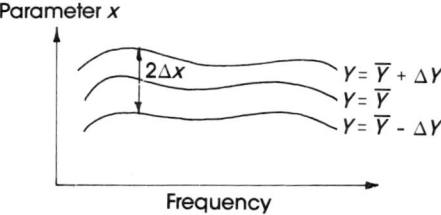

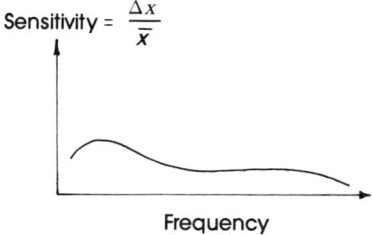

Fig. 9.21 The sensitivity, $\Delta X/\bar{X}$, of a performance parameter, X, to changes, ΔY, in an element value, Y.

9.3.4.2 Monte-Carlo Yield Predictions

Both Touchstone and SuperCompact have a Monte-Carlo yield prediction capability. Unlike the sensitivity analysis, a Monte Carlo analysis takes into account many device and process variations simultaneously and totals their combined effects. The yield is calculated as the percentage of those circuits that will meet all of the electrical performance specifications despite statistical process variations.

In Touchstone Monte-Carlo, the specified performance is inserted into the circuit file in the optimizer block. In fact, if the optimization goals for a circuit are its specifications, the optimizer block can be used as is for the Monte-Carlo analysis.

The device and process variations are entered into individual element file lines as a constrained variable. For instance, in Touchstone, the syntax for a statistical variable is

ELEMENT   1   2 PAR   #   Y1   Y2   Y3

where
ELEMENT is any Touchstone library element,
PAR is a parameter of that element,
Y1 is its mean value minus the variation ($\overline{Y} - \sigma y$),
Y2 is its mean value ($\overline{Y}$),
Y3 is its mean value plus the variation ($\overline{Y} + \sigma y$).

As many elements as the designer deems necessary may be designated as statistically variable. When Touchstone is operated in the optimizer mode, many iterations are calculated, each iteration picking random values of each variable in the range Y_1 to Y_3 and comparing the calculated performance to the goals (specifications) given in the optimizer block. A running summary is displayed of the percentage of the total number of iterations that are fully compliant with specifications. After many iterations, this percentage starts to approach a statistically significant *yield*. To calculate at least 100 iterations usually is necessary to guarantee a meaningful yield value.

Of course, a very high yield (90 to 100%) always is desirable, but, for many circuits, a yield in excess of 50% may be acceptable. If a yield under 30% is predicted, the designer must address the difficult challenge of either modifying the specifications or reconsidering the total approach to the circuit's design. Although a redesign can be extremely painful at this time, considering the work already invested in the design, to proceed and complete the layout, make a mask set, and fabricate circuits with a predicted yield of less than 30% is to invite economic ruin. Most important, an MMIC design must make economic sense.

In the case of a design predicted to have a low (under 50%) yield, the designer should review the material in Section 9.1 and do a cost estimate to answer the question of economic viability. If a circuit is judged not to be viable on the basis of yield, consider simple things that might be done to raise the yield. Perhaps, by adding a gain stage, marginal amplifier gain can be raised to the point that the minimum gain is no longer as sensitive to FET process variations. Perhaps, a particular element is unusually sensitive to process variation, and by eliminating this element the yield may improve significantly. There is no substitute for sound judgment and creative thinking when designing MMIC circuits.

9.3.5 Final Layout

The final stage of the MMIC design process is the completion of a final layout. The final layout should not be commenced until all the questions raised during layout modification and sensitivity analysis or yield prediction have been answered satisfactorily. Once the designer is convinced of the viability of the design from both the performance and cost viewpoints, work may proceed to the final layout stage.

How the final layout is done and of what it consists are strongly influenced by the foundry service used by the designer. Some foundries need only a top-level metal (first- and second-level metal) layout at this time. Top-level metal layouts can be done on various CAD layout programs such as AutoCad, or even by hand. The devices, such as FETs and capacitors, are drawn only at their topmost level, and the detailed device dimensions are supplied later at the foundry. The foundry will take all of this information, digitize it, and place it on their own CAD system (typically, CALMA GDSII). Then, the foundry will fill in all of the missing levels, such as implant, ohmic metal, dielectric, *et cetera,* necessary to produce a mask set. When the full-scale layout is completed, pen plots of each level are supplied to the designer for review. Close cooperation between the designer and the foundry is very important to ensure careful checking of the final layout for "bugs" prior to making the mask set.

Some foundries require that a designer deliver complete CALMA GDSII tapes, including all layout levels before mask-making can begin. Under this arrangement, completion of the GDSII tapes becomes the responsibility of the designer. Because few microwave companies have a CALMA system available, designers often must seek out a service company that will produce the final GDSII layout for a fee. The usual working arrangement is for the designer to take the modified layout to the CALMA service, where the final layout is digitized and the first- and second-level metal patterns are combined with standard cell information, supplied on a tape by the foundry. The designer must work very closely with the CALMA operator to ensure that all levels (there may be 11 or more) are correct in terms of both realizing the desired circuit layout and obeying the foundry's design rules. A second option is to produce a multilevel design with a layout program, such as AUTOCAD, and then translate it into GDSII using one of the software translation packages now becoming available.

REFERENCES

1. R. E. Williams, *Gallium Arsenide Processing Techniques,* Artech House, Norwood, MA, 1984.
2. S. K. Ghandi, *VLSI Fabrication Principles, Silicon and Gallium Arsenide,* John Wiley and Sons, New York, 1983.
3. J. Mun, *GaAs Integrated Circuits,* Macmillan, New York, 1988.
4. D. K. Ferry, *Gallium Arsenide Technology,* Howard W. Sams, Indianapolis, 1985.
5. R. A. Pucel, *Monolithic Microwave Integrated Circuits,* IEEE Press, New York, 1985.
6. R. E. Williams, *op. cit.*
7. *Ibid.*
8. Barna, *VHSIC* (complete citation not available).
9. M. Parisot et al., "Highly Accurate Design of Spiral Inductors for MMICs with Small Size and High Cutoff Frequency," *IEEE Trans. Microwave Theory Tech.,* Vol. MTT-32, pp. 91–95.
10. Statistical Quality Control Handbook, Western Electric Company, 1956.

Chapter 10
Final Layout of the MMIC Amplifier Examples

Creating the final layout of a MMIC circuit is a highly interactive, creative process. Chapter 9 discusses the following basic steps, which are systematically followed during this process. For the MMIC designer to create the best possible model for a circuit layout is very important, because empirical modifications are not possible.

- Obtain the foundry's design rules.
- Perform an initial layout.
- Simulate the performance of this layout using Touchstone or SuperCompact.
- Modify the layout to achieve the desired simulated performance.
- Perform a sensitivity or Monte Carlo yield analysis.
- Perform final layout.

First, obtain a set of design rules from the foundry that will be fabricating the MMIC circuits. For the purpose of these examples, a set of design rules will be assumed. The rules shown in Table 10.1 are for a hypothetical foundry. However,

Table 10.1 Foundry Process Design Role Summary

dc FET Parameters	Foundry A	Foundry B	Foundry C
I_{DSS} (mA/mm)	200.0		
g_m (mS/mm)	190.0		
V_p (V)	1.8		
V_S (V)	1.0		
V_{BGD} (V)	11.0		
V_{BGS} (V)	11.0		

Table 10.1 cont'd.

dc FET Parameters	Foundry A	Foundry B	Foundry C
RF FET Parameters			
g_M (mS/mm)	140.0		
C_{GS} (pF/mm)	1.00		
RI (Ω-mm)	4.1		
C_{DG} (pF/mm)	0.24		
R_{DS} (Ω-mm)	80.0		
C_{DS} (pF/mm)	0.36		
R_G (Ω/mm)	—		
R_S (Ω-mm)	—		
R_D (Ω-mm)	—		
Sheet Resistance	Foundry A	Foundry B	Foundry C
GaAs N$^+$ (Ω/sq)	—		
GaAs N$^-$ (Ω/sq)	300		
Thin-film resistor (Ω/sq)	50		
Ohmic Metal (Ω/sq)	—		
First Metal (Ω/sq)	.05		
Second (airbridge) Metal (Ω/sq)	.02		
Gate Metal (Ω/sq)	—		
Layer Thickness			
Ohmic Metal (μm)	—		
Gate Metal (μm)	—		
First Metal (μm)	—		
Thin-Film Resistor (μm)	—		
Second (airbridge) Metal (μm)	—		
MIM Dielectric (Å)	2000		
GaAs Substrate (mils)	4		
Maximum Safe Current per Width	Foundry A	Foundry B	Foundry C
Gate Metal (mA/μm)	2		

Table 10.1 cont'd.

Maximum Safe Current per Width	Foundry A	Foundry B	Foundry C
Thin-Film Resistor ($mA/\mu m$)	1		
First Metal ($mA/\mu m$)	3		
Second (airbridge) Metal ($mA/\mu m$)	20		
Minimum Line Width			
Ohmic Metal (μm)	12.0		
Gate Metal (μm)	.5		
First Metal (μm)	12.0		
Thin-Film Resistor (μm)	12.0		
Second (airbridge) Metal (μm)	12.0		
Minimum Line Spacing			
Ohmic Metal (μm)	6		
First Metal (μm)	6		
Thin-film Resistor (μm)	6		
Second (airbridge) Metal (μm)	6		
Capacitance/ Inductance			
MIM Capacitance ($pF/\mu m^2$)	250.00		
Substrate Via Inductance (nH)	.02		

Statistical Info Standard Deviation	Foundry A	Foundry B	Foundry C
g_M (mS/mm)	14.00		
C_{GS} (pF/mm)	.10		
RI (Ω-mm)	.40		
C_{DG} (pF/mm)	.02		
R_{DS} (Ω-mm)	8.00		
C_{DS} (Ω-mm)	.03		

Table 10.1 cont'd.

Statistical Info Standard Deviation	Foundry A	Foundry B	Foundry C
C_G (Ω-mm)	—		
R_G (Ω-mm)	—		
R_S (Ω-mm)	—		
R_D (Ω-mm)	—		
Sheet Resistance Thin-Film Resistor (Ω/sq)	2.5		
Sheet Resistance First Metal (Ω/sq)	—		
Line Width First Metal (μm)	—		
Line Width Thin-Film Resistor (μm)	—		
Line Width Second Metal (μm)	—		
MIM Capacitance (pF/mm^2)	25		
Substrate Via Inductance (nH)	—		

they are very similar to rules followed by several of the leading foundries. Note that not all of the process parameters have values. This is typical of many design rule packages, which do not communicate all of the process parameters to the designer. Often the foundries feel that certain parameters are unnecessary or proprietary knowledge. We will now follow the design process for each MMIC amplifier example.

10.1 MMIC LAYOUT FOR THE 2 TO 8 GHz TWO-STAGE FEEDBACK AMPLIFIER

The first major challenge confronting this amplifier design is converting the ideal lumped inductors to MMIC compatible spiral inductors. Most foundries have a library of standard spiral inductors for the designer to choose from. However, no such library exists for our hypothetical foundry example, so we must find a way to design these spiral inductors. We do this with the help of the HP-41C calculator programs listed in Sections 4.6.9 and 4.6.10. The layout and equivalent circuit parameters for these inductors are given in Table 10.2. The largest inductor does not conform to the inductance value in the original design, because the original value of 15 nH is too large to be physically realized as a MMIC spiral inductor.

Table 10.2 Spiral Inductors for the 2 to 8 GHz MMIC Amplifier

Inductance	No. of Turns	A-B-mils	W-mils	S-mils	$R_1\ (\Omega)$	$R_2\ (\Omega)$	$L_1\ (nH)$	$L_2\ (nH)$	$C_P\ (pF)$
.20 nH	2	5.0	.5	.8	5.87	.98	.14	.028	.021
.80 nH	2	7.0	.5	.8	9.87	1.65	.62	.12	.025
2.60 nH	3	11.0	.5	.8	22.85	3.81	2.58	.51	.049
5.6 nH	3	16.0	.5	.8	37.85	6.31	5.60	1.12	.069

Because this inductor's purpose in this circuit is to function as a bias choke, a much lower value will work just as well. We choose 5.6 nH as a good compromise between small size and high inductance. AutoCAD plots of all four inductors are shown in Figure 10.1.

The actual layout can begin by simply replacing the elements in the amplifier's schematic diagram (Figure 5.25) by the layout of the MMIC component that performs its function. The correspondence between schematic diagram symbols and layout features is as follows:

Schematic Symbol	Layout Feature
Port	Bonding Pad
Interconnects	Thin Metal Traces (modeled as microstrip lines)
Capacitor	MIM Cap
Ground	Via Hole Ground
Inductor	Spiral Inductor
Resistor	Thin-Film Resistor or GaAs Resistor
FET	Interdigitated Gate Fingers Surrounded by Source and Drain Ohmic Contacts
Cross Over	Airbridge Metal

The designer initially must decide what method of layout will be used. This decision depends critically on the kind of layout the foundry needs. If the foundry needs only a "top metal" layout, the designer is free to use any one of a number of

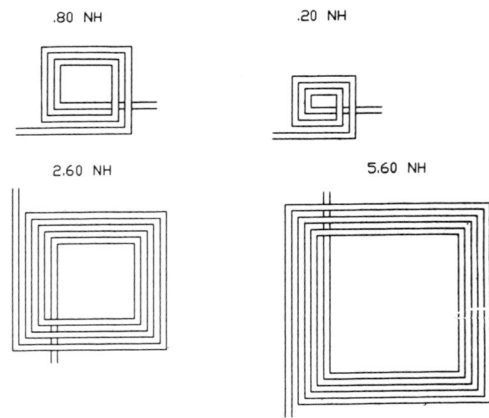

Fig. 10.1 AutoCAD plots of the four spiral inductors from Table 10.2.

generic PC-based drawing or layout CAD packages such as AutoCAD, or for that matter, a scaled, hand-drawn sketch may suffice. In this case, the foundry will add the additional masking layers needed to complete the layout. On the other hand, if the foundry requires full GDSII tapes containing all masking layers, then all masking levels become the responsibility of the designer, who may need to perform these layouts on a CALMA machine. The middle ground is held by multilevel layout performed on a PC-based CAD package such as AutoCAD, which is then translated to GDSII by one of the software translators just now becoming available.

For the purpose of these examples, we assume that our foundry accepts top-level metal layouts. Therefore, we will use AutoCAD to produce our layouts. The initial layout for the 2 to 8 GHz feedback amplifier is shown in Figure 10.2. This layout is basically the schematic diagram in Figure 5.25 with MMIC devices replacing the symbols. The inductors are all spirals. Grounds are metal pads with via holes (circles) in their centers. MIM capacitors are represented as nested rectangles, where the outer rectangle is the capacitor's bottom plate (which is in contact with the GaAs) and the inner rectangle is the top plate of the capacitor (cap top). If the bottom of the capacitor is grounded (bypass capacitor), a via hole (circle) is placed in the middle of the capacitor's layout. Small MIM capacitors are drawn as small tabs of "cap top" metal placed over slightly larger tabs of bottom plate metal. The cap top metal is understood to be always connected to an airbridge trace, even if for only a short distance. Furthermore, all crossovers also are done with airbridge metal.

Figures 10.3 to 10.5 show blowups of certain areas of the layout. These figures give details on the layout of the FET, the spiral inductors, the MIM capacitors, and the thin-film resistors (cross-hatched material).

The spiral inductor's airbridge metal must be supported periodically by posts, called *airbridge vias*. These posts (which are not shown in the layout) join the airbridge metal to the first-level metal, which is in contact with the GaAs. These posts may be thought of as analogous to the support columns beneath freeway overpasses. Their sole purpose is mechanical support.

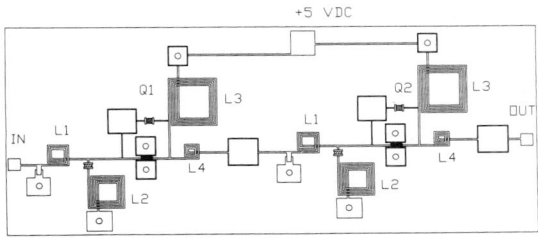

Fig. 10.2 The layout of the 2 to 8 GHz two-stage feedback amplifier, chip size is 164 × 72 mils.

338

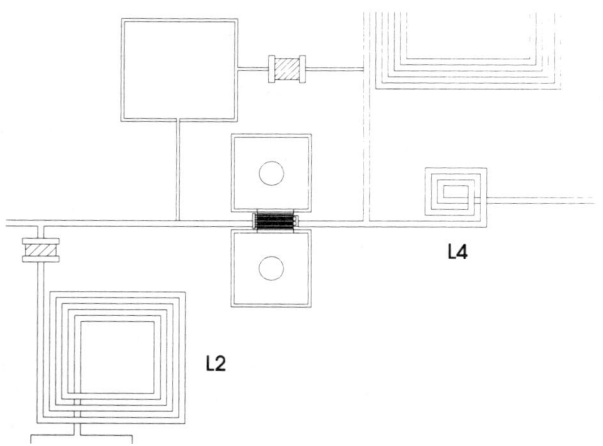

Fig. 10.3 The 2 to 8 GHz MMIC amplifier layout, FET area.

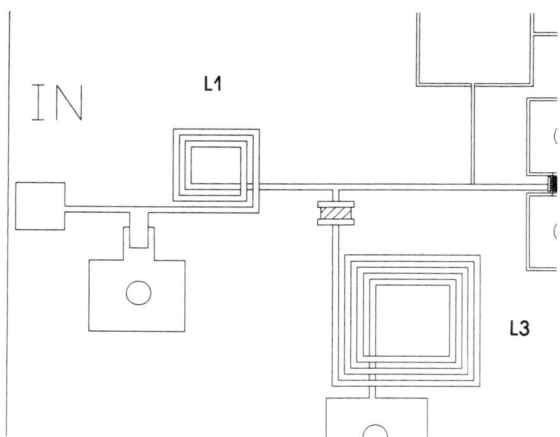

Fig. 10.4 The 2 to 8 GHz MMIC amplifier layout, input area.

Figure 10.6 gives the amplifier's modified schematic diagram based on the layout in Figure 10.2. All of the layout parasitic elements are included in this schematic diagram. These parasitic elements are

- Spiral inductor model elements (see Figure 9.14).
- Tees.
- MIM capacitor bottom plate capacitance to ground.
- Via hole inductance.
- Thin-film resistor lengths and widths.

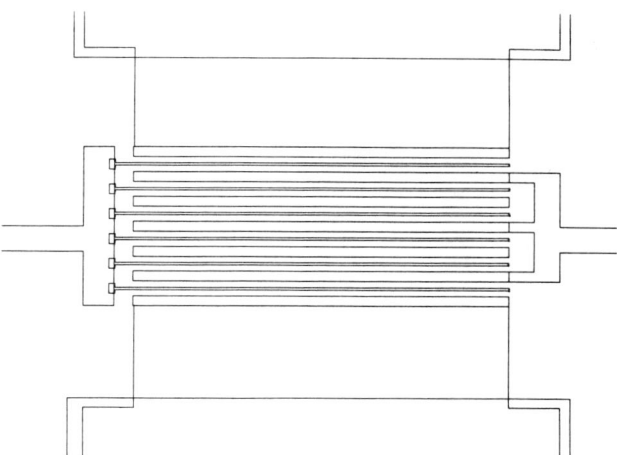

Fig. 10.5 The 2 to 8 GHz MMIC amplifier layout, FET details.

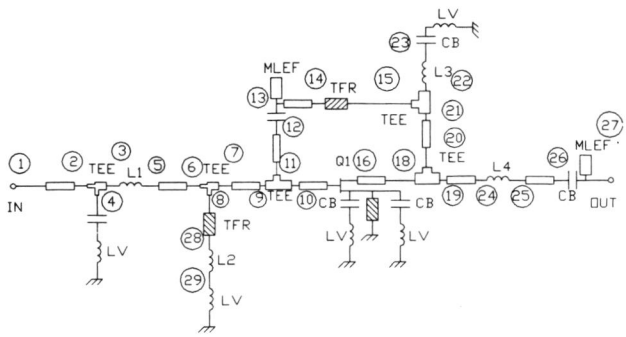

Fig. 10.6 The final schematic diagram of the MMIC 2 to 8 GHz feedback amplifier, stage 1 of two stages.

This schematic diagram, which is for the first of two identical stages, is used as the basis for a complete Touchstone file, which is given in Table 10.3. The initial layout is very conservative in terms of not physically compacting the MMIC elements; therefore, there are fairly long interconnecting lines between the elements. These line lengths may be made variables of the Touchstone optimizer to determine if shorter interconnections are a help or a detriment to performance.

The initial Touchstone simulations indicated that a fairly significant high-end gain reduction had occurred relative to the basic topology. Such a gain reduction in

Table 10.3 Touchstone—Ver (1.30-May-17-85)—Ser (15738-1598-1000)—Con (1CCC22124P) 396.ckt 02-03-90 17:05:54

```
! FINAL TOPOLOGY FOR THE 2 TO 8 GHZ TWO STAGE MMIC FEEDBACK AMPLIFER
! INCLUDING ALL LAYOUT PARASITIC ELEMENTS.
VAR

   CB=10
   LV=.013

CKT
   ! MODEL FOR L1
   CAP 1 0 C=.030
   CAP 2 0 C=.030
   CAP 1 2 C=.030
   RES 1 3 R=11.87
   RES 1 4 R=1.98
   IND 3 2 L=.879
   IND 4 3 L=.175
   DEF2P 1 2 L1

   ! MODEL FOR L2
   CAP 1 0 C=.030
   CAP 2 0 C=.030
   CAP 1 2 C=.049
   RES 1 3 R=22.85
   RES 1 4 R=3.81
   IND 3 2 L=2.59
   IND 4 3 L=.517
   DEF2P 1 2 L2

   ! MODEL FOR L3
   CAP 1 0 C=.030
   CAP 2 0 C=.030
   CAP 1 2 C=.070
   RES 1 3 R=37.85
   RES 1 4 R=6.31
   IND 3 2 L=5.6
   IND 4 3 L=1.12
   DEF2P 1 2 L3

   ! MODEL FOR L4
   CAP 1 0 C=.030
   CAP 2 0 C=.030
   CAP 1 2 C=.021
   RES 1 3 R=5.87
   RES 1 4 R=.98
   IND 3 2 L=.14
   IND 4 3 L=.028
   DEF2P 1 2 L4


   ! MODEL FOR A SINGLE FEEDBACK AMPLIFIER STAGE
   MSUB ER=12.5 H=5 T=.1 RHO=5 RGH=0
   MLIN 1 2 W=.5 L=1
   MTEE 2 3 4 W1=.5 W2=.5 W3=1.0
   SLC 4 0 L^LV C=.33
   L1 3 5
   MLIN 5 6 W=.5 L=3
   MTEE 6 7 8 W1=.5 W2=.5 W3=2.0
```

Table 10.3 cont'd.

```
    TFR 8 28 W=2 L=2.4 RS=50 F=0
    L2 28 29
    IND 29 0 L^LV
    MLIN 7 9 W=.5 L=2
    MTEE 9 10 11 W1=.5 W2=.5 W3=.5
    MLIN 11 12 W=.5 L=5
    CAP 12 13 C^CB
    MLEF 13 W=9.9 L=9.5
    MLIN 13 14 W=.5 L=4.5
    TFR 14 15 W=1 L=5 RS=50 F=0
    MLIN 10 30 W=.5 L=5
    FET 30 16 17 G=.07 T=3 F=0 CGS=.50 GGS=0 RI=8.2 CDG=.12 CDC=0 CDS=.18 RDS=160
    SLC 17 0 L^LV C^CB
    SLC 17 0 L^LV C^CB
    TFR 17 0 W=2 L=.8 RS=50 F=0
    MLIN 16 18 W=.5 L=3
    MTEE 18 19 20 W1=.5 W2=.5 W3=.5
    MLIN 20 21 W=.5 L=6
    MTEE 21 22 15 W1=.5 W2=.5 W3=.5
    L3 22 23
    SLC 23 0 L^LV C^CB
    MLIN 19 24 W=.5 L=2
    L4 24 25
    MLIN 25 26 W=.5 L=2
    CAP 26 27 C^CB
    MLEF 27 W=9.5 L=9.5
    DEF2P 1 27 AMP

    ! CASCADE TWO STAGES
    AMP 1 2
    AMP 2 3
    DEF2P 1 3 CAMP
FREQ
    SWEEP 1 10 1
OUT
    CAMP DB[S21] GR1
    CAMP DB[S11] GR1
    CAMP DB[S22] GR1
    CAMP DB[S12] GR2
GRID
    RANGE 1 10 1
    GR1 -15 15 5
    GR2 -50 0 5
OPT
    RANGE 2 7                          RANGE 7 8
    CAMP DB[S21]=10.0  50              CAMP DB[S21]=10.0  150
    CAMP DB[S11]<-10   20              CAMP DB[S11]<-10    20
    CAMP DB[S22]<-10   20              CAMP DB[S22]<-10    20
```

this type of amplifier almost certainly is due to high-frequency losses introduced by the spiral inductors. The designer has three options available at this point:

1. Redesign the spiral inductors for lower loss (i.e., widen their trace widths to reduce resistance).
2. Start the design over with smaller FETs. This will ensure lower value inductors in the matching structure, which means smaller, lower-loss inductors. Smaller FETs also will have more gain at high frequencies due to reduced C_{GS}.
3. Accept the performance as it stands, and reduce the expected bandwidth of the amplifier to include only the region of flat gain. The initial simulations of this particular amplifier indicate that 2 to 6 GHz will be its useful bandwidth.

We chose option 3, and accept that our amplifier has been reduced to a 2 to 6 GHz bandwidth by the realities of the MMIC layout. Such a decision always is complicated, based on the needs and resources of the organization doing the development. It will not always be easy for a designer to go ahead with a design whose performance has been significantly reduced by layout parasitics. This judgment must be handled on a case-by-case basis.

Once the decision has been made to go ahead with the design as it stands, the question that should be answered is, What minor modifications will improve performance? The answer to this question is given by the Touchstone optimizer, which reduces the interconnected lengths to their minimum values. Therefore, the designer needs to modify the layout in Figure 10.2 so that all interconnects are at absolute minimum lengths. It will help if the 2.60 nH inductor is "mirrored" from right to left. Mirroring the .20 nH inductor from top to bottom also may help. A very good move would be to completely eliminate the interconnect from node 7 to node 9. Other useful changes can be found by applying some creative thinking. The major cautionary note is to be sure that no two spiral inductors are extremely close to each other. Problems can develop if two spirals are close enough to experience significant mutual inductance. Mutual inductance between spiral inductors is a very difficult parasitic element to model and is best avoided if at all possible. This problem is avoided by being sure all inductors are separated by at least five trace widths.

The final values for all the interconnect lines are shown in the Touchstone file in Table 10.3. Figure 10.7 shows the final simulation, including these minor layout modifications. Performance is excellent from 2 to 6 GHz, but drops off rapidly above 7 GHz.

The remaining task before completing this design is to investigate the circuit's sensitivity to statistical variations in the fabrication process. In this example, we will check the circuit's sensitivity to various parameter variations on a one-by-one basis. In the second MMIC example (the 2 to 20 GHz distributed amplifier), we will perform a Monte-Carlo yield analysis that simultaneously takes into account all of the parameter variations.

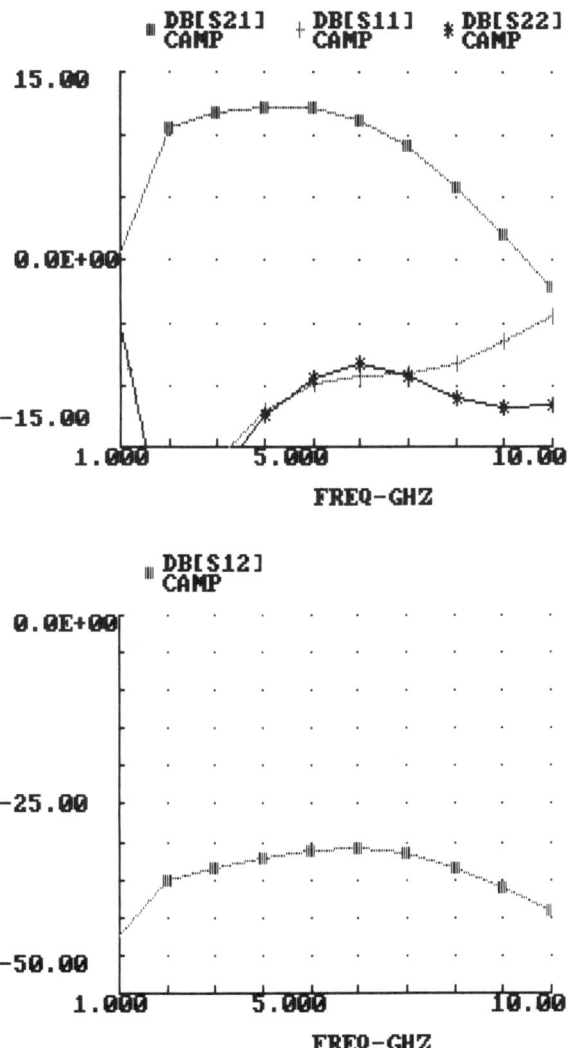

Fig. 10.7 Touchstone simulation of the 2 to 6 GHz MMIC feedback amplifier, with all layout elements in the model.

We will test the circuit's sensitivity to all of the design parameters that have statistical variations listed in the design rules. These are g_m, C_{GS}, RI, C_{DG}, C_{DS}, R_{DS}, R_f (thin-film feedback resistor), R_1 (thin-film matching resistor), and all MIM caps. The statistical variations for these parameters are listed in the design rules as standard deviations relative to the mean value of the given parameter. We will measure the circuit's sensitivity to a given parameter by running the Touchstone simulation first with the parameter's mean value *plus* the standard deviation and then with the parameter's mean value *minus* the standard deviation. Because this range of performance statistically includes 68% of all values of the given parameter (see Subsection 9.3.4), we can say that about 70% of all circuits will perform between these two simulations, if only the single parameter under consideration is varying. Strictly speaking, we should take the difference between the simulated performance levels (i.e., Δ gain) and divide it by the simulated performance at the mean value of the parameter at all frequencies. However, this very often is more work than it is worth, as the designer can tell by inspecting the two simulations whether the circuit's performance is unduly sensitive to a particular parameter over the range of frequencies.

The sensitivity simulation for each of these parameters is shown in Figures 10.8 to 10.16. The circuit's gain is most sensitive to the FET's g_m, with 4 dB of total variation occurring at midband; g_m is clearly the leading "yield driver" for this circuit. The next most important parameter in terms of sensitivity is the FET's C_{GS}, which causes about 2 dB overall variation at the high end of the band but almost no change at the low end. This is typical of C_{GS} in a wide range of circuits, because only at the high frequencies does C_{GS} become a major limitation on the circuit's ability to provide a good match. The good news is that g_m and C_{GS} are high-correlated random variables, which means that when g_m is high, C_{GS} also is high, reducing the high-end gain. When g_m is low, C_{GS} also is low, raising the high-end gain. Therefore, at least at the high end, g_m and C_{GS} variations tend to cancel each other out. However, at midband, g_m dominates, and there is still considerable variation due to g_m alone.

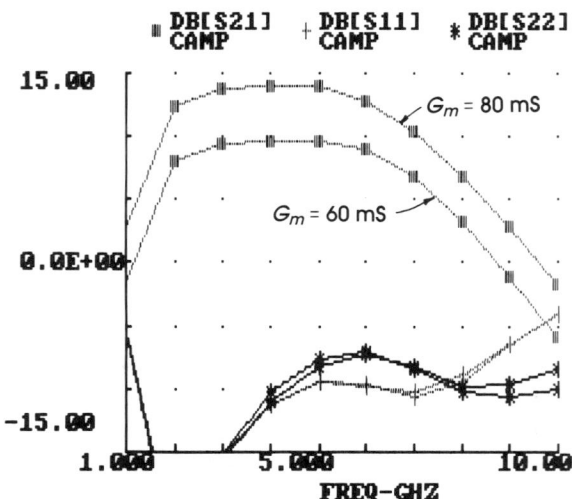

Fig. 10.8 Simulated sensitivity of the 2 to 6 GHz MMIC amplifier to g_m.

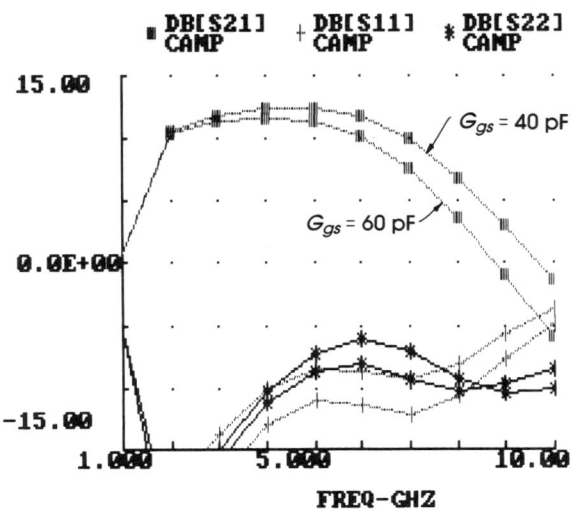

Fig. 10.9 Simulated sensitivity of the 2 to 6 GHz MMIC amplifier to C_{gs}.

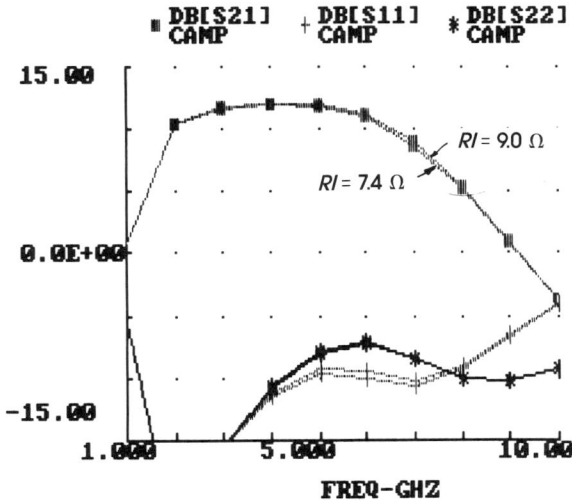

Fig. 10.10 Simulated sensitivity of the 2 to 6 GHz MMIC amplifier to RI.

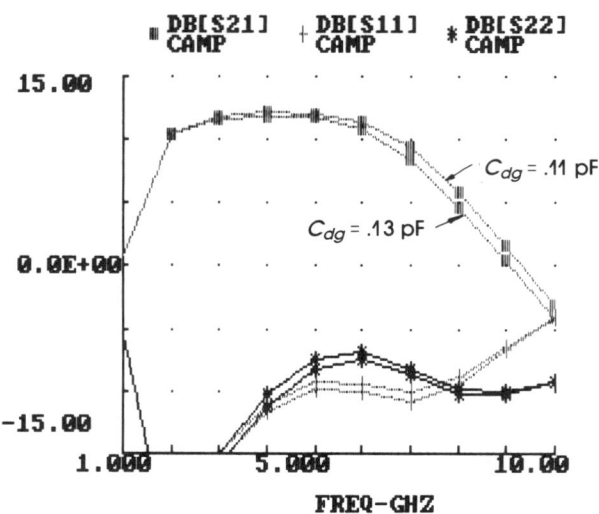

Fig. 10.11 Simulated sensitivity of the 2 to 6 GHz MMIC amplifier to C_{dg}.

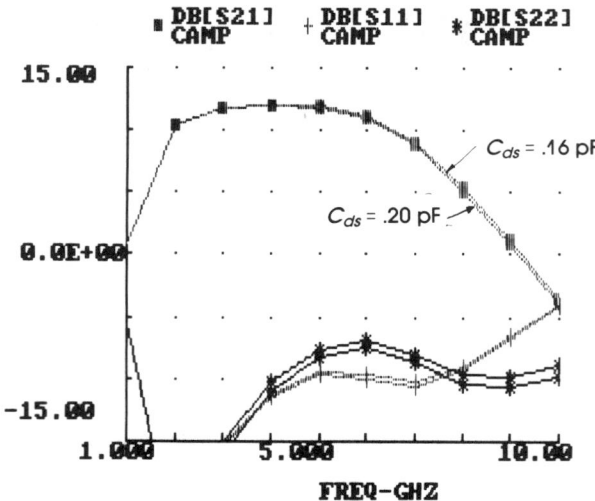

Fig. 10.12 Simulated sensitivity of the 2 to 6 GHz MMIC amplifier to C_{ds}.

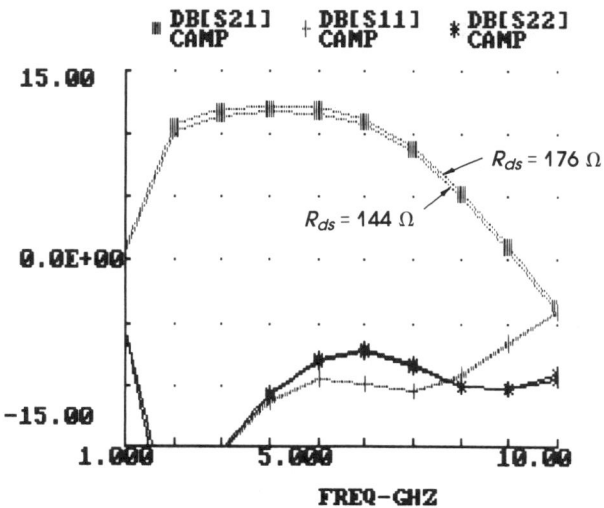

Fig. 10.13 Simulated sensitivity of the 2 to 6 GHz MMIC amplifier to R_{ds}.

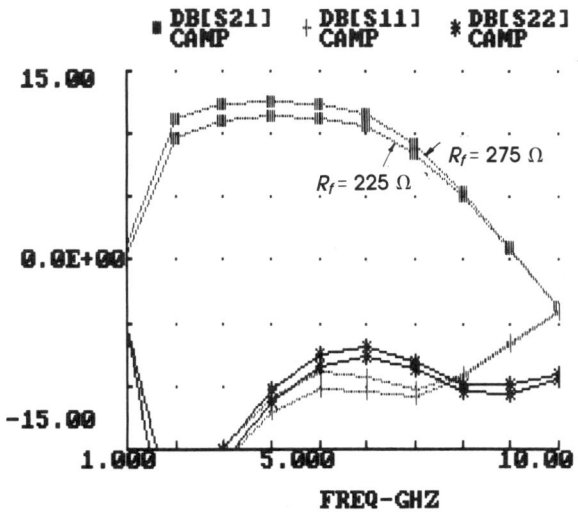

Fig. 10.14 Simulated sensitivity of the 2 to 6 GHz MMIC amplifier to R_f.

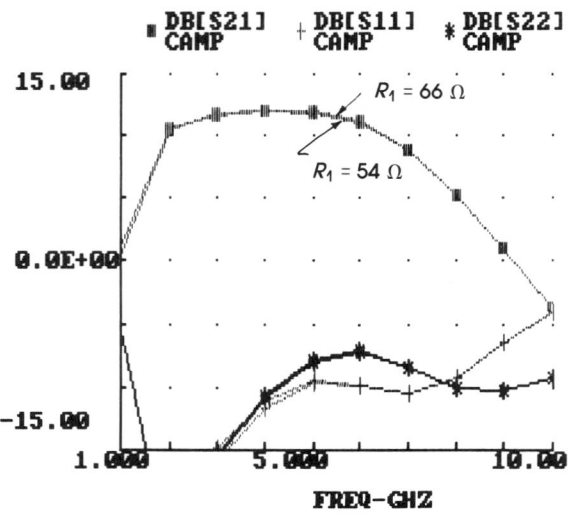

Fig. 10.15 Simulated sensitivity of the 2 to 6 GHz MMIC amplifier to R_1.

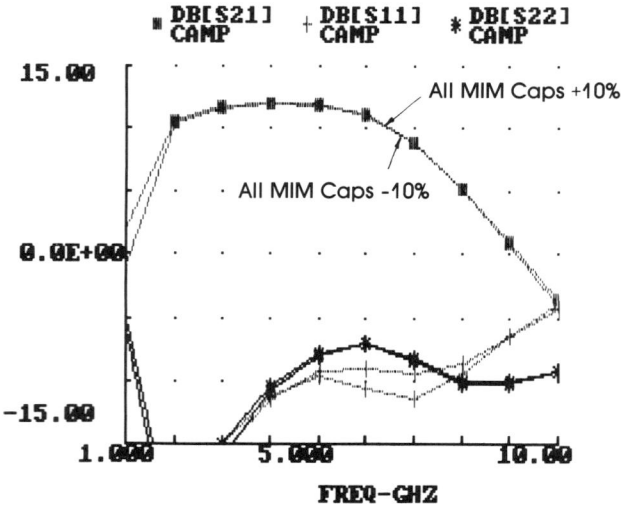

Fig. 10.16 Simulated sensitivity of the 2 to 6 GHz MMIC amplifier to all MIM capacitors.

None of the other FET parameter variations significantly affects the simulated performance, which is very fortunate and typical of most FET amplifiers. Therefore, a very good first estimate of sensitivity usually is achieved by performing the analysis for g_m and C_{GS} only.

Variations in the feedback resistor, R_f, affects the gain by about 2 dB overall at the low end, as would be expected. For this reason, to use thin-film resistors as feedback resistors rather than GaAs resistors is important because GaAs resistors have considerably more statistical variation than equivalent thin-film resistors.

Variations in R_1 and all the MIM capacitors acting together have negligible effect. Only below the band edge do performance variations due to the MIMs become apparent. This is because the large value capacitors no longer are large enough to give adequate bypassing at the very low frequencies.

If satisfied that nothing catastrophic is happening when the circuit is subjected to these statistical variations, the designer may proceed to the final step; that is, to modify the layout per the results of the final Touchstone optimization, which in this case means shorting all the interconnecting traces. This modification is advantageous for overall chip size, because reducing these trace lengths will bring the chip into line with what is normally considered an economically realistic size.

10.2 FINAL LAYOUT OF THE 2 TO 20 GHz DISTRIBUTED AMPLIFIER

The layout of the 2 to 20 GHz distributed amplifier is shown in Figure 10.17. The sources of all four FETs are connected to a common ground bus that runs along the length of the chip near its middle. Between each FET is a via hole that provides a direct ground connection for all of the FET source contacts. The gate artificial lines, which are microstrip traces as narrow as the design rules allow, meander back and forth across the bottom half of the chip, interconnecting all of the FET gates. The narrow line width is chosen for the gate lines to keep the microstrip parasitic shunt capacitance at a minimum, because shunt capacitance reduces cut-off frequency and may reduce high-frequency gain.

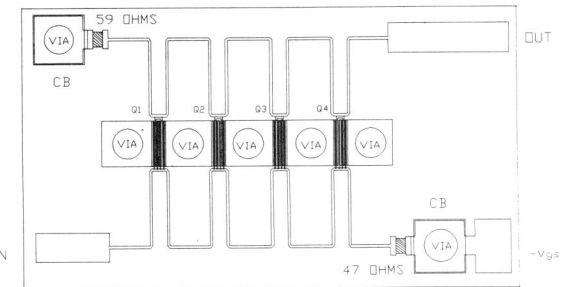

Fig. 10.17 The layout of a 2 to 20 GHz MMIC distributed amplifier. Chip size is 50 × 35 mils.

A very similar meander structure is established for the drain lines. In both drain and gate line structures, the meanders are not separated by equal amounts, but alternate between closely spaced sections and widely spaced sections. This alternating pattern is set up to get around a modeling problem. The problem is that Touchstone does not have a coupled-line model that allows coupling on both sides of a line. Thus, in our gate and drain line meander structure, the closely spaced sections are modeled as coupled line pairs, whereas the widely spaced sections are modeled as two microstrip lines, which are assumed not to couple. This simple alternating spacing strategy makes accurate modeling possible.

At the termination end of both the gate and the drain line structure is a thin-film resistor of approximately 50 Ω, which is RF grounded through an MIM bypass capacitor on top of a via hole. The gate line capacitor's top metal is connected to a bonding pad that is used to supply gate (negative) bias voltage. This biasing arrangement works because the gates do not draw current, so no voltage drop occurs across the gate line termination. The same approach cannot be used on the drain line because the full drain current of all four FETs would have to flow through the drain line termination, creating considerable voltage drop. For this reason, drain voltage must be introduced externally to the chip by a choke coil that is connected at the output terminal of the amplifier. The drain line bypass capacitor keeps the drain line dc isolated while providing RF grounding on one side of the drain termination resistor.

Because the drain and gate lines do not reach to the edge of the chip, short 50 Ω line sections are provided for inputs and outputs. Although not indicated by this layout, the designer might seriously consider providing bends in these 50 Ω lines to extend them up and down so that the input and the output of the amplifier are on a common center line. Such a configuration greatly simplifies the mechanical designs of any subsystems that use this chip, especially those with long cascaded chains.

Figures 10.18 to 10.22 show blown-up layouts of the FETs and source vias, the gate and drain line terminations and their bypass capacitors, the gate line and how it connects to the FET's gate, and the drain line and how it connects to the FET's drain. The only component that is not shown in these detailed diagrams is the supplemental drain shunt capacitors. Although these capacitors are shown in the schematic diagram in Figure 10.23, they have been omitted from the layout because, after the final optimization, their values are so small that they are no longer necessary and can be conveniently deleted. These capacitors are really present in the form of drain line microstrip parasitic shunt capacitance, which works to advantage in this instance.

The schematic diagram, based on the layout in Figure 10.17, is shown in Figure 10.23. The gate and drain line meanders are broken into basic units that are modeled per the schematic diagrams in Figures 10.24 and 10.25. The basic units are configured such that all of the coupling between parallel lines occurs within one basic unit. The meander pattern is always the same, and so the basic drain and gate line units are repeated four times and treated as *identical, noncoupled entities*.

Fig. 10.18 The 2 to 20 GHz MMIC amplifier layout, FET structure and source grounding vias.

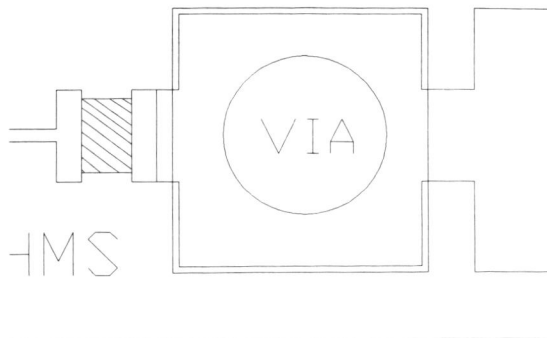

Fig. 10.19 The 2 to 20 GHz MMIC amplifier layout, gate termination and bypass capacitor.

Fig. 10.20 The 2 to 20 GHz MMIC amplifier layout, drain line termination and bypass capacitor.

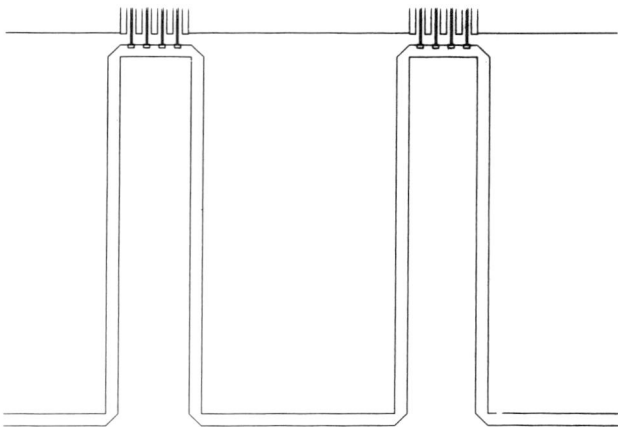

Fig. 10.21 The 2 to 20 GHz MMIC amplifier layout, gate line sections.

Fig. 10.22 The 2 to 20 GHz MMIC amplifier layout, drain line sections.

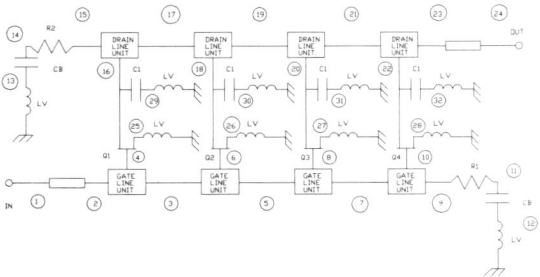

Fig. 10.23 The final schematic diagram of the MMIC 2 to 20 GHz distributed amplifier, including all layout parasitic elements.

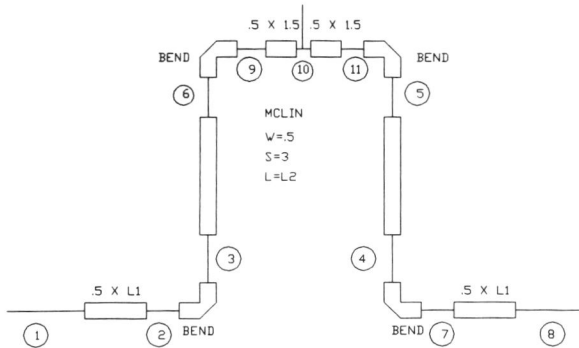

Fig. 10.24 Schematic diagram of the distributed amplifier's gate line section layout unit.

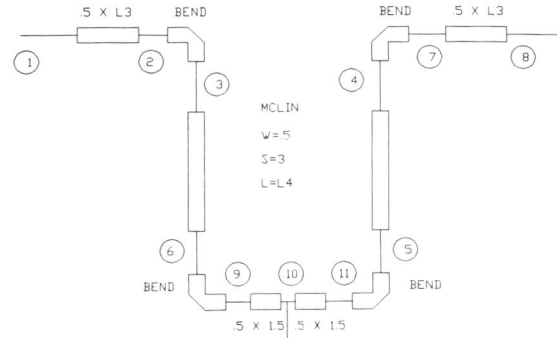

Fig. 10.25 Schematic diagram of the distributed amplifier's drain line section unit.

Within the Touchstone file, the drain line and gate line units are set up as independent three-port networks and then are nested into the top-level schematic diagram shown in Figure 10.23.

The Touchstone file for the whole amplifier including all layout parasitics (bends and coupled line sections) is given in Table 10.4. The line lengths within the gate and drain line units and the terminating resistor values remain variables of the optimizer. The final topology is optimized for flat gain from 2 to 20 GHz. The optimized lengths and resistances are given in the Touchstone file in Table 10.4. The final simulated performance is shown in Figure 10.26. The simulated gain is a flat 9.0 dB from 2 to 18 GHz and is still useful up to 23 GHz. The simulated input and output VSWR is under 2:1 from 1 to 19 GHz, and the simulated reverse isolation is greater than 34 dB over the same band. This is excellent performance for such a simple, straightforward amplifier design.

Table 10.4 Touchstone—Ver (1.30-May-17-85)—Ser (15738-1598-1000)—Con (1CCC22124P) 390.ckt
02-03-90 14:58:03

```
! THE FINAL TOPOLOGY OF THE FOUR SECTION 2 TO 20 GHZ DISTRIBUTED MMIC AMPLIFIER
! THE GaAs FETS ARE .5X230 MICRON FOUNDRY FETS
! DISTRIBUTED ELEMENT MODEL INCLUDING ALL ON CHIP PARASITIC ELEMENTS.
VAR
    L1\7.43859
    L2\12.85791
    L3\9.75344
    L4\19.68518
    CB=10
    C1\0.01219
    R1\47.32693
    R2\59.47283
    LV=.020
CKT
    ! GaAs FET MODEL
    FET 1 2 3 G=.035 T=1.7 F=0 CGS=.23 GGS=0 RI=8 CDG=.005 CDC=0 CDS=.08 RDS=1400
    DEF3P 1 2 3  TRAN

    MSUB ER=12.5 H=5 T=.2 RHO=5 RGH=0

    ! BASIC GATE LINE "UNIT"
    MLIN 1 2 W=.5 L^L1
    MBEND 2 3 W=.5 ANG=90 M=0
    MCLIN 3 4 5 6 W=.5 S=3 L^L2
    MBEND 4 7 W=.5 ANG=90 M=0
    MLIN 7 8 W=.5 L^L1
    MBEND 6 9 W=.5 ANG=90 M=0
    MLIN 9 10 W=.5 L=1.5
    MLIN 10 11 W=.5 L=1.5
    MBEND 5 11 W=.5 ANG=90 M=0
    DEF3P 1 10 8   GU

    ! BASIC DRAIN LINE "UNIT"
    MLIN 1 2 W=.5 L^L3
    MBEND 2 3 W=.5 ANG=90 M=0
    MCLIN 3 4 5 6 W=.5 S=3 L^L4
    MBEND 4 7 W=.5 ANG=90 M=0
    MLIN 7 8 W=.5 L^L3
    MBEND 6 9 W=.5 ANG=90 M=0
    MLIN 9 10 W=.5 L=1.5
    MLIN 10 11 W=.5 L=1.5
    MBEND 5 11 W=.5 ANG=90 M=0
    DEF3P 1 10 8   DU

    ! AMPLIFIER TOPOLOGY
    MLIN 1 2 W=4 L=10
    GU 2 4 3
    GU 3 6 5
    GU 5 8 7
    GU 7 10 9
    RES 9 11 R^R1
    CAP 11 12 C^CB
    IND 12 0 L^LV
    IND 13 0 L^LV
    CAP 13 14 C^CB
    RES 14 15 R^R2
    DU 15 16 17
    DU 17 18 19
    DU 19 20 21
    DU 21 22 23
    MLIN 23 24 W=4 L=10
    CAP 16 29 C^C1
    CAP 18 30 C^C1
    CAP 20 31 C^C1
    CAP 22 32 C^C1
    IND 29 0 L^LV
    IND 30 0 L^LV
    IND 31 0 L^LV
    IND 32 0 L^LV
    TRAN 4 16 25
    TRAN 6 18 26
    TRAN 8 20 27
    TRAN 10 22 28
    IND 25 0 L^LV
    IND 26 0 L^LV
    IND 27 0 L^LV
    IND 28 0 L^LV
    DEF2P 1 24  AMP

FREQ
    SWEEP 1 26 1

OUT
    AMP DB[S21] GR1
    AMP DB[S11] GR1
    AMP DB[S22] GR1
    AMP DB[S12] GR2
```

Table 10.4 cont'd.

```
GRID                          OPT
   RANGE 1  26  1                RANGE 2  20
   GR1  -20  15  5               AMP  DB[S21]=10  50
   GR2  -50   0  10              AMP  DB[S11]<-15  10
                                 AMP  DB[S22]<-15  10
```

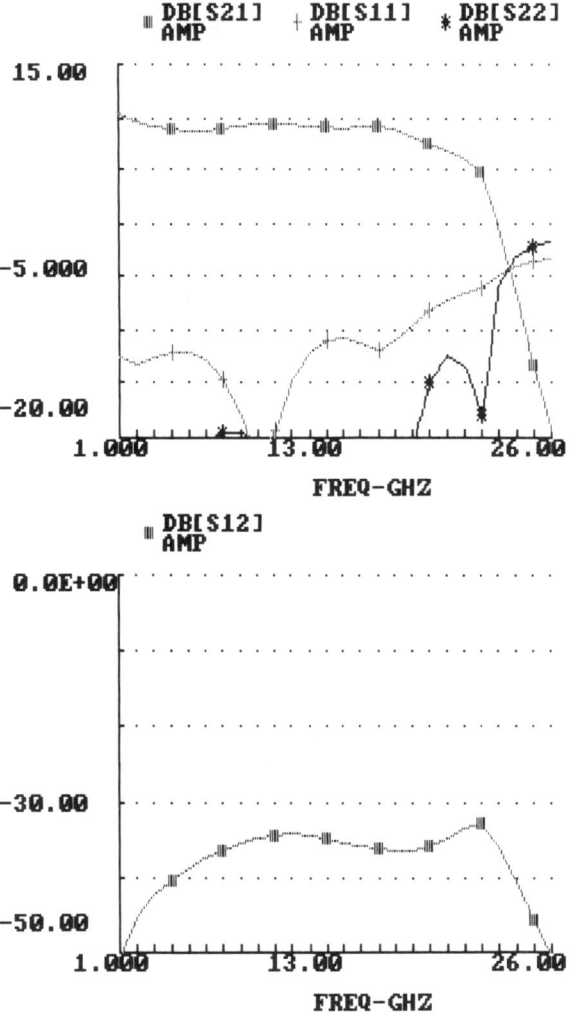

Fig. 10.26 Touchstone simulation of the 2 to 20 GHz MMIC distributed amplifier, with all layout elements in the model.

The final issue to be addressed before completing the final layout is sensitivity or yield prediction. In the case of the 2 to 20 GHz distributed amplifier, we use the Touchstone Monte-Carlo program to simulate fabrication with all of the statistical process variables operating simultaneously. The Touchstone Monte-Carlo file for this circuit, including the foundry-supplied parameter standard deviations (Table 10.1), is given in Table 10.5. Each statistically variable parameter is set up within the Touchstone Monte-Carlo file to vary *plus* or *minus* from its mean value by its standard deviation. The performance "pass" criteria is established in the OPT block. We use as passing performance, 8 dB to 11 dB gain, from 2 to 18 GHz, with 2:1 VSWR at the input and the output. The program calculates the amplifier's performance for a number of iterations with new random values for each variable parameter chosen, each iteration from within its respective range. With each iteration, the calculated performance is compared to the pass criteria in the OPT block to determine whether the amplifier passed or failed. A running tally of yield (number of passes per total number of iterations) is maintained throughout the process. As the number of iterations increases, the yield should converge to an ultimate value that is a statistically significant yield. One hundred or more iterations may be necessary to bring about convergence. Table 10.6 gives a tabulation of yield as the number of iterations increases. After 160 iterations, the yield has converged to about 57 percent, which is a very good yield from a practical, economic and manufacturability point of view.

Table 10.5 Touchstone-MC–Ver (1.30-May-17-85) Ser (15738-1598-1000)—Con (1CCC22124P) 390 OMC.CKT 09-04-89 11:26:09

```
! THE FINAL TOPOLOGY OF THE FOUR SECTION 2 TO 20 GHZ DISTRIBUTED MMIC AMPLIFIER
! THE GaAs FETS ARE .5X230 MICRON FOUNDRY FETS
! DISTRIBUTED ELEMENT MODEL INCLUDING ALL ON CHIP PARASITIC ELEMENTS.
! MONTE CARLO YIELD ANALYSIS
VAR

     GM   #.032    0.03500   .039      ! +/- 10% VARIATION IN THE FET'S GM
     CGS  #.20     0.23000   .25       ! +/- 10% VARIATION IN THE FET'S CGS
     RI   #7.2     8.00000   8.8       ! +/- 10% VARIATION IN THE FET'S RI
     RDS  #1260    1.400E+03 1540      ! +/- 10% VARIATION IN THE FET'S RDS
     CDG  #.0045   0.00500   .0055     ! +/- 10% VARIATION IN THE FET'S CDG
     CDS  #.072    0.08000   .088      ! +/- 10% VARIATION IN THE FET'S CDS
     L1=7.43859
     L2=12.85791
     L3=9.75344
     L4=19.68518
     CB  # 9       10.00000  11        ! +/- 10% VARIATION IN THE MIM CAPACITORS
     C1  # .011    0.01219   .0134     ! +/- 10% VARIATION IN THE MIM CAPACITORS
     R1  # 45      47.32693  50        ! +/-  5% VARIATION IN THE THIN FILM RESISTOR
     R2  # 56      59.47283  62        ! +/-  5% VARIATION IN THE THIN FILM RESISTOR
     LV=.020
```

Table 10.5 cont'd.

```
! GaAs FET MODEL
FET 1 2 3 G^GM    T=1.7 F=0 CGS^CGS GGS=0 RI^RI CDG^CDG   CDC=0 CDS^CDS RDS^RDS
DEF3P 1 2 3    TRAN

MSUB ER=12.5 H=5 T=.2 RHO=5 RGH=0

! BASIC GATE LINE "UNIT"                TRAN 6 18 26
MLIN 1 2 W=.5 L^L1                      TRAN 8 20 27
MBEND 2 3 W=.5 ANG=90 M=0               TRAN 10 22 28
MCLIN 3 4 5 6 W=.5 S=3 L^L2             IND 25 0 L^LV
MBEND 4 7 W=.5 ANG=90 M=0               IND 26 0 L^LV
MLIN 7 8 W=.5 L^L1                      IND 27 0 L^LV
MBEND 6 9 W=.5 ANG=90 M=0               IND 28 0 L^LV
MLIN 9 10 W=.5 L=1.5                    DEF2P 1 24   AMP
MLIN 10 11 W=.5 L=1.5
MBEND 5 11 W=.5 ANG=90 M=0           FREQ
DEF3P 1 10 8   GU
                                        SWEEP 1 26 1
! BASIC DRAIN LINE "UNIT"
MLIN 1 2 W=.5 L^L3                   OUT
MBEND 2 3 W=.5 ANG=90 M=0
MCLIN 3 4 5 6 W=.5 S=3 L^L4             AMP DB[S21] GR1
MBEND 4 7 W=.5 ANG=90 M=0               AMP DB[S11] GR1
MLIN 7 8 W=.5 L^L3                      AMP DB[S22] GR1
MBEND 6 9 W=.5 ANG=90 M=0
MLIN 9 10 W=.5 L=1.5                 GRID
MLIN 10 11 W=.5 L=1.5
MBEND 5 11 W=.5 ANG=90 M=0              RANGE 1 26 1
DEF3P 1 10 8   DU                       GR1 -20 15 5

! AMPLIFIER TOPOLOGY                 OPT
MLIN 1 2 W=4 L=10
GU 2 4 3                                RANGE 2 18
GU 3 6 5                                AMP DB[S21]<11
GU 5 8 7                                AMP DB[S21]>8.0
GU 7 10 9                               AMP DB[S11]<-9.6
RES 9 11 R^R1                           AMP DB[S22]<-9.6
CAP 11 12 C^CB
IND 12 0 L^LV
IND 13 0 L^LV
CAP 13 14 C^CB
RES 14 15 R^R2
DU 15 16 17
DU 17 18 19
DU 19 20 21
DU 21 22 23
MLIN 23 24 W=4 L=10
CAP 16 29 C^C1
CAP 18 30 C^C1
CAP 20 31 C^C1
CAP 22 32 C^C1
IND 29 0 L^LV
IND 30 0 L^LV
IND 31 0 L^LV
IND 32 0 L^LV
TRAN 4 16 25
```

Table 10.6 Yield as a Function of the Number of Iterations

No. of Trials	Yield (%)
10	55.56
20	52.23
40	48.72
80	55.70
160	57.23

INDEX

Abrupt junction, 27
ac Magnetic field, 38
ac Magnetization
    210 mode, 42
    magnetostatic mode chart, 42
Active material, 9
Active resistance, 35
Adhesion layer metalization, 240, 241
Adhesion metal, 299
Air gap, 51
Airbridge
    second-level metal, 302
    vias, 304, 337
Alumina
    microstrip transmission line substrate, 63
    substrate material, 239, 240
Aluminum
    gate metal, 299
    housing, 235
    metalization, 241
    Shottky gate metal, 7
Amplifier(s), 70
    feedback amplifier, 138, 140
    MMIC chips, 294
Anistropic field, 39
Anisotropy, 47
Artwork, 236–238
    photolithographic techniques, 217
Atomic spin, 36
AutoCAD, 79

Ball bonding, *see* Wire bonding
Barrier layer metalization, 241
Barrier metal, 299
Barrier potential, 26, 27
Base contact fingers, 23

Base resistance, 24
Base-to-collector current gain, 24
Base transit time, 25
Beryllia, *see* Substrate material
Bias current, 35
Bipolar transistors, 233, 250
    Smith chart analysis, 65
Blocking capacitors, 231
Boltzmann's constant, 33
Bonding
    ball, 249–250
    schedule, 249
    wedge, 248–249
    wire, *see* wire bonding
Brazing, 7, 245–247
    fixtures, 246
    material, 245
Built-in potential, 28
Bypass capacitors, 231

CAD optimzation, 73
CALMA, 337
    MMIC masks, 288
Capacitor(s), 67, 231
Capillary, ball wire bonding, 249
Carpenter 49 steel, 51
Carriers, 231, 245, 246
Channel, 6
Characteristic impedance, 59, 65–66
    even and odd modes, 65
    Smith chart calculations, 64–65
Chromium, *see* Gate metal
Chromium-platinum-gold, *see* First metal
Circuit analysis, 73
Circuit board,
    fabrication techniques, 217

Circuit modeling, 2
CKT
    Touchstone blocks, 74
Coil, 54
Coincidence limiting region, 47
Collector, 23-25
Collector-base junction, 23, 25
Collector depletion capacitance, 24
Compensated bends, *see* Microstrip bends
Computer-aided design (CAD), 1
Conduction channel depth, 11
Conductivity, 34
Conductor loss, 61-62
Contact resistance, 17
Coordinatograph, 236
Copper rib, 234-235
Cost factors
    as a function of volume, 291
    CALMA GDSII, 288
    chips, 289
    commercially available circuits, 293
    comparison, 292
    design rules, 287-288
    development, 287, 292
    dicing, 289
    engineering costs, 292
    inspection, 291
    make or buy decisions, 294
    packaging, 291
    photomask, 288
    wafers, 288-290
    yield, 290
Couplers, 69
    Lange, *see* Lange Coupler
Coupling loop, 39
Cross coupling, 139
Crossing mode, 42
Crosstalk, 65
Crystal
    damage, 297
    growth, 296
    YIG, 39, 41, 43
Cup-end-cap structure, 52
Curie temperature, 37
Current amplification, 23
Cut-off frequency, 30

dc Magnetic field, 38, 39, 51
dc Power input, 21

Depletion
    depth, 11
    layer, 8
    region, 26
Design procedure, 128
Design rules
    cost, 287-288
    foundry specifications, 307-318
    photolithographic techniques, 217
    standard cell library, 315
Device channels, 296-297
Dielectric
    constant, 65
    loss, 61
    material, 301-302
    pinhole damage, 301
    substrate, 63
    vias, 304
Die attach process, 231-232, 247-248
    epoxy, 247-248
    eutectic, 247-248
Diffusion constant, 33
Dimstrip, 62, 78
Diodes
    Smith chart analysis, 65
Doping, 43-44
    by ion implantation, 296-297
Drain
    current, 12
    GaAs FET structure, 6
    line, 120
    ohmic contact, 6
    resistance, 16
Drain-to-source capacitance, 17
Duroid, 63, 217, 239, 260

E-frame structure, 52
Einstein relationship, 33
Electric field, 8
Electrical loss, 61-62
Electromagnet, 50
Electron beam (e-beam) lithography, 299
    direct write, 19
Electron drift velocity, 5, 8
Electron recombination, 54
Electrons, 34
Emitter
    contact figures, 23
    depletion capacitance, 24
    resistance, 24

Emitter-base junction, 23, 25
Emulsions, 238
Encapsulation, 297
Equivalent circuit, 7-19
Etchback process, 219, 236, 242

Fano's limit, 109, 120
Feedback, 22
  amplifier, 138, 139, 331
FET, 139, 141
  gate, 297
Field-effect transistor
  GaAs, 1
Filters, 36
First metal
  chromium-platinum-gold, 300
  evaporation process, 300-301
  titanium-platinum-gold, 300
Footprint, 231, 249
Forming gas, 231
  reducing atmosphere, 246
Forward bias, 31
Forward voltage, 31
Foundry,
  design rules, 287-288, 307-318
  fabrication, 287, 307
  specifications, 307
  standard cell element library, 81, 315
FRED
  Touchstone blocks, 74
Fringing capacitance, 223

GaAs, 5
  circuit elements, 139
  microstrip transmission line substrate, 63
  MMIC chips, 2
GaAs FET(s), 5, 234, 250, 260
  Smith chart analysis, 65
  "t" gate, 8
Gain compensation, 107
Gain compression, 79
Gallium, 43
Gate, 6
  capacitance, 12-13
  line, 120
  metal, 299
  recess, 299
  resistance, 12, 14-15
  voltage, 9, 11

Gate-to-drain breakdown voltage, 20
Gate-to-drain capacitance, 17
GDSII, 337
  workstations, 81
Germanium, see Ohmic metals
Gold
  epoxy, 247
  first metal, 300
  gate metal, 300
  metalization, 241
  ohmic metals, 7
  Schottky gate metal, 7
Gold-germanium
  die attach process, 247
  ohmic contact metal, 298
Gold-silicon
  die attach process, 247
Gold-tin
  die attach process, 247
  thermal conductivity, 248
Gold germanium
  braze, 245
Gold plated copper, 246
GRID
  Touchstone blocks, 74
Grounding, 220-223
Gunn, J.B., 5
Gyromagnetic ratio, 39

Hardboard, 217
Harmonic
  balance technique, 79
  generation, 31, 78-79
Heat treatment, 244
High temperature anneal, 297
Hole(s), 33-36
Horizontal Bridgeman technique, 296
HP-41C microwave design programs, 83
Hund's rule, 37
Hybrid, 5
Hyperabrupt junction, 28

$1/f$ noise, 22
I region, 32
Impedance, see Characteristic impedance
Impurities, 296
Inductance, 67-69
  of bond wires, 229
Inductors, 67, 137

Intermetallic compounds, 231
Intermodulation product levels, 79
Intervalley scattering, 5
Ion implantation, 6, 296–297
Iron, 41

Junction diode, 26

Kovar carrier, 234, 235, 246, 274

Lange coupler, 69–70
Large-signal effects, 19–22
Libra, 78
Lifetime, 34
Lift-off process, 7, 300, 302
LineCalc, 78
Liquid encapsulated Czochralski (LEC) technique, 296
Liquid phase epitaxy, 6
Load line FET model, 19
Loop inductance
    parallel R-L-C, 46
LSA, 5
Lumped elements, 67
    matching networks, 108

MAG, 12, 18–19, 107
Magnetic
    field, 37, 38
    material, 54
    moment, 38
    return path, 51
Magnetization, 39, 41
Magnetostatic mode chart, 41
Mask, *see* Photomasks
Mass of an electron, 38
Matching elements, 140
Matching network(s), 109
Maximum available gain (*see* MAG)
Maximum channel current, I_{DSS}, 11
Maximum current, 19
Maximum power output, 20
Metalization, 240–241
MIC, xi, 1–2
    microstrip transmission line, 59
MICAD, 80
Microstrip
    bends, 220, 226

    cross junction, 228
    crosses, 220
    steps, 220
    tee(s), 220
Microstrip transmission line(s)
    characteristic impedance, 59
    conductor surface, 62
    coupler designs, 65–66
    coupling factor, 67
    crosscoupling, 65
    dielectric loss, 62
    effective dielectric constant, 59
    electrical loss, 61
    equivalent circuits, 67
    Lange couplers, 69
    MIC, 59
    MMIC, 59
    phase velocity, 59
    skin 62
    Smith chart calaculation, 64–65
    substrates, 62
    surface roughness, 62
    wavelength, 59
Microwave Harmonica, 78
MIM capacitors, 301, 302, 337, 349
Minority carriers, 28
Mismatch
    gain compensation, 108
    loss, 107
Mixing, 79
MMIC, xi, 1–2
    design, 254
    layout, 323–325
    microstrip transmission line, 59
    parasitic elements, 139
    processing summary sheet, 307
    resistors, 298
Mobility, 30
Mode, 41–42
    magnetostatic mode chart, 41
    110 mode, 41
    210 mode, 42
Molecular beam epitaxy (MBE), 6
Molybdenum, *see* Gate metal
Monolithic microwave integrated circuits, *see* MMIC
Mylar, 236

Negative differential mobility, 5

npn-Type transistors, 23
Negative bias voltage, 8
Negative differential mobility, 5
Network optimization, 73
Nickel, 241
Nickel-chromium, 241
Noise, 22
Nonlinearities, 19
Normograph, 12
Nucleus, 37

Ohmit contact
    drain, 6
    FETs and diodes, 297
    GaAs FET structure, 6
    gold-germanium, 298
    source, 6
Ohmic metals, 7
111 crystal plane, 49
110 mode, 41
OPT
    Touchstone blocks, 74
Open circuit, 35
Optical photolithographic process, 19
Optimization, 137–138
Oscillator(s), 22, 36, 41, 70
    MMIC chips, 294
Oxide layer, 244–245

Palladium, 241
Paper dolls, 323
Parallel plate capacitor, 26–27
Parallel *R-L-C* resonator, 46
Parasitic elements, 6, 12, 67, 139, 220
Parasitic
    inductance, 249
    shunt capacitors, 67
Pattern generators, 236
Photolithography, 7, 217, 240–244
    etching, 242, 299
    lift-off process, 300
    photoresist, 240, 242, 299
Photomasks, 236, 294
    chrome-glass, 238
    emulsions, 238
    film, 238
    pattern generators, 236
    step and repeat, 238
PIN diode, 32

Smith chart analysis, 65
Pinch-off voltage, 9
Planck's constant, 38
Plate up process, 219, 242, 243
Plasma enhanced chemical vapor deposition
    (PECVD), 301
Platinum, 299
Pole piece, 51
Polyimids, 302
Pop-up windows, 81
Power added efficiency, 21
Precession, 38–40, 41
Preforms, 246
Profile, 28
PUFF, 81, 83
Pure YIG (*see* YIG)

Q factor, *see* Quality factor
Quartz, 240
Quality (Q) factor, 30, 53, 202
Quantum mechanical property, 37
Quantum number, 38
Quasi-TEM transmission media, 59

R-C ladder network, 14
Rectangular coordinates, 81, 83
Rectification, 79
Rectifying Schottky barrier, 6
Reducing atmosphere, 246
Resistors, 244–245
Resonant frequency, 39, 54
Resonators, 36, 52
    parallel *R-L-C*, 46
Ribs, 246
Rubylith, 236

S-parameters, 70–72
Sapphire, 240
Saturation, 34
Sawing, 245
Scale factors, 236
Schematic capture, 81, 83
Schottky barrier(s)
    FET gate, 299
    normograph, 12
Self-alignment
    FET gate, 299
    photolithographic process, 7
Series resistance, 27

Shunt capacitor, 229
Silicon FET, 5
Silicon nitride, 301
Silk screen metalization, 240
Silver epoxy, 231
   die attach process, 247
Skin depth, 62, 301
Slope parameter, 28
Smith chart, 62–65
   analysis of GaAs FETS, 65
   design tools, 83
Softboard
   circuit boards, 217
   Duroid™, 239
   Teflon™, 239
Source,
   ohmic contact, 6
Space charge, 26
Speed of light, 38
Spheres, YIG, 43–44, 277
SPICE, 78
Spreading resistance, 235
Sputtering, 240, 242, 301
Stability, 111–112
Standard cells, 315, 319
Step and repeat technique, 245. *See also* Photomasks
Substrate,
   dielectric, 62, 65
   grounds, 221
   metallization of, 240–241
Summary sheet
   MMIC processing, 307
SuperCompact, 73, 253
   Lange coupler models, 70
   S-parameters, 70
Surface roughness, 62
Symmetric step changes, 227

Tantalum-nitride, 241
Teflon™, 239
Temperature, 47
   slope, 32
Temperature-frequency drift, 47, 53
Thermal
   compression, 248
   conductivity, 248
   considerations, 233–236
   resistance, 234–236, 248

Thick film metalization, 240
Thin film metalization, 240
Thin-film resistors
   parasitic elements, 230
   resistor stabilization, 244–245
Three-dimensional cylinder, 34
Titanium, 299
Titanium-platinum-gold, 300
Titanium-tungsten, 241
Touchstone, 73, 253
   blocks, 74
   Lange coupler models, 70
   microstrip transmission line calculation, 62
   optimizer mode, 77
   S-parameters, 70
   topology, 77
Transconductance, 9
   channel current, 8
   GaAs FET equivalent circuit, 7
   GaAs FET structure, 6
   gate voltage, 11
Transistor action, 8
Transit time, 5
   frequency, 32
Transmission media, quasi-TEM, 59
Traveling wave amplification, 120
Turn key system, 81
Two-port devices, 70–71

Ultrasonic scrubbing, 248
Uniform precession mode, 41
Unity gain cut-off frequency, 24
Unix operating systems, 80–81

Vapor phase epitaxy (VPE), 6
VAR
   Touchstone blocks, 74
Varactor diodes
   Smith chart analysis, 65
Via,
   airbridge, 304
   dielectric, 304
   substrate, 304–305
Via hole, 219, 221

Wedge bonding, 248–249
Wire bonding, 248
   ball bonding, 248

thermal compression techniques, 248
   wedge bonding, 248
Workstation, 80–83
Wrap-around grounds, 230

YIG, 36
   crystal, 39, 41, 43
   heater, 49
   limiting, 46–47
   loop, 277
   sphere, 43–44, 277
   pure, 44
   G factor of, 202
   resonance, 46
Yittrium iron garnet (*see* YIG)

About the Author

After earning a B.S. from Worcester Polytechnic Institute, Allen A. Sweet went on to receive both an M.S. and a Ph.D in Electrical Engineering from Cornell University, Ithaca, New York. He worked on the technical staff of several well-known companies in the microwave industry including Watkins-Johnson, Varian Associates, and Microwave Associates. In the late 1970's he formed his own consulting firm, which he operates today. Dr. Sweet is currently the president of Monolithic Application Inc., a design and consulting service company. He has taught various courses on microwave amplifiers and oscillators since 1981.